iOS 6 高级开发手册

（第4版）

The
Advanced iOS 6
Developer's Cookbook

［美］Erica Sadun 著

陈宗斌 译

人民邮电出版社
北京

图书在版编目（CIP）数据

iOS 6高级开发手册 : 第4版 / (美) 萨顿
(Sadun, E.) 著 ; 陈宗斌译. -- 北京 : 人民邮电出版社, 2014.4
 ISBN 978-7-115-34425-0

Ⅰ. ①i… Ⅱ. ①萨… ②陈… Ⅲ. ①移动终端－应用程序－程序设计－技术手册 Ⅳ. ①TN929.53-62

中国版本图书馆CIP数据核字(2014)第015888号

版 权 声 明

Erica Sadun: The Advanced iOS 6 Developer's Cookbook (4th Edition)
Copyright © 2013 Pearson Education, Inc.
ISBN: 978-0321884220

All rights reserved. No part of this publication may be reproduced, stored in a retrieval system, or transmitted in any form or by any means, electronic, mechanical, photocopying, recording, or otherwise without the prior consent of Addison Wesley.

版权所有。未经出版者书面许可，对本书任何部分不得以任何方式或任何手段复制和传播。

本书中文简体字版由人民邮电出版社经 Pearson Education, Inc.授权出版。版权所有，侵权必究。

本书封面贴有 Pearson Education（培生教育出版集团）激光防伪标签。无标签者不得销售。

+ 著　　　[美]Erica Sadun
+ 译　　　陈宗斌
  责任编辑　傅道坤
  责任印制　程彦红　杨林杰
+ 人民邮电出版社出版发行　北京市丰台区成寿寺路11号
  邮编　100164　电子邮件　315@ptpress.com.cn
  网址　http://www.ptpress.com.cn
  大厂聚鑫印刷有限责任公司印刷
+ 开本：800×1000　1/16
  印张：30.5
  字数：607千字　　　　　　　2014年4月第1版
  印数：1-2 500册　　　　　　2014年4月河北第1次印刷
  著作权合同登记号　图字：01-2013-5716号

定价：89.00元
读者服务热线：(010)81055410　印装质量热线：(010)81055316
反盗版热线：(010)81055315
广告经营许可证：京崇工商广字第 0021 号

内容提要

本书是市面上最畅销的 iOS 开发图书的全新升级版本，以苹果发布的 iOS 6 为基础编写而成。在本书中，资深 iOS 开发专家 Erica Sadun 与大家分享了一些用于 iOS 6 高端开发的成熟、可靠的方法，并借助大量的代码示例对这些方法进行演示讲解，从而降低了 iOS 开发的学习难度。

本书是《iOS 6 核心开发手册（第 4 版）》的姊妹篇，总共分为 13 章，先后讲解了特定设备的 iOS 开发、文档和数据的共享、Core Text 的使用方式、iOS 开发中使用到的几何学知识、应用在接入网络时需要解决的问题、图像的处理、图像捕获、音频处理、Address Book 框架在应用中的使用、地理定位、GameKit 和 StoreKit 的使用，以及如何使用推送通知等内容。

本书语言简练、内容丰富，并在配套网站上提供了完整的示例代码，适合具有一定 iOS 开发经验或其他移动开发经验的人员阅读。对 iOS 开发感兴趣的入门者，也可以从本书姊妹篇《iOS 6 核心开发手册（第 4 版）》开始起步，逐步学会、掌握 iOS 的开发。

关于作者

Erica Sadun 是数十本畅销书的著者、合著者和供稿者，这些书涉及程序设计、数字视频、数字摄影和网页设计，包括广受欢迎的 *The iOS 5 Developer's Cookbook*。她目前在 tuaw.com 上开通了博客，过去也曾在 O'Reilly 的 Mac DevCenter、Lifehacker 和 Ars Technica 上发表过博文。Erica 是数十种 iOS 原生应用程序的编写者，并拥有佐治亚理工学院图形、可视化和可用性计算中心的计算机科学博士学位。作为极客、程序员和写作者，她还没遇到过不喜欢的小工具。笔耕之余，她和她同为极客的丈夫还养育了三个小极客。这三个小极客如果不是忙着在屋子里重新布线和策划控制全球，就会按耐着困惑，盯着他们的爸爸妈妈。

献辞

谨将本书献给我的丈夫 Alberto，多年以来，他一直在默默忍受我不时提及与 iOS 开发相关的各种 SDK 和小工具，而且在本书完成之日，他依然能够保持良好的心态。

致谢

如果没有 Chuck Toporek 的努力，本书将不会问世，Chuck Toporek 是我的编辑，多年来担任过多家出版社的项目负责人。他现在在 Apple 工作，我非常想念他。如果没有他，读者将不会看到这本 Cookbook。他兼具两种优秀的技能：激励作者做他们认为自己做不到的事情，以及挥舞着巨大的"真实鳟鱼"[1]吸引作者专注于图书的主题并且不脱离现实。这样的反复鞭挞使图书得以在最后期限前问世，并且使书中的内容非常有吸引力。

还要感谢 Trina MacDonald（我的非常优秀的新编辑）、Chris Zahn（才华横溢的开发编辑）和 Olivia Basegio（忠实且富有激情的助理编辑，总能在幕后把事情办好）。此外，还要对整个 Addison-Wesley/Pearson 制作团队深表谢意，特别感谢 Kristy Hart、Jovana San Nicolas-Shirley、San Dee Phillips、Nonie Ratcliff 和 Chuti Prasertsith。还要感谢 Safari 的工作人员使我的图书初步成形，并且在出现技术性问题时进行修正。

还要感谢我多年的经纪人 Neil Salkind，以及技术评审 Oliver Drobnik、Rich Wardwell 和 Duncan Champney，他们帮助我使本书合乎情理，而不是一厢情愿地去组织内容，还要感谢我在 TUAW、Ars Technica 和 Digital Media/Inside iPhone 博客的所有现在和以前的同事。

我非常感谢人才济济的 iOS 开发人员社区，包括 Jon Bauer、Tim Burks、Matt Martel、Tim Isted、Joachim Bean、Aaron Basil、Roberto Gamboni、John Muchow、Scott Mikolaitis、Alex Schaefer、Nick Penree、James Cuff、Jay Freeman、Mark Montecalvo、August Joki、Max Weisel、Optimo、Kevin Brosius、Planetbeing、Pytey、Michael Brennan、Daniel Gard、Michael Jones、Roxfan、MuscleNerd、np101137、UnterPerro、Jonathan Watmough、Youssef Francis、Bryan Henry、William DeMuro、Jeremy Sinclair、Arshad Tayyeb、Jonathan Thompson、Dustin Voss、Daniel Peebles、ChronicProductions、Greg Hartstein、Emanuele Vulcano、Sean Heber、Josh Bleecher Snyder、Eric Chamberlain、Steven Troughton-Smith、Dustin Howett、Dick Applebaum、Kevin Ballard、Hamish Allan、Lutz Bendlin、Oliver Drobnik、Rod Strougo、Kevin McAllister、Jay Abbott、Tim Grant Davies、Maurice Sharp、Chris Samuels、Chris Greening、Jonathan Willing、Landon Fuller、Jeremy Tregunna、Christine Reindl、Wil Macaulay、Stefan Hafeneger、Scott Yelich、Mike Kale、chrallelinder、John

---

[1] 在本书的开发和制作过程中没有伤害真实的或想象的鳟鱼，但是无数罐健怡可乐就没有这么好的命运了，在编写本书原稿的过程中，我们喝了大量的可乐。

致谢

Varghese、Robert Jen、Andrea Fanfani、J. Roman、jtbandes、Artissimo、Aaron Alexander、Christopher Campbell Jensen、Nico Ameghino、Jon Moody、Julián Romero、Scott Lawrence、Evan K. Stone、Kenny Chan Ching-King、Matthias Ringwald、Jeff Tentschert、Marco Fanciulli、Neil Taylor、Sjoerd van Geffen、Absentia、Nownot、Emerson Malca、Matt Brown、Chris Foresman、Aron Trimble、Paul Griffin、Paul Robichaux、Nicolas Haunold、Anatol Ulrich (hypnocode GmbH)、Kristian Glass、Remy "psy" Demarest、Yanik Magnan、ashikase、Shane Zatezalo、Tito Ciuro、Mahipal Raythattha、Jonah Williams of Carbon Five、Joshua Weinberg、biappi、Eric Mock，以及 irc.saurik.com 和 irc.freenode.net 上的 iPhone 开发人员频道中的每一个人，还有许多其他的人，在此不能一一指出。他们的技术、建议和反馈帮助使本书成为可能。如果我忽视了任何帮助过我的人，请接受我的道歉。

在此要特别感谢我的家人和朋友，他们支持我月复一月地进行新测试版发布，并且对我莫名其妙的缺席和绝望的嚎叫表现出了极大的耐心。我感谢你们一直坚持陪伴在我身边。我还要感谢我的孩子们，当他们知道他们的母亲天天驼着背，忙着敲击键盘而无暇顾及他们时，依然表现出了坚定的支持。我的孩子们在过去几个月给我提供了非常宝贵的帮助，他们测试应用程序、提供建议，并且成长为了不起的人。我尽量每天都提醒自己，这些孩子成为我生命中的一部分我该有多幸运。

前言

欢迎阅读另一本 iOS Cookbook 图书!有了 iOS 6,Apple 的移动设备家族在使人兴奋和可能做到的事情方面达到了一个新的层次。本书可以帮助你开始开发程序。这个修订版介绍了最新的 WWDC 中公布的所有新特性,说明了如何把它们纳入应用程序中。

对于这个版本,我的出版团队明智地把 Cookbook 材料划分成册,以使之容易管理。本书关注的是公共框架(比如 StoreKit、GameKit 和 Core Location)以及一些便于使用的技术(比如图像处理排版)。它有助于构建利用专用库的应用程序并超越基本技术。本册针对的是那些精通 iOS 开发并且谋求掌握专业领域的实用知识的读者。

与之配套的一册图书是《iOS 6 核心开发手册(第 4 版)》,它提供了针对日常开发的核心的解决方案。它涵盖了使用标准 API 和接口元素创建 iOS 应用程序所需的所有类,还包含处理图形、触摸和视图以创建移动应用程序所需的秘诀。

最后,还有一本图书是 *Learning iOS 6: A Hands-on Guide to the Fundamentals of iOS Programming*,其中包含了大量教程材料,它由 Cookbook 的前几章组成。在该书中,可以找到从头开始学习 iOS 6 开发所需的所有基本知识。从 Objective-C 到 Xcode,从调试到部署,该书讲述了如何开始使用 Apple 的开发工具套件。

像过去一样,可以在 GitHub 上找到示例代码。在 https://github.com/erica/iOS-6-Cookbook 上可以找到这本 Cookbook 的知识库,在 WWDC 2012 发布之后,它全都针对 iOS 6 进行了更新。

如果你有什么建议、错误修正方法和校正意见,或者可以为将来的版本做贡献的其他任何想法,可以给我发送电子邮件: erica@ericasadun.com。让我提前感谢你。我会感谢所有的反馈,它们有助于使本书变得更好、更优秀。

——Erica Sadun,2012 年 9 月

用户需要什么

如果计划构建 iOS 应用程序,将至少需要一个 iOS 设备,用于测试应用程序,首选新款 iPhone 或平板电脑。下面的列表涵盖了开始时必须准备的一些物品。

- Apple 的 iOS SDK——可以从 Apple 的 iOS Dev Center（http://developer.apple.com/ios）下载 iOS SDK 的最新版本。如果计划通过 App Store 销售应用程序，可以变成一名付费的 iOS 开发者。对于个人开发者，其费用是 99 美元/年，对于企业（即公司）开发者，费用则是 299 美元/年。注册的开发者将收到证书，允许他们"登录"并把应用程序下载到他们的 iPhone/iPod Touch 上，以进行测试和调试，并且及早访问 iOS 的预发行版本。免费程序开发者可以在基于 Mac 的模拟器上测试他们的软件，但是不能部署到设备上或者提交给 App Store。

> **大学计划**
> Apple 还为学生和教师提供了一个大学计划（University Program）。如果你是一名计算机学科的学生，并且学习的是大学级别的课程，就要与教员核实一下，看看你所在的学校是否是 University Program 的一部分。有关 iPhone 开发者大学计划（iPhone Developer University Program）的更多信息，参见 http://developer.apple.com/support/iphone/university。

- 一台现代的 Mac 机，运行 Mac OS X Lion（版本 10.7），最好是运行 Mac OS X Mountain Lion（版本 10.8）——需要大量的磁盘空间用于开发，并且 Mac 应该具有尽可能多的 RAM。
- iOS 设备——尽管 iOS SDK 包括一个模拟器用于测试应用程序，但是确实需要拥有 iOS 硬件，以便为平台开发程序。可以把单元连接到计算机上并安装你构建的软件。对于真实的 App Store 部署，手边有多个单元是有益的，用以表示几代不同的硬件和固件，以便可以在目标受众将使用的相同平台上执行测试。
- Internet 连接——该连接允许利用真实的 Wi-Fi 连接以及 EDGE 或 3G 服务测试程序。
- 熟悉 Objective-C——要为 iPhone 编写程序，需要知道 Objective-C 2.0。该语言基于具有面向对象扩展的 ANSI C，这意味着还需要知道一点 C 语言的知识。如果利用 Java 或 C++编写过程序并且熟悉 C 语言，就应该能够比较容易地转向 Objective-C。

Mac/iOS 开发的路标

一本书不可能做到面面俱到。如果尝试把你需要知道的所有知识都打包进这本书中，那么你将学不会它（事实是，本书提供了一个优秀的智力开发工具，不要让自己过于紧张）。为 Mac 和 iOS 平台开发程序的确需要知道许多知识。如果你只是刚刚起步并且没有任何编程经验，那么第一步应该是学习 C 程序设计语言的大学级别的课程。尽管字母表可能从字母 A 开始，但是大多数程序设计语言都起源于 C，当然你通往开发者的道路也是如此。

一旦知晓了 C 语言以及如何使用编译器（在基本的 C 语言课程中将会学到它），余下的应该就简单了。从此，可以直接跳到 Objective-C，并且学习如何利用该语言以及 Cocoa 框架编程。

图 0-1 所示的流程图显示了 Pearson Education 提供的关键图书，它们可以提供必要的培训，而这是你为了变成熟练的 iOS 开发者所需要接受的。

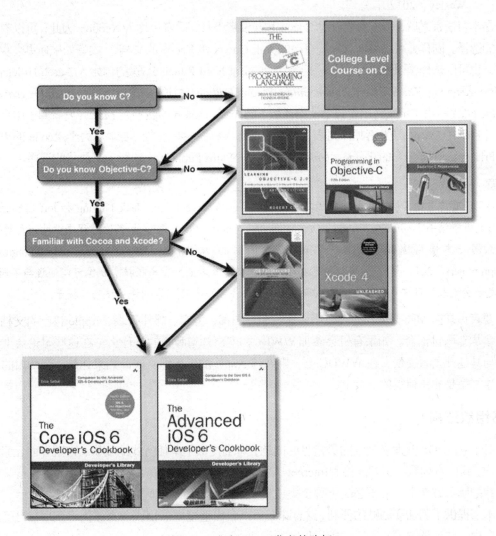

图 0-1　成为 iOS 开发者的路标

一旦知道了 C 语言，就可以选择学习如何利用 Objective-C 编程。如果你想深入了解该语言，可以阅读 Apple 自己的文档，或者从下面这些关于 Objective-C 的图书中选择一本来阅读。

- Aaron Hillegass 编写的 *Objective-C Programming: The Big Nerd Ranch Guide* 一书（Big Nerd Ranch，2012 年）。
- Robert Clair 编写的 *Learning Objective-C: A Hands-on Guide to Objective-C for Mac and*

- *iOS Developers* 一书（Addison-Wesley，2011 年）。
- Stephen Kochan 编写的 *Programming in Objective-C 2.0, Fourth Edition* 一书（Addison-Wesley，2012 年）。

在学习语言之后，接下来要应对 Cocoa 和开发者工具，后者也称为 Xcode。为此，可以有几个不同的选择。同样，可以参考 Apple 自己的关于 Cocoa 和 Xcode 的文档[1]，或者如果你更喜欢阅读图书，也可以选择学习最好的图书。亚特兰大的 Big Nerd Ranch 公司[2]的创始人 Aaron Hillegass 是 *iOS Programming: The Big Nerd Ranch Guide, Second Edition* 一书的合著者，也是 *Cocoa Programming for Mac OS X* 一书（不久将出现该书的第 4 版）的作者。Aaron 的图书在 Mac 开发者圈子中享有盛誉，并且是在 Cocoa 开发者邮件列表上看到的推荐次数最多的图书。要学习关于 Xcode 的更多知识，Fritz Anderson 编写的 *Xcode 4 Unleashed* 一书（Sams Publishing 出版）是最佳的选择。

> **注意：**
> 市场上还有许多由其他出版社出版的其他类图书，包括 Dave Mark、Jack Nutting 和 Jeff LaMarche 编写的最畅销书 *Beginning iPhone 4 Development*（Apress，2011 年）。如果你对程序设计一无所知，那么另一本值得选择的图书是 Tim Isted 编写的 *Beginning Mac Programming*（Pragmatic Programmers，2011 年）。不要把自己局限于某一本书或者某一家出版社。就像可以通过与不同的开发者交流来学习许多知识一样，从市场上的其他图书中也可以学到许多技巧与提示。

要真正掌握 Mac 开发，需要查看多种材料：图书、博客、邮件列表、Apple 自己的文档，最好是参加专题讨论会。如果有机会参加 WWDC，就会知道我说的是什么。在这些讨论会上要花时间与其他开发者交谈，在 WWDC 上，则要与 Apple 的工程师交谈，如果你是一名严肃的开发者，这样做是非常值得的。

本书组织结构

对于新的 iOS 开发者面临的最常见的问题，本书提供了单任务式的秘诀：布置界面元素、响应用户、访问本地数据源和连接到 Internet。每一章都把相关的任务组织在一起，允许直接跳转到所寻找的解决方案上，而不必决定哪个类或框架最匹配所讨论的问题。

本书提供了剪切和粘贴的便利，这意味着可以自由地把本书秘诀中的源代码重用在自己的应用程序中，然后调整代码以适应应用程序的需要。

下面简要描述了本书各章的内容。

---

[1] 要从头开始学习 Cocoa，可以参阅 *Cocoa Fundamentals Guide*（http://developer.apple.com/mac/library/documentation/Cocoa/Conceptual/CocoaFundamentals/CocoaFundamentals.pdf）；要学习 Xcode，可以参阅 *A Tour of Xcode*（http://developer.apple.com/mac/library/documentation/DeveloperTools/Conceptual/A_Tour_of_Xcode/A_Tour_of_Xcode.pdf）。

[2] Big Nerd Ranch 公司：www.bignerdranch.com。

- 第 1 章,"特定于设备的开发",每个 iOS 设备都代表独特、共享、短暂和持久属性的融合。这些属性包括设备的当前物理方位、它的型号名称、它的电池状态以及它对机载硬件的访问。本章从设备的构建配置到它的活动机载传感器进行了广泛探讨,并且提供了一些秘诀,用于返回关于正在使用的单元的各类信息项。

- 第 2 章,"文档和数据共享",在 iOS 下,应用程序可以使用系统提供的多个特性共享信息和数据,以及把控制从一个应用程序转移给另一个应用程序。本章介绍可以在应用程序之间集成文档和进行数据共享的方式。你将看到如何把这些特性添加到自己的应用程序中,并且聪明地使用它们,使应用程序成为 iOS 生态系统中具有协作精神的一员。

- 第 3 章,"Core Text",本章介绍了属性化文本处理,并且探索了如何将文本特性构建到应用程序中。你将学到把属性化字符串添加到公共 UIKit 元素中,如何创建 Core Text 强化的视图,以及如何为自由的文本排版突破更大的界限。在学完本章后,你将发现 Core Text 带给 iOS 的强大能力。

- 第 4 章,"几何学",尽管与 Core Animation 或 Open GL 相比,UIKit 不太需要应用数学,但是在处理贝塞尔路径和视图变换时,几何学仍然扮演着重要的作用。为什么需要几何学呢?它有助于以非标准的方式操纵视图,包括沿着自定义的路径布置文本,以及沿着路径运行动画。如果你在一提到贝塞尔曲线、凸包和样条时就目光呆滞,本章可以帮助澄清这些术语,使你能够向工具箱中添加一些功能强大的自定义选项。

- 第 5 章,"联网",Apple 以各种网络计算及其支持技术给 iOS 打下了坚实的基础。本书姊妹篇《iOS 核心开发手册(第 4 版)》中有关联网的章节介绍了网络状态检查、同步和异步下载、JSON 和 XML 解析。本书中的这一章继续讨论这个主题,并且介绍了更高级的技术。其中包括身份验证质询、使用系统密钥链(keychain)、处理 OAuth 等。这里介绍了一些方便的方法,应该可以帮助你从事开发工作。

- 第 6 章,"图像",图像是抽象的表示,存储构成图片的数据。本章介绍了 Cocoa Touch 图像,特别是 UIImage 类,并且讲述了在 iOS 上处理图像数据所需的全部专业知识。在本章中,你将学习如何在应用程序中加载、存储和修改图像数据。你将发现如何处理图像数据以创建特殊效果,如何逐个字节地访问图像,等等。

- 第 7 章,"照相机",照相机把图像抬高到了下一个层次。它们使你能够把实时馈送和用户指示的快照集成到应用程序中,并且提供源于现实世界的原始数据。在本章中将会学习图像捕获。你将发现如何使用 Apple 提供的类来获取图片,以及如何从零开始操纵图像。你将学习控制图像元数据,以及如何把实时馈送与高级过滤集成起来。本章从硬件的角度重点介绍了图像捕获。无论你是在打开照相机闪光灯,还是在检测面孔,本章都会介绍 iOS 图像捕获技术的方方面面的知识。

- 第 8 章,"音频",iOS 设备是一个媒体大师。其内置的 iPod 特性可以专业地处理音频和视频。iOS SDK 向开发者展示了这种功能。一套丰富的类通过回放、搜索和录制简化了媒体处理。本章介绍了一些秘诀，使用那些类操纵音频、把媒体展示给用户，以及允许用户与媒体交互。你将看到如何构建音频播放器和录音机，还将发现如何浏览 iPod 库以及如何选择要播放的项目。

- 第 9 章，"连接到 Address Book"，本章将介绍 Address Book 并将演示如何在应用程序中使用它的框架。你将学到如何访问各个联系人的信息，如何修改和更新联系人信息，以及如何使用谓词只查找感兴趣的联系人。本章还将介绍 GUI 类，它们提供了交互式解决方案，用于挑选、查看和修改联系人。

- 第 10 章，"位置"，快速计算正变得与计算的方式和计算的内容同样重要。iOS 终日忙个不停，整天都与它的用户一起旅行。Core Location 给 iOS 注入了按需进行地理定位的功能。MapKit 添加了交互式的应用程序内的地图绘制功能，使用户能够查看和操纵加注释的地图。利用 Core Location 和 MapKit，可以开发应用程序，帮助用户与朋友会面、搜索本地资源，或者提供基于位置的个人信息流。本章介绍了这些可以感知位置的框架，并且说明了如何把它们集成进 iOS 应用程序中。

- 第 11 章，"GameKit"，本章介绍了可以通过 GameKit 创建相联系的游戏玩法的多种方式。GameKit 提供了一些特性，使应用程序能够超越单玩家/单设备场景，并过渡到使用 Game Center（游戏中心）和设备与设备之间的联网。Apple 的 Game Center 添加了一种集中式服务，使游戏能够提供共享的排行榜和基于 Internet 的配对。GameKit 还为对等连接提供了即席的联网解决方案。

- 第 12 章，"StoreKit"，StoreKit 提供了应用程序中的采购功能，可以把它集成进你的软件中。利用 StoreKit，最终用户可以使用他们的 iTunes 凭证从应用程序内购买可以解锁的特性、媒体订阅或可消费的资产，比如鱼食或阳光。本章介绍了 StoreKit，并且说明了如何使用 StoreKit API 为用户创建采购选项。

- 第 13 章，"推送通知"，当脱离设备的服务需要直接与用户通信时，推送通知提供了一种解决方案。就像本地通知允许应用程序在预定的时间联系用户一样，推送通知可以从基于 Web 的系统递送消息。推送通知可以让设备显示一条提醒、播放一段自定义的声音，或者更新应用程序图标。脱离设备的服务可以用这种方式联系基于 iOS 的客户，使他们能够知道新的数据或更新。本章将介绍你需要知道的关于推送通知的所有基本的知识。

关于示例代码

出于教学的目的，本书的示例代码使用单个 main.m 文件。人们通常不这样开发 iPhone 或 Cocoa 应用程序，或者诚实地讲，不应该这样开发它们，但它提供了一种极佳的方式来表达一个好主意。当读者必须同时查看 5 个、7 个或 9 个单独的文件时，将很难讲述一个故事。提供单个文件可以把注意力集中在那个故事上,允许在单个代码块中访问那个主意。

这些示例并不打算作为独立的应用程序。它们只是为了演示单个秘诀和单个主意。一个具有中心表达方式的 main.m 文件可以在一个位置呈现实现故事。读者可以研究这些集中的主意,并且使用标准的文件结构和布局把它们转换成正常的应用程序结构。本书中的表达方式不会以标准的日常最佳实践方法产生代码。作为替代,它会反映一种教学方法,提供可以根据需要纳入到工作中的简明的解决方案。

相比之下,在 Apple 的标准示例代码中,必须梳理许多文件来构建将要演示的概念的心智模型(mental model)。这些示例被构建为完整的应用程序,它们所执行的任务通常与你需要解决的问题相关,但并非是必要的。只找出那些相关的部分就要做许多的工作,可能会导致得不偿失。

在本书中,将会发现这种用一个文件讲故事的规则的例外情况:当类实现就是秘诀时,Cookbook 将提供标准的类和头文件。一些秘诀不是为了强调某种技巧,而是提供这些类和类别(即对预先存在的类(而不是新类)的扩展)。对于这些秘诀,除了封闭故事余下部分的骨架式 main.m 文件之外,还要寻找单独的.m 和.h 文件。

一般来讲,本书中的示例都使用单个应用程序标识符:com.sadun.helloworld。本书使用一个标识符来避免数十个示例同时塞满 iOS 设备。每个示例将替换前一个示例,确保主屏幕保持相对整洁。如果想同时安装多个示例,只需简单地编辑标识符,添加独特的后缀,比如 com.sadun.helloworld.table-edits。也可以编辑自定义的显示名称,使应用程序看上去有所区别。团队配置文件(Team Provisioning Profile)将匹配每个应用程序标识符,包括 com.sadun.helloworld。这允许把编译过的代码安装到设备上,而不必更改标识符,只需确保在每个项目的构建设置中更新签名身份即可。

获取示例代码

在开源 GitHub 托管站点上的 github.com/erica/iOS-6-Cookbook 中可以找到本书的源代码。其中,可以找到按章划分的源代码集合,它们提供了本书中所介绍材料的工作示例。秘诀是按书中那样编号的。例如,第 5 章中的秘诀 6 出现在 06 子文件夹中的 C05 文件夹中。

任何编号为 00 或者带有后缀(比如 05b 或 02c)的项目都指示用于创建大量文本和图形的材料。通常,我会删除这些额外的项目。本书原稿的早期读者要求我在这个版本中包括进它们。你将发现其中有 6 个左右的额外示例散布在代码库周围。

如果你感觉直接使用 git 不那么得心应手,GitHub 提供了一个下载按钮。在编写本书时,该按钮位于站点主页的右边,大约在首页下面一半的位置。它使你能够将整个代码库作一个 ZIP 存档文件或压缩包下载。

投稿

示例代码永远不是一个固定的目标。随着 Apple 更新其 SDK 和 Cocoa Touch 库,它也会继续演化。参与进来吧!可以通过建议错误修正方式和校正措施以及扩展代码介入这个过程。GitHub 允许分解代码库,并利用自己的调整和特性扩充它们,并把它们共享回主代码库中。如果提出新的主意或方法,请让我知道。我的团队和我都很高兴在代码库和本书的下一个版本中吸纳非常好的建议。

获取 git

可以使用 git 版本控制系统下载本书的源代码。在 http://code.google.com/p/git-osx-installer 上提供了 git 的 OS X 实现。OS X git 实现包括命令行和 GUI 解决方案,因此要搜索最适合开发需求的版本。

获取 GitHub

GitHub(http://github.com)是最大的 git 托管站点,具有超过 15 万个公共代码库。它既为公共项目提供了免费托管,也为私有项目提供了付费选项。利用一个自定义的 Web 界面,其中包括 wiki(维基)托管、问题跟踪以及对项目开发者的社交网络的强调,很容易发现新代码或者在现有库上开展合作。可以在它们的网站上注册一个免费账户,这样就能够复制和修改 Cookbook 代码库,或者创建自己的开源 iOS 项目以与其他人共享。

联系作者

如果你对本书有任何建议或问题,请给我发送电子邮件,地址是:erica@ericasadun.com,或者访问 GitHub 代码库,并在那里联系我。

目录

第1章 特定于设备的开发 1
1.1 访问基本的设备信息 1
1.2 添加设备能力限制 2
    1.2.1 用户权限描述 3
    1.2.2 其他常用的 Info.plist 键 4
1.3 秘诀：检查设备接近度和电池状态 ... 4
    1.3.1 启用和禁用接近度传感器 5
    1.3.2 监测电池状态 5
    1.3.3 检测 Retina 支持 8
1.4 秘诀：取回额外的设备信息 8
1.5 秘诀：使用加速能力"向上"定位 10
1.6 处理基本的方向 12
1.7 同步获取当前的加速计角度 13
    1.7.1 通过加速计计算方向 14
    1.7.2 计算相对角度 14
1.8 使用加速度移动屏幕上的对象 15
1.9 秘诀：基于加速计的滚动视图 18
1.10 秘诀：Core Motion 基础 21
    1.10.1 测试传感器 22
    1.10.2 处理程序块 22
1.11 秘诀：获取和使用设备姿势 26
1.12 使用运动事件检测晃动 27
1.13 使用外部屏幕 29
    1.13.1 检测屏幕 29
    1.13.2 获取屏幕分辨率 30
    1.13.3 设置 Video Out 30
    1.13.4 添加显示器链接 31
    1.13.5 过扫描补偿 31
    1.13.6 VIDEOkit 31
1.14 跟踪用户 35
1.15 还有一件事：检查可用的磁盘空间 ... 36
1.16 小结 37

第2章 文档和数据共享 39
2.1 秘诀：处理统一类型标识符 39
    2.1.1 通过文件扩展名确定 UTI 40
    2.1.2 从 UTI 转向扩展名或 MIME 类型 41
    2.1.3 测试顺应性 42
    2.1.4 获取顺应性列表 43
2.2 秘诀：访问系统粘贴板 45
    2.2.1 存储数据 46
    2.2.2 存储公共类型 46
    2.2.3 获取数据 47
    2.2.4 被动更新粘贴板 47
2.3 秘诀：监测 Documents 文件夹 48
    2.3.1 支持文档文件共享 48
    2.3.2 用户控制 49
    2.3.3 Xcode 访问 49
    2.3.4 扫描新文档 50
2.4 秘诀：展示活动视图控制器 54
    2.4.1 展示活动视图控制器 55

2 目录

- 2.4.2 活动项目源 55
- 2.4.3 项目提供者 55
- 2.4.4 项目源回调 56
- 2.4.5 添加服务 57
- 2.4.6 项目和服务 62
- 2.4.7 支持 HTML 电子邮件 63
- 2.4.8 排除活动 63
- 2.5 秘诀：Quick Look 预览控制器 ... 63
- 2.6 秘诀：添加 QuickLook 动作 66
- 2.7 秘诀：使用文档交互控制器 68
  - 2.7.1 创建文档交互控制器实例 ... 69
  - 2.7.2 文档交互控制器属性 70
  - 2.7.3 提供文档 Quick Look 支持 ... 70
  - 2.7.4 检查打开菜单 71
- 2.8 秘诀：声明文档支持 75
  - 2.8.1 创建自定义的文档类型 76
  - 2.8.2 实现文档支持 77
- 2.9 秘诀：创建基于 URL 的服务 ... 81
  - 2.9.1 声明模式 82
  - 2.9.2 测试 URL 83
  - 2.9.3 添加处理程序方法 83
- 2.10 小结 84

第 3 章 Core Text 85
- 3.1 Core Text 和 iOS 85
  - 3.1.1 属性 85
  - 3.1.2 C 语言与 Objective-C 86
  - 3.1.3 UIKit 87
- 3.2 属性化字符串 87
- 3.3 秘诀：基本的属性化字符串 90
- 3.4 秘诀：可变的属性化字符串 91
- 3.5 响应者样式的神秘性 94
- 3.6 秘诀：属性栈 96

- 3.7 秘诀：使用伪 HTML 创建属性化文本 101
- 3.8 利用 Core Text 绘图 105
- 3.9 创建图像图案 108
- 3.10 秘诀：在滚动视图上绘制 Core Text 110
- 3.11 秘诀：探讨字体 112
- 3.12 向应用程序中添加自定义的字体 114
- 3.13 秘诀：把 Core Text 进行分页处理 114
- 3.14 秘诀：把属性化文本绘制到 PDF 中 116
- 3.15 秘诀：大电话文本 117
- 3.16 小结 121

第 4 章 几何学 123
- 4.1 秘诀：获取贝塞尔路径中的点 ... 123
- 4.2 稀释点 125
- 4.3 秘诀：平滑绘图 128
- 4.4 秘诀：基于速度的笔画 130
- 4.5 秘诀：限制贝塞尔路径 132
- 4.6 秘诀：放入路径 137
- 4.7 处理曲线 139
- 4.8 秘诀：沿着贝塞尔路径移动项目 ... 143
- 4.9 秘诀：沿着贝塞尔路径绘制属性化文本 145
- 4.10 秘诀：视图变换 148
  - 4.10.1 基本变换 149
  - 4.10.2 揭密 149
  - 4.10.3 获取变换值 150
  - 4.10.4 设置变换值 151
  - 4.10.5 获取视图点的位置 152

4.11 秘诀：测试视图相交 154
4.12 小结 159

第 5 章 联网 161
5.1 秘诀：安全凭证存储 161
5.2 秘诀：输入凭证 165
5.3 秘诀：处理身份验证质询 170
5.4 秘诀：上传数据 172
5.5 秘诀：构建简单的 Web 服务器 176
5.6 秘诀：OAuth 实用程序 180
5.7 秘诀：OAuth 过程 184
    5.7.1 第 1 步：从 API 提供商请求令牌 184
    5.7.2 第 2 步：获取和存储令牌 184
    5.7.3 第 3 步：请求用户访问 185
    5.7.4 第 4 步：获取一个 OAuth 验证者令牌 185
    5.7.5 第 5 步：利用访问令牌进行身份验证 187
5.8 小结 191

第 6 章 图像 193
6.1 图像源 193
6.2 读取图像数据 195
    6.2.1 UIImage 类的便捷方法 195
    6.2.2 查找沙盒中的图像 197
    6.2.3 从 URL 加载图像 198
    6.2.4 从资产库中读取数据 198
6.3 秘诀：放入和填充图像 199
6.4 秘诀：旋转图像 203
6.5 秘诀：处理位图表示 206
    6.5.1 坐标系统之间的转换 206
    6.5.2 查找数据 207
    6.5.3 在图像数据与位图数据之间转换 208
6.6 秘诀：基本的图像处理 210
6.7 秘诀：图像卷积 212
6.8 秘诀：基本的 Core Image 处理 215
6.9 抓取基于视图的截屏图 217
6.10 绘制到 PDF 文件中 218
6.11 秘诀：倒影 219
6.12 秘诀：发射器 222
6.13 小结 224

第 7 章 照相机 225
7.1 秘诀：拍摄照片 225
    7.1.1 设置选择器 225
    7.1.2 显示图像 226
    7.1.3 把图像保存到相册 227
7.2 秘诀：启用闪光灯 229
7.3 秘诀：访问 AVFoundation 照相机 231
    7.3.1 需要照相机 231
    7.3.2 查询和获取照相机 232
    7.3.3 建立照相机会话 233
    7.3.4 切换照相机 235
    7.3.5 照相机预览 236
    7.3.6 布置照相机预览 236
    7.3.7 照相机图像助手 237
7.4 秘诀：EXIF 238
    7.4.1 ImageIO 238
    7.4.2 查询元数据 239
    7.4.3 包装 UIImage 240
7.5 图像方位 243
7.6 秘诀：Core Image 过滤 245
7.7 秘诀：Core Image 人脸检测 247
7.8 秘诀：对实时馈送进行抽样 253

7.9 小结 257

第 8 章 音频 259
8.1 秘诀：利用 AVAudioPlayer 播放音频 259
    8.1.1 初始化音频播放器 259
    8.1.2 监测音频级别 260
    8.1.3 播放进度和擦除 261
    8.1.4 捕获播放的结束 262
8.2 秘诀：循环音频 267
8.3 秘诀：处理音频中断 270
8.4 秘诀：录制音频 273
8.5 秘诀：利用 Audio Queues 录制音频 ... 279
8.6 秘诀：利用 MPMediaPickerController 选择音频 286
8.7 创建媒体查询 288
    8.7.1 构建查询 288
    8.7.2 使用谓词 289
8.8 秘诀：使用 MPMusicPlayerController 290
8.9 小结 294

第 9 章 连接到 Address Book 295
9.1 AddressBook 框架 295
    9.1.1 AddressBookUI 295
    9.1.2 AddressBook 及其数据库 ... 296
    9.1.3 记录 296
    9.1.4 自定义的 ABStandin 类 297
    9.1.5 查询 Address Book 301
    9.1.6 包装 AddressBook 框架 301
    9.1.7 使用记录函数 302
    9.1.8 获取和设置字符串 302
    9.1.9 处理日期属性 304
    9.1.10 多值记录属性 305
    9.1.11 存储多值数据 309
    9.1.12 处理多值项目 310
    9.1.13 地址、社会概况和即时消息属性 311
    9.1.14 Address Book 中的图像 313
    9.1.15 创建、添加和删除记录 314
    9.1.16 搜索联系人 316
    9.1.17 对联系人排序 317
    9.1.18 处理群组 317
    9.1.19 ABContact、ABGroup 和 ABContactsHelper 320
9.2 秘诀：搜索 Address Book 321
9.3 秘诀：访问联系人图像数据 323
9.4 秘诀：选择人员 325
9.5 秘诀：限制联系人选择器属性 327
9.6 秘诀：添加和删除联系人 329
9.7 修改和查看单独的联系人 332
    9.7.1 用户编辑 332
    9.7.2 委托方法 333
9.8 秘诀："未知的"人员控制器 333
9.9 小结 336

第 10 章 位置 337
10.1 授权 Core Location 337
    10.1.1 测试位置服务 337
    10.1.2 重置位置和隐私 338
    10.1.3 检查用户权限 339
    10.1.4 测试 Core Location 特性 ... 340
10.2 秘诀：Core Location 简介 341
    10.2.1 位置属性 343
    10.2.2 跟踪速度 344
10.3 秘诀：地理围栏 344

| 10.4 | 秘诀：使用行进方向值跟踪"North" 347
| 10.5 | 秘诀：前向和反向地理编码 349
| 10.6 | 秘诀：查看位置 352
| 10.7 | 秘诀：用户位置注释 356
| 10.8 | 创建地图注释 359
  10.8.1 创建、添加和删除注释 360
  10.8.2 注释视图 360
  10.8.3 自定义注释视图 361
  10.8.4 响应注释按钮的点按动作 362
| 10.9 | 小结 365

第 11 章 GameKit 367

| 11.1 | 启用 Game Center 367
| 11.2 | 秘诀：登录到 Game Center 369
| 11.3 | 设计排行榜和成绩 370
  11.3.1 构建排行榜 370
  11.3.2 创建成绩 372
| 11.4 | 秘诀：访问排行榜 373
| 11.5 | 秘诀：显示 Game Center 视图控制器 375
| 11.6 | 秘诀：提交分数 377
| 11.7 | 秘诀：检查成绩 378
| 11.8 | 秘诀：把成绩报告给 Game Center 379
| 11.9 | 秘诀：多玩家配对安排 381
| 11.10 | 秘诀：响应配对安排者 382
| 11.11 | 秘诀：创建邀请处理程序 384
| 11.12 | 管理配对状态 385
| 11.13 | 秘诀：处理玩家状态改变 386
| 11.14 | 秘诀：获取玩家名字 387
  11.14.1 比较玩家 388
  11.14.2 获取本地玩家 388

| 11.15 | 游戏玩法 389
| 11.16 | 序列化数据 389
| 11.17 | 秘诀：同步数据 393
| 11.18 | 秘诀：逐回合地安排配对 395
| 11.19 | 秘诀：响应基于回合的邀请 397
| 11.20 | 秘诀：加载配对 398
| 11.21 | 秘诀：响应玩法 400
| 11.22 | 秘诀：结束游戏玩法 404
| 11.23 | 删除配对 407
| 11.24 | 秘诀：Game Center 语音 409
  11.24.1 测试聊天可用性 409
  11.24.2 建立播放和录制音频会话 409
  11.24.3 创建语音聊天 411
  11.24.4 开始和停止聊天 411
  11.24.5 聊天状态监测 411
  11.24.6 实现聊天按钮 412
  11.24.7 控制音量 413
| 11.25 | GameKit 对等服务 413
  11.25.1 GameKit Bluetooth 的局限性 414
  11.25.2 Bonjour 会话 414
  11.25.3 服务器、客户和对等方 415
  11.25.4 对等连接过程 415
  11.25.5 发送和接收数据 418
  11.25.6 状态改变 419
  11.25.7 创建 GameKit 助手 420
  11.25.8 对等语音聊天 420
  11.25.9 实现语音聊天 420
  11.25.10 创建"联机" GameKit 连接 421
| 11.26 | 小结 423

第 12 章 StoreKit 425

12.1 初识 StoreKit 425
　12.1.1 履约 425
　12.1.2 StoreKit 开发悖论 426
　12.1.3 开发和测试 427
　12.1.4 提交 427
12.2 创建测试账户 427
12.3 创建新的应用程序中的购买项目..... 428
　12.3.1 填写细节区域 429
　12.3.2 添加本地化描述 430
　12.3.3 填写定价区域 431
　12.3.4 提供购买 GUI 截屏图 431
　12.3.5 提交应用程序中的购买产品以
　　　　 进行评审 432
12.4 构建店面 GUI 432
12.5 购买项目 434
　12.5.1 签出 iTunes 账户以进行测试
　　　　 435
　12.5.2 在购买后重新获得程序控制
　　　　 435
　12.5.3 注册购买 438
　12.5.4 恢复购买 438
　12.5.5 购买多个项目 439
　12.5.6 处理注册购买中的延迟 439
12.6 验证收据 439
12.7 小结 442

第 13 章 推送通知 443
13.1 推送通知简介 443
　13.1.1 推送的工作原理 444
　13.1.2 多个提供者支持 444
　13.1.3 安全 445
　13.1.4 推送限制 445
　13.1.5 推送通知与本地通知 446
13.2 配置推送 446
　13.2.1 生成新的应用程序标识符 446
　13.2.2 生成 SSL 证书 447
　13.2.3 特定于推送的配置 448
　13.2.4 创建推送兼容的应用程序 449
13.3 注册应用程序 449
　13.3.1 获取设备令牌 449
　13.3.2 处理令牌请求错误 450
　13.3.3 响应通知 450
13.4 秘诀：推送客户骨架 452
13.5 构建通知有效载荷 457
　13.5.1 本地化的提醒 458
　13.5.2 从字典转换为 JSON 458
　13.5.3 自定义的数据 458
　13.5.4 在启动时接收数据 458
13.6 秘诀：发送通知 459
　13.6.1 沙盒和生产 460
13.7 反馈服务 464
13.8 设计推送 465
13.9 小结 466

第 1 章
特定于设备的开发

每个 iOS 设备都代表独特、共享、短暂和持久属性的融合。这些属性包括设备的当前物理方位、它的型号名称、它的电池状态以及它对机载硬件的访问。本章从设备的构建配置到它的活动机载传感器进行了广泛探讨，并且提供了一些秘诀，用于返回正在使用的单元的各类信息项。你将学到在运行时测试硬件的前提条件，并在应用程序的 Info.plist 文件中指定那些前提条件。你将发现如何通过 Core Motion 请求传感器反馈，以及订阅通知，在传感器状态改变时创建回调。你将学到添加屏幕镜像和第二屏输出，以及请求特定于设备的详细信息以便进行跟踪。本章将介绍 iPhone 设备上可用的硬件、文件系统和传感器，并且帮助你以编程方式利用那些特性。

1.1 访问基本的设备信息

UIDevice 类展示了一些关键的特定于设备的属性，包括使用的 iPhone、iPad 或 iPod Touch 型号、设备名称，以及 OS 名称和版本。它是一种一站式解决方案，用于提取出某些系统详细信息。每个方法都是一个实例方法，它们是使用 UIDevice 单例通过[UIDevice currentDevice]调用的。

可以通过 UIDevice 获取的系统信息包括下面这些项。

- **systemName**：它用于返回当前使用的操作系统的名称。对于目前这一代 iOS 设备，在平台上只运行一种 OS：iPhone OS。Apple 还没有更新这个名称，以匹配一般性的 iOS 品牌重塑举动。
- **systemVersion**：这个值将列出单元上目前安装的固件版本，例如，4.3、5.1.1、6.0 等。
- **model**：iPhone 型号返回一个描述其平台的字符串，即 iPhone、iPad 和 iPod Touch。如果将 iOS 扩展到新设备上，将使用额外的字符串描述那些型号。localizedModel 提供了该属性的本地化版本。
- **userInterfaceIdiom**：这个属性表示当前设备上使用的界面风格，即 iPhone（用于 iPhone 和 iPod Touch）或 iPad。当 Apple 提供另外的平台风格时，可能会引入其他的用语。

- **name**：这个字符串表示由 iTunes 中的用户指定的 iPhone 名称，比如 "Joe's iPhone" 或 "Binky"。这个名称也用于创建设备的本地主机名。

下面给出了几个使用这些属性的示例：

```
UIDevice *device = [UIDevice currentDevice];
NSLog(@"System name: %@", device.systemName);
NSLog(@"Model: %@", device.model);
NSLog(@"Name: %@", device.name);
```

对于当前的 iOS 版本，可以利用一个简单的布尔测试进行风格检查。下面给出了一个示例，说明如何实现 iPad 检查。它用于测试选择器一致性，如果可能，将会返回[UIDevice currentDevice].userInterfaceIdiom，否则，将返回 UIUserInterfaceIdiomPhone：

```
#define IS_IPAD (UI_USER_INTERFACE_IDIOM() == UIUserInterfaceIdiomPad)
```

万一这个测试失败，目前可以假定使用的是 iPhone/iPod Touch。当 Apple 发布新的设备家族时，将需要根据更细致的测试更新代码。

1.2 添加设备能力限制

应用程序的 Info.plist 属性列表使你能够在向 iTunes 提交应用程序时指定应用程序的要求。这些限制允许告诉 iTunes 应用程序需要哪些设备特性。

每个 iOS 单元都会提供一个独特的特性集。一些设备会提供照相机和 GPS 能力，另外一些则不会。一些设备具有机载陀螺仪、自动聚焦，以及其他强大的选项。你可以指定在设备上运行应用程序时需要哪些特性。

在 Info.plist 文件中包括 UIRequiredDeviceCapabilities 键时，iTunes 将限制把应用程序安装到提供必需能力的设备。把这个列表作为一个字符串数组或者字典提供。

数组指定每个必需的能力；该数组中的每一项都必须存在于设备上。字典允许显式要求或禁止某个特性，字典键就是能力，字典值用于设置特性是必须存在（布尔值 true）还是必须省略（布尔值 false）。

表 1-1 中详细说明了当前的键。其中只包括应用程序绝对需要或者不能支持的那些特性。如果应用程序可以提供解决办法，就不要以这种方式添加限制。表 1-1 讨论了每个特性。当使用禁令而不是需求时，意义就颠倒了，例如，不能机载自动聚焦照相机或陀螺仪，或者不支持游戏中心（Game Center）访问。

表 1-1　必需的设备能力

键	使用
telephony	应用程序需要 Phone 应用程序或者使用 tel://URL
wifi	应用程序需要基于本地 802.11 的网络访问。如果在应用程序运行时 iOS 必须维持该 Wi-Fi 连接，可以添加 UIRequiresPersistentWiFi 作为顶级属性列表键
sms	应用程序需要 Messages 应用程序或者使用 sms://URL
still-camera	应用程序需要机载静物照相机，并且可以使用图像拾取器界面从该静物照相机捕获照片
auto-focus-camera	应用程序需要额外的聚焦能力以进行微距摄影，或者拍摄特别清晰的图像以进行图像内的数据检测
front-facing-camera	应用程序需要在设备上前置摄像头
camera-flash	应用程序需要闪光灯特性
video-camera	应用程序需要能够录制视频的照相机
accelerometer	应用程序需要特定于加速计的反馈，而不止是简单的 UIViewController 定向事件
gyroscope	应用程序需要设备上的机载陀螺仪
location-services	应用程序需要使用任意类型的 Core Location
gps	应用程序需要使用 Core Location，并且需要更为精确的 GPS 定位
magnetometer	应用程序需要使用 Core Location，并且需要与前进方向相关的事件，即行进的方向（磁力计是内置的罗盘）
gamekit	应用程序需要访问游戏中心（Game Center）(iOS 4.1 及更高版本)
microphone	应用程序需要使用内置的麦克风或者可以提供麦克风的（被认可的）附件
opengles-1	应用程序需要 OpenGL ES 1.1
opengles-2	应用程序需要 OpenGL ES 2.0
armv6	应用程序仅针对 armv6 指令集（3.1 或更高版本）进行编译
armv7	应用程序仅针对 armv7 指令集（3.1 或更高版本）进行编译
peer-peer	应用程序通过蓝牙技术（3.1 或更高版本）使用 GameKit 对等连接
bluetooth-le	应用程序需要蓝牙技术的低功耗支持（5.0 及更高版本）

例如，考虑一个应用程序，当在备有照相机的设备上运行时，它将提供一个选项用于拍摄图片。如果应用程序是在前置摄像头的 iPod Touch 单元上工作，就不要包括进静物照相机限制。可代之以从应用程序内检查照相机兼容性，并在合适时展示照相机选项。添加静物照相机限制将从潜在的顾客池中排除掉许多早期的 iPod Touch（第 1~3 代）和 iPad（第 1 代）所有者。

1.2.1　用户权限描述

为了保护隐私，最终用户必须明确地允许应用程序访问提醒信号、照片、位置、联系人和日历数据。为了说服用户接受，它有助于解释应用程序可以怎样使用这类数据，并且说明访问它的原因。给位于 Info.plist 文件顶层的以下键分配字符串值。当 iOS 提示用户有关特定于资源的权限时，它将显示这些字符串，作为它的标准对话框的一部分：

- **NSRemindersUsageDescription**
- **NSPhotoLibraryUsageDescription**
- **NSLocationUsageDescription**
- **NSContactsUsageDescription**
- **NSCalendarsUsageDescription**

1.2.2 其他常用的 Info.plist 键

下面给出了你可能想在属性列表中分配的另外几个常用键，以及有关它们可以做什么的描述。

- **UIFileSharingEnabled**（Boolean 型，默认为关）：允许用户从 iTunes 中访问应用程序的 Documents 文件夹的内容。这个文件夹出现在应用程序沙盒的顶级。
- **UIAppFonts**（Array 型，字体名称（包括其扩展）的字符串）：指定在软件包中提供的自定义 TTF 字体。在添加字体时，可以使用标准的 UIFont 调用访问它们。
- **UIApplicationExitsOnSuspend**（Boolean 型，默认为关）：当用户单击 Home 按钮时使应用程序能够终止，而不是转移到后台。当启用这个键时，iOS 将会终止应用程序，并从内存中清除它。
- **UIRequiresPersistentWifi**（Boolean 型，默认为关）：指示 iOS 在应用程序活动时维持一条 Wi-Fi 连接。
- **UIStatusBarHidden**（Boolean 型，默认为关）：如果启用这个键，则会在应用程序启动时隐藏状态栏。
- **UIStatusBarStyle**（String 型，默认为 UIStatusBarStyleDefault）：指定应用程序启动时的状态栏的风格。

1.3 秘诀：检查设备接近度和电池状态

UIDevice 类提供了一些 API，使你能够跟踪设备的特征，包括电池的状态和接近度传感器。秘诀 1-1 演示了如何启用和查询对这两种技术的监测。它们二者都以通知的形式提供更新，可以订阅它们，以便在有重要的更新时通知你的应用程序。

1.3.1 启用和禁用接近度传感器

接近度在此时是一个特定于 iPhone 的特性。iPod Touch 和 iPad 没有提供接近度传感器。除非具有相对身体部位握持 iPhone 的某个迫切的理由（或者反之亦然），否则使用接近度传感器获益甚少。

当启用接近度传感器时，它具有一项主要的任务。它会检测正前方是否有较大的物体。如果是，它将会关闭屏幕，并发送一个普通的通知。把阻挡的物体移开，将会再次打开屏幕。在你打电话时，这可以防止耳朵接触屏幕导致按键或者拨号。一些设计不佳的保护性外壳将会阻止 iPhone 的接近度传感器正确地工作。

Siri 使用了这个特性，当把手机抬高到耳朵附近时，它会记录你的询问，发送它以进行解释。Siri 的语音接口在工作时不依赖于可视化的 GUI。

秘诀 1-1 还演示了在 iPhone 上如何处理接近度传感。它的代码使用 UIDevice 类切换接近度监测，并且订阅 UIDeviceProximityStateDidChangeNotification 以捕获状态改变。两种状态是开和关。当 UIDevice proximityState 属性返回 YES 时，就激活了接近度传感器。

1.3.2 监测电池状态

可以以编程方式跟踪电池和设备状态。这些 API 使你能够知道电池充电的程度，以及设备是否插入到了充电电源中。电池电量是一个范围在 1.0（完全充电）~0.0（完全放电）之间的浮点值。它提供了一个近似的放电水平，在执行将给设备施加罕见重负的操作之前，可以查询它。

例如，在用户执行一系列大型的数学计算之前，你可能想提醒它，并且建议插入电源。可以通过下面这个 UIDevice 调用获取电池电量，返回的值是以 5%的增量产生的：

```
NSLog(@"Battery level: %0.2f%%",
    [UIDevice currentDevice].batteryLevel * 100);
```

充电状态具有 4 个可能的值：正在充电（即连接到电源）、充满、拔掉电源插头和笼统的"未知状态"。可以使用 UIDevice batteryState 属性取回这些状态：

```
NSArray *stateArray = @[
    @"Battery state is unknown",
    @"Battery is not plugged into a charging source",
    @"Battery is charging",
    @"Battery state is full"];

NSLog(@"Battery state: %@",
    stateArray[[UIDevice currentDevice].batteryState]);
```

不要把这些选择视作持久的状态。可代之以把它们视作对设备上实际发生的事情的短暂反应。它们不是标志，不能用"或"把它们连接起来构成一般性的电池描述。相反，这些值反映了最近的状态改变。

可以通过响应电池状态改变的通知，轻松地监测状态改变。这样，就可以捕获瞬时事件，比如当电池最终充满电时，当用户插入电源充电时，以及当用户断开与电源的连接时。

要开始监测，可以把 batteryMonitoringEnabled 属性设置为 YES。在监测期间，当电池状态或电量改变时，UIDevice 类将产生通知。秘诀 1-1 订阅了这两种通知。请注意：也可以直接检查这些值，而不必等待通知。Apple 对于电量改变更新的频率没有提供任何保证，但是可以通过测试这个秘诀来断定，它们是以相当规则的方式发生的。

秘诀 1-1　监测接近度和电池

```objc
// View the current battery level and state
- (void) peekAtBatteryState
{
    NSArray *stateArray = [NSArray arrayWithObjects:
                            @"Battery state is unknown",
                            @"Battery is not plugged into a charging source",
                            @"Battery is charging",
                            @"Battery state is full", nil];
    
    NSString *status = [NSString stringWithFormat:
        @"Battery state: %@, Battery level: %0.2f%%",
        [stateArray objectAtIndex:[UIDevice currentDevice].batteryState],
        [UIDevice currentDevice].batteryLevel * 100];
    
    NSLog(@"%@", status);
}
// Show whether proximity is being monitored
- (void) updateTitle
{
    self.title = [NSString stringWithFormat:@"Proximity %@",
        [UIDevice currentDevice].proximityMonitoringEnabled ? @"On" : @"Off"];
}
// Toggle proximity monitoring off and on
- (void) toggle: (id) sender
{
    // Determine the current proximity monitoring and toggle it
    BOOL isEnabled = [UIDevice currentDevice].proximityMonitoringEnabled;
```

```objc
        [UIDevice currentDevice].proximityMonitoringEnabled = !isEnabled;
        [self updateTitle];
}
- (void) loadView
{
        [super loadView];
        // Enable toggling and initialize title
        self.navigationItem.rightBarButtonItem =
            BARBUTTON(@"Toggle", @selector(toggle:));
        [self updateTitle];
        // Add proximity state checker
        [[NSNotificationCenter defaultCenter]
            addObserverForName:UIDeviceProximityStateDidChangeNotification
            object:nil queue:[NSOperationQueue mainQueue]
            usingBlock:^(NSNotification *notification) {
                // Sensor has triggered either on or off
                NSLog(@"The proximity sensor %@",
                    [UIDevice currentDevice].proximityState ?
                    @"will now blank the screen" : @"will now restore the screen");
        }];
        // Enable battery monitoring
        [[UIDevice currentDevice] setBatteryMonitoringEnabled:YES];
        // Add observers for battery state and level changes
        [[NSNotificationCenter defaultCenter]
            addObserverForName:UIDeviceBatteryStateDidChangeNotification
            object:nil queue:[NSOperationQueue mainQueue]
            usingBlock:^(NSNotification *notification) {
                // State has changed
                NSLog(@"Battery State Change");
                [self peekAtBatteryState];
        }];
        [[NSNotificationCenter defaultCenter]
            addObserverForName:UIDeviceBatteryLevelDidChangeNotification
            object:nil queue:[NSOperationQueue mainQueue]
            usingBlock:^(NSNotification *notification) {
                // Level has changed
                NSLog(@"Battery Level Change");
                [self peekAtBatteryState];
        }];
}
```

> **获取这个秘诀的代码**
> 要查找这个秘诀的完整示例项目，可以浏览 https://github.com/erica/iOS-6-Advanced-Cookbook，并进入第 1 章的文件夹。

1.3.3 检测 Retina 支持

近年来，Apple 在其旗舰设备上引入了 Retina 显示屏。根据 Apple 的说法，它的像素密度非常高，足以使人眼无法区分单独的像素。带有更高分辨率的艺术的应用程序可以利用这种改进的显示质量。

UIScreen 类提供了一种容易的方式，用于检查当前设备是否提供了内置的 Retina 显示屏。检查屏幕的 scale 属性，它提供了从逻辑坐标空间（磅，大约是 1/160 英寸）转换为设备坐标空间（像素）的转换因子。对于标准显示屏，转换因子是 1.0，因此 1 点对应为 1 像素。对于 Retina 显示屏，它是 2.0（4 像素/磅）：

```
- (BOOL) hasRetinaDisplay
{
    return ([UIScreen mainScreen].scale == 2.0f);
}
```

UIScreen 类还提供了两个有用的显示屏尺寸属性。bounds 返回屏幕的边界矩形，以磅为单位。无论屏幕上有任何元素（比如状态栏、导航栏或标签栏），这都会提供屏幕的完全尺寸。

applicationFrame 属性（同样以磅为单位）把状态栏排除在外，提供了应用程序的初始窗口尺寸的框架。

1.4 秘诀：取回额外的设备信息

sysctl()和 sysctlbyname()允许获取系统信息。这些标准的 UNIX 函数用于询问操作系统有关硬件和 OS 的详细信息。看一眼 Macintosh 上的/usr/include/sys/sysctl.h 包括文件，就能对所提供的范围类型有一个感觉。在那里，能够找到一份可以用作这些函数的参数常量的详尽列表。

这些常量使你能够检查核心信息，比如系统的 CPU 频率、可用的内存量等。秘诀 1-2 演示了这种功能。它引入了一个 UIDevice 类，用于收集系统信息，并通过一系列方法调用返回它。

你可能想知道：当标准的 UIDevice 类可以根据需要返回设备型号时，为什么这个类还要包括一个平台方法。答案在于区分不同的单元类型。

iPhone 3GS 的型号只是"iPhone"，iPhone 4S 也是一样。与之相反，这个秘诀为 3GS 返回的平台值是"iPhone2,1"，为 iPhone 4S 返回的是"iPhone 4,1"。这允许以编程方式把 3GS 单元与第一代 iPhone（"iPhone1,1"）或 iPhone 3G（"iPhone1,2"）区分开。

每种型号都提供了独特的内置能力。准确知道你正在处理哪款 iPhone 有助于确定那个单元是否有可能支持诸如可访问性、GPS 和磁力计之类的特性。

秘诀 1-2　扩展设备信息收集

```
@implementation UIDevice (Hardware)
+ (NSString *) getSysInfoByName:(char *)typeSpecifier
{
    // Recover sysctl information by name
    size_t size;
    sysctlbyname(typeSpecifier, NULL, &size, NULL, 0);
    
    char *answer = malloc(size);
    sysctlbyname(typeSpecifier, answer, &size, NULL, 0);
    
    NSString *results = [NSString stringWithCString:answer
        encoding: NSUTF8StringEncoding];
    free(answer);
    
    return results;
}
- (NSString *) platform
{
    return [UIDevice getSysInfoByName:"hw.machine"];
}
- (NSUInteger) getSysInfo: (uint) typeSpecifier
{
    size_t size = sizeof(int);
    int results;
    int mib[2] = {CTL_HW, typeSpecifier};
    sysctl(mib, 2, &results, &size, NULL, 0);
    return (NSUInteger) results;
}


- (NSUInteger) cpuFrequency
{
    return [UIDevice getSysInfo:HW_CPU_FREQ];
}

- (NSUInteger) busFrequency
```

```
    {
        return [UIDevice getSysInfo:HW_BUS_FREQ];
    }

    - (NSUInteger) totalMemory
    {
        return [UIDevice getSysInfo:HW_PHYSMEM];
    }

    - (NSUInteger) userMemory
    {
        return [UIDevice getSysInfo:HW_USERMEM];
    }

    - (NSUInteger) maxSocketBufferSize
    {
        return [UIDevice getSysInfo:KIPC_MAXSOCKBUF];
    }
    @end
```

> **获取这个秘诀的代码**
> 要查找这个秘诀的完整示例项目，可以浏览 https://github.com/erica/iOS-6-Advanced-Cookbook，并进入第 1 章的文件夹。

1.5 秘诀：使用加速能力"向上"定位

iPhone 提供了 3 个机载传感器，用于沿着 iPhone 的 3 根相互垂直的轴（左/右（x 轴）、上/下（y 轴）和前/后（z 轴））度量加速能力。这些值指示作用于 iPhone 的力，它们来自重力和用户移动。可以通过在脑袋周围晃动 iPhone（向心力）或者把它从高楼上投下（自由落体）来获得某种净力反馈。当然，如果不幸摔坏了，它也许不能取回这类数据。

要向 iPhone 加速计更新订阅某个对象，可把它设置委托。设置为委托的对象必须实现 UIAccelerometerDelegate 协议：

```
[[UIAccelerometer sharedAccelerometer] setDelegate:self]
```

在指定时，委托将会接收 accelerometer:didAccelerate:回调消息，可以跟踪并对其做出响应。发送给委托方法的 UIAcceleration 结构包含 x 轴、y 轴和 z 轴的浮点值，每个值的范围为 −1.0~1.0：

```
float x = acceleration.x;
```

```
float y = acceleration.y;
float z = acceleration.z;
```

秘诀 1-3 使用这些值来帮助确定"向上"的方向。它会计算 x 和 y 加速度向量之间的反正切值，返回垂直向上的偏移角度。当接收到新的加速消息时，秘诀将会利用其箭头图片（在图 1-1 中可以看到它）旋转 UIImageView 实例，以指向上方。对用户动作的实时响应确保箭头会继续指向上方，而无论用户怎样改变手机的方向。

秘诀 1–3　捕获加速事件

```
- (void)accelerometer:(UIAccelerometer *)accelerometer
    didAccelerate:(UIAcceleration *)acceleration
{
    // Determine up from the x and y acceleration components
    float xx = -acceleration.x;
    float yy = acceleration.y;
    float angle = atan2(yy, xx);
    [arrow setTransform:
        CGAffineTransformMakeRotation(angle)];
}

- (void) viewDidLoad
{
    // Initialize the delegate to start catching accelerometer events
    [UIAccelerometer sharedAccelerometer].delegate = self;
}
```

图 1-1　使用 x 和 y 方向的力向量，通过执行一个反正切函数，利用一点数学计算即可恢复"向上"的方向。在这个示例中，无论用户怎样改变 iPhone 的方向，箭头总会指向上方

> **获取这个秘诀的代码**
>
> 要查找这个秘诀的完整示例项目，可以浏览 https://github.com/erica/iOS-6-Advanced-Cookbook，并进入第 1 章的文件夹。

1.6 处理基本的方向

UIDevice 类使用内置的 orientation 属性获取设备的物理方向。iOS 设备支持这个属性的 7 个可能的值。

- **UIDeviceOrientationUnknown**：方向目前未知。
- **UIDeviceOrientationPortrait**：Home 键按下。
- **UIDeviceOrientationPortraitUpsideDown**：Home 键升起。
- **UIDeviceOrientationLandscapeLeft**：Home 键在左边。
- **UIDeviceOrientationLandscapeRight**：Home 键在右边。
- **UIDeviceOrientationFaceUp**：设备正面朝上。
- **UIDeviceOrientationFaceDown**：设备正面朝下。

设备在典型的应用程序会话期间可能经历其中任何或所有的方向。尽管创建的方向要与机载加速计保持一致，但是不会以任何方式把这些方向绑定到内置的角度值。

iOS 提供了两个内置宏，帮助确定设备方向枚举值是纵向还是横向的：即 UIDeviceOrientationIsPortrait()和 UIDeviceOrientationIsLandscape()。能够很方便地扩展 UIDevice 类，提供这些测试作为内置的设备属性：

```
@property (nonatomic, readonly) BOOL isLandscape;
@property (nonatomic, readonly) BOOL isPortrait;

- (BOOL) isLandscape
{
    return UIDeviceOrientationIsLandscape(self.orientation);
}

- (BOOL) isPortrait
{
    return UIDeviceOrientationIsPortrait(self.orientation);
}
```

你的代码可以直接订阅设备重定向通知。为此，可以把 beginGeneratingDeviceOrientationNotifications 发送给 currentDevice 单例。然后添加一个观察者，捕获随后发生的 UIDevice-

OrientationDidChangeNotification 更新。如你所期望的，可以通过调用 UIDeviceOrientationDid-ChangeNotificationOrientationNotification 来完成侦听。

1.7 同步获取当前的加速计角度

有时，你可能想在不用把自己设定为完全委托的情况下查询加速计。下面的方法打算在 UIDevice 类别内使用，允许与 x/y 平面（iOS 设备的正面）一起同步返回当前的设备角度。为此，可输入一个新的运行循环，等待加速计事件，从那个回调中获取当前的角度，然后让运行循环返回那个角度：

```
- (void)accelerometer:(UIAccelerometer *)accelerometer
    didAccelerate:(UIAcceleration *)acceleration
{
    float xx = acceleration.x;
    float yy = -acceleration.y;
    device_angle = M_PI / 2.0f - atan2(yy, xx);
    if (device_angle > M_PI)
        device_angle -= 2 * M_PI;
    CFRunLoopStop(CFRunLoopGetCurrent());
}

- (float) orientationAngle
{
    // Supercede current delegate
    id priorDelegate = [UIAccelerometer sharedAccelerometer].delegate;
    [UIAccelerometer sharedAccelerometer].delegate = self;

    // Wait for a reading
    CFRunLoopRun();

    // Restore delegate
    [UIAccelerometer sharedAccelerometer].delegate = priorDelegate;

    return device_angle;
}
```

这不是用于持续轮询的方法，可为此直接使用回调。但是，对于偶尔的角度查询，这些方法提供了对当前屏幕角度的简单、直接的访问。

1.7.1 通过加速计计算方向

在第一次启动应用程序时，UIDevice 类不会报告正确的方向。仅当设备移到一个新位置或者 UIViewController 方法起作用之后，它才会更新方向。

直到用户把设备移开然后再移回正确的方向之后，才可能会把纵向启动的应用程序视作"纵向"的。在模拟器和 iPhone 设备上存在这种状况，并且很容易测试（由于对特性进行了更新，使之像设计的那样工作，因此关闭了针对这个问题的无线电探测器）。

对于解决办法，可以考虑像刚才所示的那样直接恢复角度方向。然后，在确定了设备的角度之后，从基于加速计的角度转换为设备方向。下面用代码说明了它的工作方式：

```
// Limited to the four portrait/landscape options
- (UIDeviceOrientation) acceleratorBasedOrientation
{
    CGFloat baseAngle = self.orientationAngle;
    if ((baseAngle > -M_PI_4) && (baseAngle < M_PI_4))
        return UIDeviceOrientationPortrait;
    if ((baseAngle < -M_PI_4) && (baseAngle > -3 * M_PI_4))
        return UIDeviceOrientationLandscapeLeft;
    if ((baseAngle > M_PI_4) && (baseAngle < 3 * M_PI_4))
        return UIDeviceOrientationLandscapeRight;
    return UIDeviceOrientationPortraitUpsideDown;
}
```

要知道的是，这个示例只考虑了 *x-y* 平面，而大多数用户界面决策都需要在这里做出。这个代码段完全忽略了 *z* 轴，这意味着最终将会得到模糊的随机结果：即正面朝上和正面朝下的方向。可以修改这段代码，以根据需要提供这种细微差别。

UIViewController 类的 interfaceOrientation 实例方法报告了视图控制器的界面的方向。尽管这不能代替加速计读数，但是许多界面布局问题都基于底层的视图方向，而不是设备特征。

还要知道的是，尤其是在 iPad 上，子视图控制器使用的布局方向可能不同于设备方向。例如，嵌入式控制器可能在一个横向划分的视图控制器内展示一种纵向布局。即便如此，也可以考虑底层的界面方向是否可以满足方向检测代码。它可能比设备方向更可靠，尤其是在应用程序启动时。

1.7.2 计算相对角度

屏幕重定向支持意味着必须在象限中支持界面对于给定设备角度的关系，其中每个象限都用于一个可能的面向前方的屏幕方向。由于 UIViewController 可以自动旋转其屏幕上的视图，需要了解一些数学知识以解释那些重定向。

下面的方法（编写它是为了在 UIDevice 类别中使用）可以计算角度，以使得角度与设备方向保持同步。这将在垂直方向上产生简单的偏移，以匹配当前展示 GUI 的方式：

```
- (float) orientationAngleRelativeToOrientation:
    (UIDeviceOrientation) someOrientation
{
    float dOrientation = 0.0f;
    switch (someOrientation)
    {
        case UIDeviceOrientationPortraitUpsideDown:
            {dOrientation = M_PI; break;}
        case UIDeviceOrientationLandscapeLeft:
            {dOrientation = -(M_PI/2.0f); break;}
        case UIDeviceOrientationLandscapeRight:
            {dOrientation = (M_PI/2.0f); break;}
        default: break;
    }
    float adjustedAngle =
        fmod(self.orientationAngle - dOrientation, 2.0f * M_PI);
    if (adjustedAngle > (M_PI + 0.01f))
        adjustedAngle = (adjustedAngle - 2.0f * M_PI);
    return adjustedAngle;
}
```

这个方法使用一个浮点模数获取实际的屏幕角度与界面方向角度偏移量之间的差值，返回非常重要的垂直角度偏移量。

> **注意：**
> 在 iOS 6 中，可以使用 Info.plist 代替 shouldAutorotateToInterfaceOrientation:，以允许和禁止方向改变。

1.8 使用加速度移动屏幕上的对象

借助一点编程工作，iPhone 的机载加速计就可以使对象在屏幕上四处"移动"，实时响应用户倾斜手机的方式。秘诀 1-4 创建了一只动画式的蝴蝶，用户可以使之快速移过屏幕。

使之工作的秘密在于：向程序中添加一个所谓的"物理计时器"。它不是直接响应加速中的变化，而是像秘诀 1-3 所做的那样，加速计回调用于测量当前的力。它取决于计时器例程随着时间的推移通过改变它的画面对蝴蝶应用那些力。下面列出了一些要记住的关键点。

- 只要力的方向仍然保持相同，蝴蝶就会加速。它的速度会依据加速力在 x 或 y 方向上的量度成比例地提高。
- 由计时器调用的 tick 例程将通过向蝴蝶的原点添加速度向量来移动蝴蝶。
- 蝴蝶移动的范围是有界限的。因此，当它撞到某个边缘时，将会停止在那个方向上移动。这可以一直把蝴蝶保留在屏幕上。tick 方法将会检查界限条件。例如，如果蝴蝶撞到垂直边缘，那它仍然可以在水平方向上移动。
- 蝴蝶会改变它自身的方向，使之总是"下落"。可以在 tick 方法中应用一个简单的旋转变换来实现这一点。在使用变换时，还要关注画面或中心偏移。在应用偏移之前，总是要重置数学处理，然后重新应用任何角度改变。不这样做的话，可能导致画面出人意料地放大、收缩或扭曲。

> **注意：**
> 计时器在自然状态下不会处理块。如果你愿意使用基于块的设计，可以查询 github，找到它的解决办法。

秘诀 1-4　基于加速计的反馈移动屏幕上的对象

```
- (void)accelerometer:(UIAccelerometer *)accelerometer
    didAccelerate:(UIAcceleration *)acceleration
{
    // Extract the acceleration components
    float xx = -acceleration.x;
    float yy = acceleration.y;

    // Store the most recent angular offset
    mostRecentAngle = atan2(yy, xx);

    // Has the direction changed?
    float accelDirX = SIGN(xvelocity) * -1.0f;
    float newDirX = SIGN(xx);
    float accelDirY = SIGN(yvelocity) * -1.0f;
    float newDirY = SIGN(yy);

    // Accelerate. To increase viscosity lower the additive value
    if (accelDirX == newDirX) xaccel =
        (abs(xaccel) + 0.85f) * SIGN(xaccel);
    if (accelDirY == newDirY) yaccel =
```

```
            (abs(yaccel) + 0.85f) * SIGN(yaccel);

    // Apply acceleration changes to the current velocity
    xvelocity = -xaccel * xx;
    yvelocity = -yaccel * yy;
}

- (void) tick
{
    // Reset the transform before changing position
    butterfly.transform = CGAffineTransformIdentity;

        // Move the butterfly according to the current velocity vector
    CGRect rect = CGRectOffset(butterfly.frame, xvelocity, 0.0f);
    if (CGRectContainsRect(self.view.bounds, rect))
        butterfly.frame = rect;
    rect = CGRectOffset(butterfly.frame, 0.0f, yvelocity);
    if (CGRectContainsRect(self.view.bounds, rect))
        butterfly.frame = rect;

    // Rotate the butterfly independently of position
    butterfly.transform =
        CGAffineTransformMakeRotation(mostRecentAngle + M_PI_2);
}


- (void) initButterfly
{
    CGSize size;

    // Load the animation cells
    NSMutableArray *butterflies = [NSMutableArray array];
    for (int i = 1; i <= 17; i++)
    {
        NSString *fileName = [NSString stringWithFormat:@"bf_%d.png", i];
        UIImage *image = [UIImage imageNamed:fileName];
        size = image.size;
        [butterflies addObject:image];
    }

    // Begin the animation
```

```
    butterfly = [[UIImageView alloc]
        initWithFrame:(CGRect){.size=size}];
    [butterfly setAnimationImages:butterflies];
    butterfly.animationDuration = 0.75f;
    [butterfly startAnimating];

    // Set the butterfly's initial speed and acceleration
    xaccel = 2.0f;
    yaccel = 2.0f;
    xvelocity = 0.0f;
    yvelocity = 0.0f;

    // Add the butterfly
    butterfly.center = RECTCENTER(self.view.bounds);
    [self.view addSubview:butterfly];

    // Activate the accelerometer
    [[UIAccelerometer sharedAccelerometer] setDelegate:self];

    // Start the physics timer
    [NSTimer scheduledTimerWithTimeInterval: 0.03f
    target: self selector: @selector(tick)
    userInfo: nil repeats: YES];
}
```

获取这个秘诀的代码

要查找这个秘诀的完整示例项目，可以浏览 https://github.com/erica/iOS-6-Advanced-Cookbook，并进入第 1 章的文件夹。

1.9 秘诀：基于加速计的滚动视图

好几位读者要求我在本书这一版中包括进一个倾斜滚轮秘诀。倾斜滚轮使用设备的内置加速计来控制在 UIScrollView 的内容周围的移动。当用户调整设备时，材料会相应地"下落"。它不会把视图定位在屏幕上，而是把内容视图滚动到一个新的偏移位置。

创建这个界面的挑战在于：确定设备在什么地方应该具有它的静止轴（resting axis）。大多数人最初建议当显示屏靠着它的背部时应该是稳定的，并且 z 轴方向笔直地指向上方。事实证明：这实际上是一种相当糟糕的设计选择。要使用那根轴，就意味着在导航期间屏幕必须实际地偏离观看者。随着设备旋转离开视图，用户将不能完全看到屏幕上所发生的事情，尤其是在固定的位置使用设备时，站在高处查看设备有时也会产生这种效果。

作为替代，秘诀 1-5 假定稳定的位置是通过 z 轴指向大约 45° 的方向，即用户把 iPhone 或 iPad 握在手中的自然位置，这处于正面朝上和正面朝前方的中间位置。对秘诀 1-5 中的数学运算做了相应的调整。从这个歪斜的位置来回倾斜，使屏幕在调整期间保持最大的可见性。

与秘诀 1-4 相比，这个秘诀中的另一处改变是低得多的加速常量。这使屏幕上的运动能够更慢地发生，让用户更容易降低速度并恢复导航。

秘诀 1-5　倾斜滚轮

```
- (void)accelerometer:(UIAccelerometer *)accelerometer
    didAccelerate:(UIAcceleration *)acceleration
{
    // extract the acceleration components
    float xx = -acceleration.x;
    float yy = (acceleration.z + 0.5f) * 2.0f; // between face-up and face-forward

    // Has the direction changed?
    float accelDirX = SIGN(xvelocity) * -1.0f;
    float newDirX = SIGN(xx);
    float accelDirY = SIGN(yvelocity) * -1.0f;
    float newDirY = SIGN(yy);

    // Accelerate. To increase viscosity lower the additive value
    if (accelDirX == newDirX) xaccel = (abs(xaccel) + 0.005f) * SIGN(xaccel);
    if (accelDirY == newDirY) yaccel = (abs(yaccel) + 0.005f) * SIGN(yaccel);

    // Apply acceleration changes to the current velocity
    xvelocity = -xaccel * xx;
    yvelocity = -yaccel * yy;
}

- (void) tick
{
    xoff += xvelocity;
    xoff = MIN(xoff, 1.0f);
    xoff = MAX(xoff, 0.0f);

    yoff += yvelocity;
```

```
        yoff = MIN(yoff, 1.0f);
        yoff = MAX(yoff, 0.0f);

        // update the content offset based on the current velocities
        CGFloat xsize = sv.contentSize.width - sv.frame.size.width;
        CGFloat ysize = sv.contentSize.height - sv.frame.size.height;
        sv.contentOffset = CGPointMake(xoff * xsize, yoff * ysize);
    }

- (void) viewDidAppear:(BOOL)animated
{
    NSString *map = @"http://maps.weather.com/images/\
        maps/current/curwx_720x486.jpg";
    NSOperationQueue *queue = [[NSOperationQueue alloc] init];
    [queue addOperationWithBlock:
    ^{
        // Load the weather data
        NSURL *weatherURL = [NSURL URLWithString:map];
        NSData *imageData = [NSData dataWithContentsOfURL:weatherURL];

        // Update the image on the main thread using the main queue
        [[NSOperationQueue mainQueue] addOperationWithBlock:^{
            UIImage *weatherImage = [UIImage imageWithData:imageData];
            UIImageView *imageView =
                [[UIImageView alloc] initWithImage:weatherImage];
            CGSize initSize = weatherImage.size;
            CGSize destSize = weatherImage.size;
            // Ensure that the content size is significantly bigger
            // than the screen can show at once
            while ((destSize.width < (self.view.frame.size.width * 4)) ||
                (destSize.height < (self.view.frame.size.height * 4)))
            {
            destSize.width += initSize.width;
            destSize.height += initSize.height;
            }

            imageView.userInteractionEnabled = NO;
            imageView.frame = (CGRect){.size = destSize};
            sv.contentSize = destSize;
```

```
        [sv addSubview:imageView];
        
        // Activate the accelerometer
        [[UIAccelerometer sharedAccelerometer] setDelegate:self];
        
        // Start the physics timer
        [NSTimer scheduledTimerWithTimeInterval: 0.03f
            target: self selector: @selector(tick)
            userInfo: nil repeats: YES];
    }];
}];
}
```

获取这个秘诀的代码

要查找这个秘诀的完整示例项目,可以浏览 https://github.com/erica/iOS-6-Advanced-Cookbook,并进入第 1 章的文件夹。

1.10 秘诀:Core Motion 基础

Core Motion 框架集中了运动数据处理。该框架是在 iOS 4 SDK 中引入的,用于取代你刚才阅读到的直接加速计访问。它提供了对 3 个关键的机载传感器的集中式监测。这些传感器由陀螺仪、磁力计和加速计组成,其中陀螺仪用于测量设备的旋转,磁力计提供了一种测量罗盘方位的方式,加速计用于检测沿着 3 根轴的重力变化。第四个入口点称为**设备移动**(device motion),它把全部 3 种传感器都结合进单个监测系统中。

Core Motion 使用来自这些传感器原始值创建可度的测量结果,主要表现为力向量的形式。可测量的项包括以下属性。

- **设备姿势**(`attitude`):设备相对于某个参照画面的方向。姿势被表示为摇晃、前倾和左右摇摆的角度,它们都以弧度为单位。
- **旋转速率**(`rotationRate`):设备围绕它的 3 根轴中的每一根轴旋转的速率。旋转包括 x、y 和 z 角速度值,它们以弧度/秒为单位。
- **重力**(`gravity`):设备当前的加速度向量,由正常的重力场提供。重力的单位是 g's,分别沿着 x、y 和 z 轴来测量。每个单位代表由地球提供的标准重力加速度(9.8 米/秒^2)。
- **用户加速度**(`userAcceleration`):用户提供的加速度向量。像重力一样,用户加速度的单位也是 g's,分别沿着 x、y 和 z 轴来测量。当把它们加到一起时,用户向量和重力向量代表给设备提供的总加速度。

- **磁场（`magneticField`）**：代表设备邻近区域里的总磁场的向量。磁场是沿着 x、y 和 z 轴以微特斯拉（microtesla）为单位测量的。还提供了校准精度，通知应用程序有关磁场测量的质量。

1.10.1 测试传感器

正如在本章前面所学到的，可以使用应用程序的 Info.plist 文件要求使用或排除机载传感器。也可以在应用程序内测试每种可能的 Core Motion 支持：

```
if (motionManager.gyroAvailable)
    [motionManager startGyroUpdates];

if (motionManager.magnetometerAvailable)
    [motionManager startMagnetometerUpdates];

if (motionManager.accelerometerAvailable)
    [motionManager startAccelerometerUpdates];

if (motionManager.deviceMotionAvailable)
    [motionManager startDeviceMotionUpdates];
```

开始更新不会产生像使用 UIAccelerometer 时遇到的委托回调机制。作为替代，你将负责轮询每个值，或者可以使用基于块的更新机制，执行在每次更新时提供的一个块（例如，startAccelerometerUpdatesToQueue:withHandler:)。

1.10.2 处理程序块

秘诀 1-6 修改了秘诀 1-4，以便使用 Core Motion。把加速度回调移入一个处理程序块中，并从数据的加速度属性中读取 x 和 y 值。否则，代码将保持不变。在这里，将看到 Core Motion 的一些基本的方面：创建一个新的运动管理器，它用于测试加速计可用性。然后，它将使用一个新的操作队列开始更新，该队列在应用程序运行期间将持续存在。

establishMotionManager 和 shutDownMotionManager 方法使应用程序能够根据需要启动和关闭运动管理器。这些方法是在应用程序变成活动时和挂起时从应用程序委托内调用的：

```
- (void) applicationWillResignActive:(UIApplication *)application
{
    [tbvc shutDownMotionManager];
}

- (void) applicationDidBecomeActive:(UIApplication *)application
```

```
{
    [tbvc establishMotionManager];
}
```

这些方法提供了一种干净的方式，用于关闭和恢复运动服务，以响应当前的应用程序状态。

秘诀 1-6　基本的 Core Motion

```
@implementation TestBedViewController
- (void) tick
{
    butterfly.transform = CGAffineTransformIdentity;
    
    // Move the butterfly according to the current velocity vector
    CGRect rect = CGRectOffset(butterfly.frame, xvelocity, 0.0f);
    if (CGRectContainsRect(self.view.bounds, rect))
        butterfly.frame = rect;
    
    rect = CGRectOffset(butterfly.frame, 0.0f, yvelocity);
    if (CGRectContainsRect(self.view.bounds, rect))
        butterfly.frame = rect;
    
    butterfly.transform = 
        CGAffineTransformMakeRotation(mostRecentAngle + M_PI_2);
}
- (void) shutDownMotionManager
{
    NSLog(@"Shutting down motion manager");
    [motionManager stopAccelerometerUpdates];
    motionManager = nil;
    
    [timer invalidate];
    timer = nil;
}

- (void) establishMotionManager
{
    if (motionManager)
        [self shutDownMotionManager];
    
    NSLog(@"Establishing motion manager");
```

```
    // Establish the motion manager
    motionManager = [[CMMotionManager alloc] init];
    if (motionManager.accelerometerAvailable)
        [motionManager
            startAccelerometerUpdatesToQueue:
                [[NSOperationQueue alloc] init]
            withHandler:^(CMAccelerometerData *data, NSError *error)
            {
                // Extract the acceleration components
                float xx = -data.acceleration.x;
                float yy = data.acceleration.y;
                mostRecentAngle = atan2(yy, xx);

                // Has the direction changed?
                float accelDirX = SIGN(xvelocity) * -1.0f;
                float newDirX = SIGN(xx);
                float accelDirY = SIGN(yvelocity) * -1.0f;
                float newDirY = SIGN(yy);

                // Accelerate. To increase viscosity,
                // lower the additive value
                if (accelDirX == newDirX)
                    xaccel = (abs(xaccel) + 0.85f) * SIGN(xaccel);
                if (accelDirY == newDirY)
                    yaccel = (abs(yaccel) + 0.85f) * SIGN(yaccel);

                // Apply acceleration changes to the current velocity
                xvelocity = -xaccel * xx;
                yvelocity = -yaccel * yy;
            }];

    // Start the physics timer
    timer = [NSTimer scheduledTimerWithTimeInterval: 0.03f
        target: self selector: @selector(tick)
        userInfo: nil repeats: YES];
}

- (void) initButterfly
{
```

```objc
    CGSize size;
    
    // Load the animation cells
    NSMutableArray *butterflies = [NSMutableArray array];
    for (int i = 1; i <= 17; i++)
    {
        NSString *fileName = 
            [NSString stringWithFormat:@"bf_%d.png", i];
        UIImage *image = [UIImage imageNamed:fileName];
        size = image.size;
        [butterflies addObject:image];
    }
    
    // Begin the animation
    butterfly = [[UIImageView alloc]
        initWithFrame:(CGRect){.size=size}];
    [butterfly setAnimationImages:butterflies];
    butterfly.animationDuration = 0.75f;
    [butterfly startAnimating];
    
    // Set the butterfly's initial speed and acceleration
    xaccel = 2.0f;
    yaccel = 2.0f;
    xvelocity = 0.0f;
    yvelocity = 0.0f;
    
    // Add the butterfly
    butterfly.center = RECTCENTER(self.view.bounds);
    [self.view addSubview:butterfly];
}

- (void) loadView
{
    [super loadView];
    self.view.backgroundColor = [UIColor whiteColor];
    [self initButterfly];
}
@end
```

> **获取这个秘诀的代码**
>
> 要查找这个秘诀的完整示例项目,可以浏览 https://github.com/erica/iOS-6-Advanced-Cookbook,并进入第 1 章的文件夹。

1.11 秘诀:获取和使用设备姿势

设想有一部 iPad 放在桌子上。iPad 上显示了一幅图像,可以弯曲并查看它。现在,设想旋转那个 iPad,就像它平放在桌子上一样,但是当 iPad 移动时,图像不会移动。它保持与周围的世界完美对齐。无论怎样旋转 iPad,图像都不会随着视图更新而"移动",以平衡物理运动。这就是秘诀 1-7 的工作方式,利用设备的机载陀螺仪(这是必需的),使这个秘诀工作。

无论怎样握持设备,图像都会调整。除了这种水平操作,还可以拾起设备并在空间中定位它的方向。如果在手中翻转设备并查看它,就会看到图像的颠倒的"底部"。还可以沿着两根轴倾斜它:一根是从 Home 按钮指向照相机,另一根则从照相机与 Home 按钮的中点开始穿过 iPad 的表面。还有一根轴,它是你最先探讨过的,从设备的中间开始,指向设备上方的空间,并且穿过它下面的中点。在操纵设备时,图像会做出响应,在那个 iPad 内创建一个可视化的静态世界。

秘诀 1-7 显示了如何利用少量简单的几何变换来执行该操作。它建立了一个运动管理器,订阅设备运动更新,然后基于运动管理器返回的摇晃、前倾和左右摇摆的角度应用图像变换。

秘诀 1-7 使用设备运动更新修正空间里的图像

```
- (void) shutDownMotionManager
{
    NSLog(@"Shutting down motion manager");
        [motionManager stopDeviceMotionUpdates];
    motionManager = nil;
}

- (void) establishMotionManager
{
    if (motionManager)
        [self shutDownMotionManager];

    NSLog(@"Establishing motion manager");

    // Establish the motion manager
    motionManager = [[CMMotionManager alloc] init];
```

```
    if (motionManager.deviceMotionAvailable)
        [motionManager
         startDeviceMotionUpdatesToQueue:
            [NSOperationQueue currentQueue]
        withHandler: ^(CMDeviceMotion *motion, NSError *error) {
            CATransform3D transform;
            transform = CATransform3DMakeRotation(
                motion.attitude.pitch, 1, 0, 0);
            transform = CATransform3DRotate(transform,
                motion.attitude.roll, 0, 1, 0);
            transform = CATransform3DRotate(transform,
                motion.attitude.yaw, 0, 0, 1);
            imageView.layer.transform = transform;
        }];
}
```

> **获取这个秘诀的代码**
>
> 要查找这个秘诀的完整示例项目，可以浏览 https://github.com/erica/iOS-6-Advanced-Cookbook，并进入第 1 章的文件夹。

1.12 使用运动事件检测晃动

当 iPhone 检测到一个运动事件时，它会把该事件传递给当前的第一个响应者，即响应者链中的主对象。响应者是可以处理事件的对象，所有的视图和窗口都是响应者，因此也是应用程序对象。

响应者链提供了一种对象层次结构，所有的对象都可以响应事件。当朝向链开始处的对象接收到一个事件时，不会进一步传递那个事件。对象会处理它。如果它不能处理，可以把该事件转移给下一个响应者。

对象通常可以通过把它们自己声明为第一个响应者来获得这种身份，这是通过 becomeFirstResponder 实现的。在这个代码段中，UIViewController 确保它会变成第一个响应者，只要它的视图出现在屏幕上即可。一旦消失，它将放弃第一个响应者的身份：

```
- (BOOL)canBecomeFirstResponder {
    return YES;
}

// Become first responder whenever the view appears
- (void)viewDidAppear:(BOOL)animated {
    [super viewDidAppear:animated];
```

```
    [self becomeFirstResponder];
}

// Resign first responder whenever the view disappears
- (void)viewWillDisappear:(BOOL)animated {
    [super viewWillDisappear:animated];
    [self resignFirstResponder];
}
```

第一个响应者将接收所有的触摸和运动事件。运动回调反映了 UIView 触摸回调阶段。回调方法如下。

- **motionBegan:withEvent::** 这个回调指示运动事件的开始。在编写本书时，只能识别一类运动事件：晃动。将来可能不是这样，因此你可能想在代码中检查运动类型。
- **motionEnded:withEvent::** 在第一个响应者在运动事件结束时接收这个回调。
- **motionCancelled:withEvent::** 与触摸一样，可以通过打入的电话和其他系统事件取消运动。Apple 建议在代码中实现全部 3 个运动事件回调（类似地，还要实现全部 4 个触摸事件回调）。

下面的代码段显示了一对运动回调的示例。如果在设备上测试它们，可以注意到几件事。第一，从用户的角度看，开始和结束事件几乎是同时发生的，为这两类事件播放声音有些小题大做。第二，它偏向于进行从一侧到另一侧的晃动检测，与前后和上下晃动相比，iPhone 更擅长检测从一侧到另一侧的晃动。最后，Apple 的运动实现使用了一种轻微锁定的方法。直到生成了另一个运动事件或者在处理了前一个运动事件之后，才能生成一个新的运动事件。Shake to Shuffle 和 Shake to Undo 事件使用了相同的锁定机制：

```
- (void)motionBegan:(UIEventSubtype)motion
    withEvent:(UIEvent *)event {

    // Play a sound whenever a shake motion starts
    if (motion != UIEventSubtypeMotionShake) return;
    [self playSound:startSound];
}

- (void)motionEnded:(UIEventSubtype)motion withEvent:(UIEvent *)event
{
    // Play a sound whenever a shake motion ends
    if (motion != UIEventSubtypeMotionShake) return;
    [self playSound:endSound];
}
```

1.13 使用外部屏幕

可以用许多方式使用外部屏幕。例如，采取最新款的 iPad。第二代和第三代型号提供了内置的屏幕监测。连接 VGA 或 HDMI 电缆，就可以把内容显示在外部显示器和内置屏幕上。某些设备允许使用 AirPlay（Apple 的专有无线缆空中下载视频解决方案）把屏幕以无线方式镜像到 Apple TV。这些镜像特性极其方便，但是并不仅限于在 iOS 中简单地把一个屏幕上的内容复制到另一个屏幕上。

UIScreen 类允许独立地检测并写到外部屏幕上。可以把任何连接的显示器视作一个新窗口，并为该显示器创建内容，使之独立于主设备屏幕上显示的任何视图。可以为任何有线屏幕执行该操作，并且从 iPad 2（及更高型号）和 iPhone 4S（及更高型号）开始，可以使用 AirPlay to Apple TV 2（及更高型号）以无线方式执行该操作。名为 Reflector 的第三方应用程序允许使用 AirPlay 把显示器镜像到 Mac 或 Windows 计算机。

几何学很重要。为什么呢？iOS 设备目前包括 320 像素×480 像素的老式 iPhone 显示器、640 像素×960 像素的 Retina 显示器单元和 1024 像素×768 像素的 iPad。典型的复合/分量输出是在 720 像素×480 像素（480i 和 480p）、1024 像素×768 像素和 1280 像素×720 像素（720p）下的 VGA 上产生的，然后还有更高质量的 HDMI 输出可用。

除此之外，还有过扫描的问题及其他目标显示器的局限性，并且 Video Out 会迅速变成一个几何挑战。幸运的是，Apple 利用一些方便的现实修改来响应这种挑战。无需尝试在输出屏幕与内置的设备屏幕之间创建一种一对一的关系，而可以基于输出显示器的可用属性构建内容。只需创建一个窗口，填充它，并显示它。

如果打算开发 Video Out 应用程序，不要想当然地认为用户会严格地使用 AirPlay。许多用户仍然使用老式的电缆接头连接到显示器和投影仪。确保每种电缆至少具有一根（复合、分量、VGA 和 HDMI），还要具有准备好使用 AirPlay 的 iPhone 和 iPad，以便可以彻底地测试每种输出配置。第三方电缆（通常是从远东进口的，没有打上 Made for iPhone/iPad 的标签）将不会工作，因此要确保购买具有 Apple 品牌的物品。

1.13.1 检测屏幕

UIScreen 类可以报告连接了多少个屏幕。你知道无论何时这个计数大于 1，都会连接有外部屏幕。screens 数组中的第一项总是主设备屏幕：

```
#define SCREEN_CONNECTED ([UIScreen screens].count > 1)
```

每个屏幕都可以报告它的边界（即它的物理尺寸，以磅为单位）以及它的屏幕比例（将磅与像素关联起来）。两个标准的通知使你能够观察屏幕何时连接到设备以及与设备断开。

```
// Register for connect/disconnect notifications
[[NSNotificationCenter defaultCenter]
    addObserver:self selector:@selector(screenDidConnect:)
    name:UIScreenDidConnectNotification object:nil];
[[NSNotificationCenter defaultCenter]
    addObserver:self selector:@selector(screenDidDisconnect:)
    name:UIScreenDidDisconnectNotification object:nil];
```

连接意指任意类型的连接，无论是通过电缆还是通过 AirPlay。无论何时接收到这种类型的更新，都一定要统计屏幕数量，并调整用户界面，以匹配新的情况。

你的职责是：无论何时加入新屏幕，都要建立一些窗口，一旦发生分离事件，就要清除它们。每个屏幕都应该具有它自己的窗口，为输出显示器管理内容。对于分离的屏幕，不要抓住它们的窗口不放。释放它们，然后当新屏幕出现时重新创建它们。

> 注意：
> 在 screens 数组中不会表示出镜像的屏幕，而代之以将镜像存储在主屏幕的 mirroredScreen 属性中。当禁用、未连接或者设备的能力简直不支持镜像时，这个属性为空。
> 创建一个新屏幕并把它用于独立的外部显示器总会撤销镜像。因此，即使用户启用了镜像，当应用程序开始写到并创建外部显示器，它将具有优先级。

1.13.2 获取屏幕分辨率

每个屏幕都提供了一个 availableModes 属性。这是一个分辨率对象的数组，其中的元素按从最低分辨率到最高分辨率排序。每种模式都有一个 size 属性，指示一个目标像素大小的分辨率。许多屏幕都支持多种模式。例如，VGA 显示器可能具有多达 6 种模式，或者具有比它提供的更多不同的分辨率。支持的分辨率数量因硬件而异。总是至少有一种分辨率可用，但是当具有多种分辨率时，应该给用户提供选择的机会。

1.13.3 设置 Video Out

在从[UIScreens screens]数组中获取了外部屏幕对象之后，可以查询可用的模式并选择要使用的尺寸。一般说来，可以放弃选择列表中的最后一种模式，总是使用尽可能高的分辨率，或者放弃使用最低分辨率的第一种模式。

要开始一个 Video Out（视频输出）流，可以创建一个新的 UIWindow，并将其尺寸调整为所选的模式。给那个窗口添加一个新视图，以便在其上绘图。然后把窗口分配给外部屏幕，并使之成为关键的和可见的窗口。这命令窗口显示出来，并准备好使用它。之后，再次使原始窗口成为关键窗口。这允许用户继续与主屏幕交互。不要跳过这一步。对于最终用户来说，什么也不会比发现他们昂贵的设备不再响应他们的触摸更令他们暴躁：

```
self.outputWindow = [[UIWindow alloc] initWithFrame:theFrame];
outputWindow.screen = secondaryScreen;
[outputWindow makeKeyAndVisible];
[delegate.view.window makeKeyAndVisible];
```

1.13.4 添加显示器链接

显示器链接是一种计时器，以便把绘图与显示器的刷新率进行同步。可以通过更改显示器链接的 frameInterval 属性，来调整这个画面刷新时间。该属性默认为 1，更高的数字会降低刷新率。如果把它设置为 2，则会把画面刷新率减半。当屏幕连接到设备时，就会创建显示器链接。UIScreen 类实现了一个方法，用于返回它的屏幕的显示器链接对象。可以为显示器链接指定一个目标以及要调用的选择器。

显示器链接定期触发，以让你知道何时更新 Video Out 屏幕。如果 CPU 负载较小，可以把时间间隔调长一些，但是这将会获得较低的画面刷新率。这是一个重要的折衷，尤其是对于在设备端需要具有高级 CPU 响应的直接操作界面更是如此。

秘诀 1-8 中的代码为运行循环使用了常见的模式，提供最少的等待时间。在处理显示器链接时，如果使之无效（invalidate），就将它从运行循环中删除。

1.13.5 过扫描补偿

UIScreen 类允许通过给 overscanCompensation 属性赋值，补偿显示器屏幕边缘的像素损失。在 Apple 的文档中描述了可以指派的技术，但是它们基本上相当于你是想剪辑内容还是想用空格填充它。

1.13.6 VIDEOkit

秘诀 1-8 介绍了 VIDEOkit，它是一个基本的外部屏幕客户。该秘诀演示了配置和使用有线与无线外部屏幕所需的所有特性。通过调用 startupWithDelegate:来建立屏幕监测。把主视图控制器传递给它，该控制器的职责将是创建外部内容。

内部的 init 方法开始侦听屏幕连接和分离事件，并根据需要构建和删除窗口。每次显示器链接触发时，都会调用一个非正式的委托方法（updateExternalView:）。它将传递外部窗口上的视图，委托可以根据需要在该窗口上绘图。

在这个秘诀的配套示例代码中，视图控制器委托存储一个本地颜色值，并使用它给外部显示器着色：

```
- (void) updateExternalView: (UIImageView *) aView
```

```
{
    aView.backgroundColor = color;
}

- (void) action: (id) sender
{
    color = [UIColor randomColor];
}
```

每次按下动作按钮时,视图控制器都会生成一种新颜色。当 VIDEOkit 查询控制器以更新外部视图时,将把该颜色设置为背景色。可以看到外部屏幕即时更新为新的随机颜色。

> **注意:**
> Reflector App(单一许可的费用是 15 美元,5 台计算机的许可费用是 50 美元,reflectorapp.com)为 AirPlay 提供了一个优秀的调试伙伴,它提供了一种可以在 Mac 和 Windows 计算机上工作的无线/无 Apple TV 的解决方案。它模拟 Apple TV AirPlay 接收器,允许从 iOS 直接广播到桌面,并记录该输出。

秘诀 1-8 VIDEOkit

```
@interface VIDEOkit : NSObject
{
    UIImageView *baseView;
}
@property (nonatomic, weak) UIViewController *delegate;
@property (nonatomic, strong) UIWindow *outputWindow;
@property (nonatomic, strong) CADisplayLink *displayLink;
+ (void) startupWithDelegate: (id) aDelegate;
@end

@implementation VIDEOkit
static VIDEOkit *sharedInstance = nil;

- (void) setupExternalScreen
{
    // Check for missing screen
    if (!SCREEN_CONNECTED) return;

    // Set up external screen
    UIScreen *secondaryScreen = [UIScreen screens][1];
    UIScreenMode *screenMode =
```

```objc
        [[secondaryScreen availableModes] lastObject];
    CGRect rect = (CGRect){.size = screenMode.size};
    NSLog(@"Extscreen size: %@", NSStringFromCGSize(rect.size));

    // Create new outputWindow
    self.outputWindow = [[UIWindow alloc] initWithFrame:CGRectZero];
    _outputWindow.screen = secondaryScreen;
    _outputWindow.screen.currentMode = screenMode;
    [_outputWindow makeKeyAndVisible];
    _outputWindow.frame = rect;

    // Add base video view to outputWindow
    baseView = [[UIImageView alloc] initWithFrame:rect];
    baseView.backgroundColor = [UIColor darkGrayColor];
    [_outputWindow addSubview:baseView];

    // Restore primacy of main window
    [_delegate.view.window makeKeyAndVisible];
}


- (void) updateScreen
{
    // Abort if the screen has been disconnected
    if (!SCREEN_CONNECTED && _outputWindow)
        self.outputWindow = nil;

    // (Re)initialize if there's no output window
    if (SCREEN_CONNECTED && !_outputWindow)
        [self setupExternalScreen];

    // Abort if encountered some weird error
    if (!self.outputWindow) return;

    // Go ahead and update
    SAFE_PERFORM_WITH_ARG(_delegate,
        @selector(updateExternalView:), baseView);
}

- (void) screenDidConnect: (NSNotification *) notification
{
```

```objc
    NSLog(@"Screen connected");
    UIScreen *screen = [[UIScreen screens] lastObject];

    if (_displayLink)
    {
        [_displayLink removeFromRunLoop:[NSRunLoop currentRunLoop]
            forMode:NSRunLoopCommonModes];
        [_displayLink invalidate];
        _displayLink = nil;
    }

    self.displayLink = [screen displayLinkWithTarget:self
        selector:@selector(updateScreen)];
    [_displayLink addToRunLoop:[NSRunLoop currentRunLoop]
        forMode:NSRunLoopCommonModes];
}


- (void) screenDidDisconnect: (NSNotification *) notification
{
    NSLog(@"Screen disconnected.");
    if (_displayLink)
    {
        [_displayLink removeFromRunLoop:[NSRunLoop currentRunLoop]
            forMode:NSRunLoopCommonModes];
        [_displayLink invalidate];
        self.displayLink = nil;
    }
}


- (id) init
{
    if (!(self = [super init])) return self;

    // Handle output window creation
    if (SCREEN_CONNECTED)
        [self screenDidConnect:nil];

    // Register for connect/disconnect notifications
    [[NSNotificationCenter defaultCenter]
        addObserver:self selector:@selector(screenDidConnect:)
```

```
            name:UIScreenDidConnectNotification object:nil];
    [[NSNotificationCenter defaultCenter] addObserver:self
            selector:@selector(screenDidDisconnect:)
            name:UIScreenDidDisconnectNotification object:nil];

    return self;
}

- (void) dealloc
{
    [self screenDidDisconnect:nil];
    self.outputWindow = nil;
}
+ (VIDEOkit *) sharedInstance
{
    if (!sharedInstance)
        sharedInstance = [[self alloc] init];
    return sharedInstance;
}

+ (void) startupWithDelegate: (id) aDelegate
{
    [[self sharedInstance] setDelegate:aDelegate];
}
@end
```

获取这个秘诀的代码

要查找这个秘诀的完整示例项目，可以浏览 https://github.com/erica/iOS-6-Advanced-Cookbook，并进入第 1 章的文件夹。

1.14 跟踪用户

跟踪是开发者的一种不幸的现实生活。Apple 不赞成使用 UIDevice 属性，该属性提供了绑定到设备硬件的唯一标识符。Apple 利用两个标识符属性取代 UIDevice 属性。它使用 identifierForAdvertising 属性返回当前设备所独有的一个特定于设备的字符串，并使用 identifierForVendor 属性提供一个绑定到每位应用程序供应商的字符串。无论使用的是哪个应用程序，这都应该会返回相同的唯一字符串，它不是顾客 id。不同设备上的相同应用程序可以返回不同的字符串，就像应用程序可以来自不同的供应商一样。

这些标识符是使用新的 NSUUID 类构建的。可以在跟踪场景之外使用这个类，创建保证全球唯一的 UUID 字符串。Apple 写道："UUID（Universally Unique Identifier，通用唯一标识符），也称为 GUID（Globally Unique Identifier，全局唯一标识符）或 IID（Interface Identifier，接口标识符），是 128 位的值。UUID 在空间和时间上都是唯一的，这是由于它结合了两个值，第一个值是生成它的计算机上所特有的，第二个值代表从 1582 年 10 月 15 日 00:00:00 起所经过的 100 纳秒数。"

UUID 类方法可以根据需要生成一个新的 RFC 4122v4 UUID。使用[NSUUID UUID]返回一个新实例（附带的好处是：它全都是大写的）。从此，可以获取 UUIDString 表示，或者通过 getUUIDBytes:直接请求字节。

1.15 还有一件事：检查可用的磁盘空间

NSFileManager 类可让你确定 iPhone 上有多少空闲空间，以及设备上总共提供了多少空间。程序清单 1-1 演示了如何检查这些值，并且使用友好的、逗号格式化的字符串显示结果。返回的值表示空闲空间（以字节为单位）。

程序清单 1-1　获取文件系统大小和文件系统空闲大小

```
- (NSString *) commaFormattedStringWithLongLong: (long long) num
{
    // Produce a properly formatted number string
    // Alternatively use NSNumberFormatter
    if (num < 1000)
        return [NSString stringWithFormat:@"%d", num];
    return    [[self commasForNumber:num/1000]
        stringByAppendingFormat:@",%03d", (num % 1000)];
}

- (void) action: (UIBarButtonItem *) bbi
{
    NSFileManager *fm = [NSFileManager defaultManager];
    NSDictionary *fattributes = 
        [fm fileSystemAttributesAtPath:NSHomeDirectory()];
    NSLog(@"System space: %@",
        [self commaFormattedStringWithLongLong:[[fattributes
        objectForKey:NSFileSystemSize] longLongValue]]);
```

```
    NSLog(@"System free space: %@",
        [self commasForNumber:[[fattributes
        objectForKey:NSFileSystemFreeSize] longLongValue]]);
}
```

1.16 小结

本章介绍了一些与 iPhone 设备交互的核心方式。你看到了如何获取设备信息，检查电池状态，以及订阅接近事件。你学习了如何将 iPod Touch 与 iPhone 和 iPad 区分开，以及确定正在使用的是哪种型号。你了解了加速计，并通过几个示例看到了它的应用，从简单的"向上"定位到更复杂的晃动检测算法。你进入到 Core Motion 中，并且学习了如何创建更新块，实时响应设备事件。最后，你看到了如何给应用程序添加外部屏幕支持。下面对你刚才遇到的秘诀列出了几条总结性考虑。

- iPhone 的加速计提供了一种新颖的方式，用于补充其基于触摸的界面。使用加速度数据把用户交互扩展到"触摸此处"基本操作之外，并且引入了知道倾斜的反馈。
- 低级调用可能是 App Store 友好的。它们不依赖于可能基于当前的固件版本而改变的 Apple API。UNIX 系统调用似乎令人畏缩不前，但是其中许多都受到 iOS 设备家族完全支持。
- 记住设备的限制。你可能希望在执行文件密集型的工作之前检查空闲的磁盘空间，以及在全速运行 CPU 之前检查电池的充电状态。
- 深入研究 Core Motion。它提供的实时设备反馈是把 iOS 设备集成进现实体验中的基础。
- 既然 AirPlay 打破了外部显示器的束缚，就可以为比你以前所想的多得多的令人兴奋的项目使用 Video Out。AirPlay 和外部视频屏幕意味着可以把 iOS 设备转变成游戏和实用程序的远程控制装置，这样就可以在大屏幕上显示并在小屏幕上控制它们。
- 在提交给 iTunes 之前，使用 Info.plist 文件确定哪些设备能力是必需的。iTunes 使用这个必需能力的列表确定是否可以把某个应用程序下载到给定的设备上，并在该设备上正确地运行。

第 2 章
文档和数据共享

在 iOS 下，应用程序可以使用多种系统特性共享信息和数据，以及把控制权从一个应用程序转移给另一个应用程序。每个应用程序都可以访问公共的系统粘贴板，允许跨应用程序进行复制和粘贴。用户可以把文档从一个应用程序传输给另一个支持该格式的应用程序。它们可以请求把系统提供的许多"动作"应用于文档，比如打印、发微博，或者贴到 Facebook 上。应用程序可以声明自定义的 URL 模式，并把它们嵌入在文本和 Web 页面中。本章介绍可以在应用程序之间集成文档和进行数据共享的方式。你将看到如何把这些特性添加到自己的应用程序中，并且聪明地使用它们，使应用程序成为 iOS 系统中具有协作精神的一员。

2.1 秘诀：处理统一类型标识符

统一类型标识符（Uniform Type Identifier，UTI）代表 iOS 信息共享的中心组件。可以把它们视作下一代 MIME 类型。UTI 是标识资源类型（比如图像和文本）的字符串，它们指定哪些类型的信息将用于公共数据对象。它们不需要为此依赖于老式的指示符，比如文件扩展名、MIME 类型，或者文件类型的元数据（比如 OSType）。UTI 利用更新、更灵活的技术替代了这些项目。

UTI 使用反向域名风格的命名约定。常见的源于 Apple 的标识符看起来如下：public.html 和 public.jpeg，它们分别指 HTML 源文本和 JPEG 图像，二者都是专用的文件信息类型。

继承对于 UTI 起着重要作用。UTI 使用类似于 OO 的继承系统，其中子 UTI 具有对父 UTI 的 "is-a"（是一个）关系。子 UTI 会继承它们的父 UTI 的所有属性，但是会添加它们表示的数据种类的更多特征。这是由于每个 UTI 都可以根据需要承担更一般或者更特定的角色。例如，以 JPEG UTI 为例。JPEG 图像（public.jpeg）是一幅图像（public.image），而图像反过来又是一种数据（public.data），数据反过来又是一种用户可查看（或者可收听）的内容（public.content），它是一种项目（public.item），即用于 UTI 的普通的基本类型。这种层次结构被称为顺应性，其中子 UTI 顺应父 UTI。例如，更具体的 jpeg UTI 顺应更一般的图像或数据 UTI。

图 2-1 显示了 Apple 的基本顺应性树的一部分。这棵树上位于较低位置的任何项目都必须顺

应其所有父数据属性。声明一个父 UTI 意味着支持它的所有子 UTI。因此，可以打开 public.data 的应用程序必须能够服务于文本、电影、图像文件等。

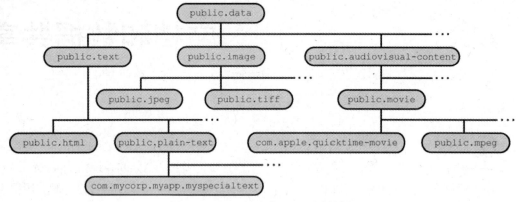

图 2-1　Apple 的公共 UTI 顺应性树

UTI 支持多重继承。一个项目可以顺应多个父 UTI 。因此，你可能设想一种同时提供文本和图像容器的数据类型，它声明了对二者的顺应性。

没有用于 UTI 项目的中心注册表，尽管每个 UTI 都应该遵守约定。public 域是为特定于 iOS 的类型预留的，是大多数应用程序所共有的。Apple 生成了公共项目的一个完整的家族层次结构。可以使用标准的预留域命名方式添加任何特定于第三方公司的名称（例如，com.sadun.myCustomType 和 com.apple.quicktime-movie）。

2.1.1　通过文件扩展名确定 UTI

Mobile Core Services 框架提供了一些实用程序，允许基于文件扩展名获取 UTI 信息。在使用这些基于 C 语言的函数时，一定要包括头文件，并把应用程序连接到框架。当给下面的函数传递一个路径扩展字符串时，它将返回一个首选的 UTI。首选的标识符是单个 UTI 字符串：

```
#import <MobileCoreServices/MobileCoreServices.h>

NSString *preferredUTIForExtension(NSString *ext)
{
    // Request the UTI via the file extension
    NSString *theUTI = (__bridge_transfer NSString *)
        UTTypeCreatePreferredIdentifierForTag(
            kUTTagClassFilenameExtension,
            (__bridge CFStringRef) ext, NULL);
    return theUTI;
}
```

可以使用 kUTTagClassMIMEType 作为第一个参数，给 UTTypeCreatePreferredIdentifierForTag() 传递一种 MIME 类型代替文件扩展名。该函数将为给定的 MIME 类型返回首选的 UTI：

```
NSString *preferredUTIForMIMEType(NSString *mime)
{
    // Request the UTI via the file extension
    NSString *theUTI = (__bridge_transfer NSString *)
        UTTypeCreatePreferredIdentifierForTag(
            kUTTagClassMIMEType,
            (__bridge CFStringRef) mime, NULL);
    return theUTI;
}
```

结合使用这些函数，将允许你从文件扩展名和 MIME 类型转向用于现代文件访问 UTI 类型。

2.1.2 从 UTI 转向扩展名或 MIME 类型

也可以另辟蹊径，从 UTI 产生首选的扩展名或 MIME 类型，这要使用 UTTypeCopyPreferredTagWithClass()。当给下面的函数传递 public.jpeg 时，它们将分别返回 jpeg 和 image/jpeg：

```
NSString *extensionForUTI(NSString *aUTI)
{
    CFStringRef theUTI = (__bridge CFStringRef) aUTI;
    CFStringRef results =
        UTTypeCopyPreferredTagWithClass(
            theUTI, kUTTagClassFilenameExtension);
    return (__bridge_transfer NSString *)results;
}

NSString *mimeTypeForUTI(NSString *aUTI)
{
    CFStringRef theUTI = (__bridge CFStringRef) aUTI;
    CFStringRef results =
        UTTypeCopyPreferredTagWithClass(
            theUTI, kUTTagClassMIMEType);
    return (__bridge_transfer NSString *)results;
}
```

必须在叶子层使用这些函数，这意味着直接在该层级声明类型扩展名。扩展名声明在属性列表中，其中将描述像文件扩展名和默认图标这样的特性。因此，例如，给扩展名函数传递 public.text 或 public.movie 将返回 nil，而 public.plain-text 和 public.mpeg 则将分别返回扩展名 txt 和 mpg。

前面的项目将在顺应性树中处于较高的位置，从而提供一种抽象类型，而不是特定的实现。对于目前为应用程序定义的给定类，没有现成的 API 函数用于寻找从它传承下来的项目。你可能想在 bugreport.apple.com 上归档一个增强请求。诚然，所有的扩展名和 MIME 类型都会注册在某个位置（否则，UTTypeCopyPreferredTagWithClass()将怎样从一开始就执行向上查找的工作），因此把扩展名映射到更一般的 UTI 的能力应该是可能存在的。

1. MIME 助手

尽管从扩展名到 UTI 的服务是很详尽的，可以为几乎任何扩展名返回 UTI，但是从 UTI 到 MIME 的结果则很随意，有些漫无目标。通常可以为任何公共项目生成正确的 MIME 表示；公共程度较低的项目则很少见。

下面的代码行显示了各种各样的扩展名、它们的 UTI（通过首选的 UTIForExtension()获取），以及从每个 UTI 生成的 MIME 类型（通过 mimeTypeForUTI()）。可以看到，其中有相当多的空白。当这些函数不能找到匹配时，它们将返回 nil：

```
xlv: dyn.age81u5d0 / (null)
xlw: com.microsoft.excel.xlw / application/vnd.ms-excel
xm: dyn.age81u5k / (null)
xml: public.xml / application/xml
z: public.z-archive / application/x-compress
zip: public.zip-archive / application/zip
zoo: dyn.age81y55t / (null)
zsh: public.zsh-script / (null)
```

为了解决这个问题，用于这个秘诀的示例代码包括了一个额外的 **MIMEHelper** 类。它定义了一个函数，返回所提供的扩展名的 MIME 类型：

```
NSString *mimeForExtension(NSString *extension);
```

它的扩展名和 MIME 类型源于 Apache Software Foundation，它将其列表放入公共域中。对于用于这个秘诀的示例代码中的 450 个扩展名，iOS 返回了全部 450 个 UTI，但是只会返回 88 个 MIME 类型。Apache 列表把这个数字增加到 230 个可识别的 MIME 类型。

2.1.3 测试顺应性

使用 UTTypeConformsTo()函数测试顺应性。该函数接受两个参数：一个源 UTI 和一个要比较的 UTI，如果第一个 UTI 顺应第二个 UTI，就返回 True。可以使用这个函数来测试一个更具体的项目是否顺应一个更一般的项目。相等性测试则使用 UTTypeEqual()。下面显示了一个示例，说明了可能如何使用顺应性测试，确定文件路径是否可能指向图像资源：

```
BOOL pathPointsToLikelyUTIMatch(NSString *path, CFStringRef theUTI)
```

```
{
    NSString *extension = path.pathExtension;
    NSString *preferredUTI = preferredUTIForExtension(extension);
    return (UTTypeConformsTo(
        (__bridge CFStringRef) preferredUTI, theUTI));
}

BOOL pathPointsToLikelyImage(NSString *path)
{
    return pathPointsToLikelyUTIMatch(path, CFSTR("public.image"));
}

BOOL pathPointsToLikelyAudio(NSString *path)
{
    return pathPointsToLikelyUTIMatch(path, CFSTR("public.audio"));
}
```

2.1.4　获取顺应性列表

UTTypeCopyDeclaration() 是 iOS API 中的所有 UTI 函数中最一般（并且最有用）的函数，它返回包含以下键的字典。

- **kUTTypeIdentifierKey**：UTI 名称，它将被传递给函数（例如，public.mpeg）。
- **kUTTypeConformsToKey**：类型顺应的任何父项目（例如，public.mpeg 顺应 public.movie）。
- **kUTTypeDescriptionKey**：正在考虑的类型（如果存在的话）的现实描述（例如，"MPEG movie"）。
- **kUTTypeTagSpecificationKey**：给定 UTI 的等价 OSType（例如，MPG 和 MPEG）、文件扩展名（mpg、mpeg、mpe、m75 和 m15）和 MIME 类型（视频/mpeg、视频/mpg、视频/x-mpeg 和视频/x-mpg）的字典。

除了这些公共项目之外，还会遇到更多的键，它们指定了导入和导出的 UTI 描述（kUTImportedTypeDeclarationsKey 和 kUTExportedTypeDeclarationsKey）、与 UTI 相关联的图标资源（kUTTypeIconFileKey）、指向描述类型的页面的 URL（kUTTypeReferenceURLKey），以及为 UTI 提供版本字符串的版本键（kUTTypeVersionKey）。

使用返回的字典向上通过顺应性树来构建一个数组，表示给定 UTI 顺应的所有项目。例如，public.mpeg 类型顺应 public.movie、public.audiovisual-content、public.data、public.item 和 public.content。通过 conformanceArray 函数将这些项目返回为一个数组，如秘诀 2-1 所示。

秘诀 2-1　测试顺应性

```objc
// Build a declaration dictionary for the given type
NSDictionary *utiDictionary(NSString *aUTI)
{
    NSDictionary *dictionary =
        (__bridge_transfer NSDictionary *)
            UTTypeCopyDeclaration((__bridge CFStringRef) aUTI);
    return dictionary;
}


// Return an array where each member is guaranteed unique
// but that preserves the original ordering wherever possible
NSArray *uniqueArray(NSArray *anArray)
{
    NSMutableArray *copiedArray =
        [NSMutableArray arrayWithArray:anArray];

    for (id object in anArray)
    {
        [copiedArray removeObjectIdenticalTo:object];
        [copiedArray addObject:object];
    }

    return copiedArray;
}


// Return an array representing all UTIs that a given UTI conforms to
NSArray *conformanceArray(NSString *aUTI)
{
    NSMutableArray *results =
        [NSMutableArray arrayWithObject:aUTI];
    NSDictionary *dictionary = utiDictionary(aUTI);
    id conforms = [dictionary objectForKey:
        (__bridge NSString *)kUTTypeConformsToKey];

    // No conformance
    if (!conforms) return results;

    // Single conformance
```

```
    if ([conforms isKindOfClass:[NSString class]])
    {
        [results addObjectsFromArray:conformanceArray(conforms)];
        return uniqueArray(results);
    }
    // Iterate through multiple conformance
    if ([conforms isKindOfClass:[NSArray class]])
    {
        for (NSString *eachUTI in (NSArray *) conforms)
            [results addObjectsFromArray:conformanceArray(eachUTI)];
        return uniqueArray(results);
    }
    // Just return the one-item array
    return results;
}
```

获取这个秘诀的代码

要查找这个秘诀的完整示例项目，可以浏览 https://github.com/erica/iOS-6-Advanced-Cookbook，并进入第 2 章的文件夹。

2.2 秘诀：访问系统粘贴板

粘贴板（在某些系统上也称为剪贴板）提供了一种核心 OS 特性，用于跨应用程序共享数据。用户可以在一个应用程序中把数据复制到粘贴板上，切换任务，然后把该数据复制到另一个应用程序中。剪切/复制/粘贴特性类似于在大多数操作系统中的那些特性。当用户在文本框或视图之间切换时，也可以在单个应用程序内执行复制和粘贴；开发人员也可以为特定于应用程序的数据建立私人粘贴板，它们将不会被其他应用程序所使用。

UIPasteboard 类允许访问共享的设备粘贴板及其内容。下面这个代码段返回一般的系统粘贴板，它适合于大多数一般的复制/粘贴应用：

```
UIPasteboard *pb = [UIPasteboard generalPasteboard];
```

除了一般的共享式系统粘贴板之外，iOS 还提供了特定于应用程序的粘贴板，以更好地确保数据隐私，它不会扩展到应用程序之外，并且自定义名称的粘贴板可以跨应用程序使用，但是仅限于那些知道并且使用粘贴板名称键的应用程序。使用 pasteboardWithUniqueName 创建特定于应用程序的粘贴板，它返回一个应用程序粘贴板对象，该对象将在应用程序退出前持续存在。

使用 pasteboardWithName:create: 创建自定义的粘贴板，它返回一个具有指定名称的粘贴板。为粘贴板使用反向 DNS 命名方式（例如，com.sadun.shared-application-pasteboard）。如果粘贴板还不存在，创建参数将指定系统是否应该创建它。这种类型的粘贴板可以超越单个应用程序的运行而持续存在；可以在创建后把持久属性设置为 YES。可以使用 removePasteboardWithName:销毁粘贴板，并释放被它使用的资源。

2.2.1 存储数据

粘贴板一次可以存储一个或多个条目。每个条目都具有一种关联的类型，可以使用 UTI 指定存储的是哪种类型的数据。例如，你可能发现 public.text（更确切地讲是 public.utf8-plain-text）存储文本数据，public.url 用于 URL 地址，public.jpeg 则用于图像数据。除此之外，iOS 上还使用了许多其他的公共数据类型。存储类型的数据的字典被称为 item，可以通过粘贴板的 items 属性获取所有可用项目的数组。

可以确定当前利用简单的消息存储的项目的种类。可以给粘贴板发送 pasteboardTypes 消息，查询粘贴板的可用类型。这将返回当前存储在粘贴板上的类型的数组：

```
NSArray *types = [pb pasteboardTypes];
```

可以在粘贴板上设置数据，并通过传递一个 NSData 对象和一个描述数据所顺应的类型的 UTI，来关联一种类型。此外，对于属性列表对象（即字符串、日期、数组、字典、数字或 URL），可以通过 setValue:forPasteboardType:设置一个 NSValue。这些属性列表对象在内部的存储方式稍微不同于它们的原始数据版本，从而导致了方法上的差异：

```
[[UIPasteboard generalPasteboard]
    setData:theData forPasteboardType:theUTI];
```

2.2.2 存储公共类型

粘贴板可以进一步专用于几种数据类型，它们代表最常用的粘贴板项目。它们是颜色（不是一个属性列表"值"对象）、图像（也不是一个属性列表"值"对象）、字符串和 URL。UIPasteboard 类提供了专用的获取器和设置器，使得更容易处理这些项目。可以把其中每个项目都视作粘贴板的属性，因此可以使用点表示法设置和获取它们。更重要的是，每个属性都具有一种复数形式，允许把这些项目作为对象的数组来访问它们。

粘贴板属性极大地简化了在最常用的情况下使用系统粘贴板。属性访问器包括如下一些。

- **string**：设置或获取粘贴板上的第一个字符串。
- **strings**：设置或获取粘贴板上的所有字符串的数组。
- **image**：设置或获取粘贴板上的第一幅图像。
- **images**：设置或获取粘贴板上的所有图像的数组。
- **URL**：设置或获取粘贴板上的第一个 URL。
- **URLs**：设置或获取粘贴板上的所有 URL 的数组。

- **color**：设置或获取粘贴板上的第一种颜色。
- **colors**：设置或获取粘贴板上的所有颜色的数组。

2.2.3 获取数据

当使用 4 个特殊类之一时，只需使用关联的属性从粘贴板中获取数据即可，否则，可以使用 dataForPasteboardType:方法取出数据。该方法返回其类型与作为参数发送的 UTI 匹配的第一个项目中的数据。粘贴板中任何其他的匹配项目都将被忽略。

如果需要获取所有匹配的数据，可以取回 itemSetWithPasteboardTypes:，然后遍历集合，以获取每个字典。可以从单个字典键中取回每个项目的数据类型，以及从其值中取回数据。

如前所述，UIPasteboard 提供了两种方法，用于粘贴到粘贴板上，这依赖于要粘贴的信息是一个属性列表对象还是原始的数据。对于属性列表对象（包括字符串、日期、数字、字典、数组和 URL），可以使用 setValueForPasteboardType:方法；对于一般的数据，则使用 setData:forPasteboardType:方法。

当粘贴板改变时，它们将发出一个 UIPasteboardChangedNotification，可以通过默认的 NSNotificationCenter 观察者侦听它。也可以监视自定义的粘贴板，并通过 UIPasteboardRemovedNotification 侦听它们的删除操作。

> **注意：**
> 如果你想成功地把文本数据粘贴到 Notes 或 Mail 中，可以在把信息存储到粘贴板上时使用 public.utf8-plain-text 作为所选的 UTI。使用 string 或 strings 属性可以自动增强这个 UTI。

2.2.4 被动更新粘贴板

坦率地讲，iOS 的选择和复制界面并不是操作系统的最高效的元素。有时，你希望为用户简化操作，同时又准备好打算与其他应用程序共享的内容。

考虑秘诀 2-2。它允许用户使用文本视图输入和编辑文本，同时自动执行更新粘贴板的过程。当观察者处于活动状态时（通过简单地点按按钮来切换），每一次编辑都会使文本更新粘贴板。这是通过实现一个文本视图委托方法（textViewDidChange:）来完成的，该方法通过自动把更改赋予粘贴板来响应编辑（updatePasteboard）。

这个秘诀演示了访问和更新粘贴板中涉及的相对简单性。

秘诀 2-2　创建自动将文本输入到粘贴板中的解决方案

```
- (void) updatePasteboard
{
    // Copy the text to the pasteboard when the watcher is enabled
```

```
    if (enableWatcher)
        [UIPasteboard generalPasteboard].string = textView.text;
}

- (void) textViewDidChange: (UITextView *) textView
{
    // Delegate method calls for an update
    [self updatePasteboard];
}

- (void) toggle: (UIBarButtonItem *) bbi
{
    // switch between standard and auto-copy modes
    enableWatcher = !enableWatcher;
    bbi.title = enableWatcher ? @"Stop Watching" : @"Watch";
}
```

> **获取这个秘诀的代码**
>
> 要查找这个秘诀的完整示例项目，可以浏览 https://github.com/erica/iOS-6-Advanced-Cookbook，并进入第 2 章的文件夹。

2.3 秘诀：监测 Documents 文件夹

　　iOS 文档并没有受困在它们的沙盒中，你可以并且应该与用户共享它们。应该允许用户直接控制他们的文档，以及访问他们可能在设备上创建的任何资料。一个简单的 Info.plist 设置将使 iTunes 能够显示用户的 Documents 文件夹的内容，并使那些用户能够根据需要添加和删除资料。

　　在将来某个时间，你可能使用一个简单的 NSMetadataQuery 监测器来监视 Documents 文件夹并报告更新。在编写本书时，元数据监视还没有扩展到 iCloud 之外以用于其他的文件夹。从 OS X 导出的代码无法像期望的那样在 iOS 上工作。目前，准确地讲，有两个搜索域可供 iOS 使用：即普遍存在的数据范围和普遍存在的文档范围（即 iCloud 和 iCloud）。

　　直到 iOS 中出现了一般的功能之后，才能使用 kqueue。这种老式技术提供了可伸缩的事件通知。利用 kqueue，可以监测添加和清除事件。这粗略地等同于寻找要添加和删除的文件，它们是你想做出反应的主要更新类型。秘诀 2-3 展示了一个用于监视 Documents 文件夹的 kqueue 实现。

2.3.1 支持文档文件共享

　　要支持文件共享，可以向应用程序的 Info.plist 中添加一个 UIFileSharingEnabled 键，并把它的值设置为 YES，如图 2-2 所示。在处理非原始的键和值时，这个项目被称为支持 iTunes 文件共享的 Application。iTunes 将在每个设备的 Apps 选项卡中列出所有声明文件共享支持的应用程序，如图 2-3 所示。

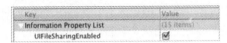

图 2-2　启用 UIFileSharingEnabled 以允许用户通过 iTunes 访问 Documents 文件夹

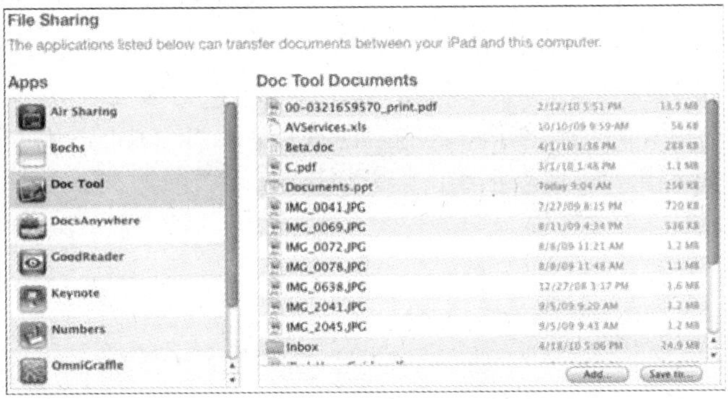

图 2-3　在 iTunes 中，将在设备的 Apps 选项卡中列出每个安装的声明了 UIFileSharingEnabled 的应用程序

2.3.2　用户控制

不能指定在 Documents 文件夹中允许存放哪些类型的项目。用户可以添加他们喜欢的任何项目，以及删除他们希望删除的任何项目。不过，他们不能做的是使用 iTunes 界面导航子文件夹。注意图 2-3 中的 Inbox 文件夹，这是一个从应用程序之间的文档共享中遗留下来的工件，但它不应该出现在那里。用户不能直接管理数据，不应该把子文件夹留在那里以使他们混淆。

用户在 iTunes 中不能像删除其他文件和文件夹那样删除 Inbox，应用程序应该也不能直接把文件写到 Inbox 中。尊重 Inbox 的角色，它用于捕获从其他应用程序传入的任何数据。在实现文件共享支持时，总是要检查 Inbox 以恢复活动状态，并且处理该数据以清空 Inbox，以及无论何时应用程序启动和恢复运行时都要删除它。在本章后面将讨论处理传入的文档的最佳实践。

2.3.3　Xcode 访问

作为一位开发人员，你不仅能够访问 Documents 文件夹，而且能够访问整个应用程序沙盒。使用 Xcode Organizer (Command-2) > Devices 选项卡>"设备">Applications >"应用程序名称"可以浏览沙盒，以及从中上传和下载文件。

通过启用应用程序的 UIFileSharingEnabled 属性，可以测试基本的文件共享，以及把数据加载到 Documents 文件夹中。在创建了那些文件之后，可以使用 Xcode 和 iTunes 检查、下载和删除它们。

2.3.4 扫描新文档

秘诀 2-3 通过在其 beginGeneratingDocumentNotificationsInPath:方法中请求 kqueue 通知来工作。在这里，它获取一个用于你所提供的路径（在这里是 Documents 文件夹）的文件描述符，并请求用于添加和清除事件的通知。它将把这个功能添加到当前的运行循环中，无论何时监测的文件夹更新，都会启用通知。

一旦接收到那个回调，它将发布一条通知（我自定义的 kDocumentChanged，在 kqueueFired 方法中），并且继承监视新事件。在主线程上的主运行循环中都会运行它，因此一旦接收到通知，GUI 就可以响应并更新它自身。

下面的代码段演示了如何使用秘诀 2-3 的监视器来更新 GUI 中的文件列表。无论何时内容改变了，更新通知都允许应用程序刷新那些目录内容清单：

```
- (void) scanDocuments
{
    NSString *path = [NSHomeDirectory()
        stringByAppendingPathComponent:@"Documents"];
    items = [[NSFileManager defaultManager]
        contentsOfDirectoryAtPath:path error:nil];
    [self.tableView reloadData];
}

- (void) loadView
{
    [self.tableView registerClass:[UITableViewCell class]
        forCellReuseIdentifier:@"cell"];
    [self scanDocuments];
    // React to content changes
    [[NSNotificationCenter defaultCenter]
        addObserverForName:kDocumentChanged
        object:nil queue:[NSOperationQueue mainQueue]
        usingBlock:^(NSNotification *notification){
            [self scanDocuments];
        }];

    // Start the watcher
    NSString *path = [NSHomeDirectory()
        stringByAppendingPathComponent:@"Documents"];
    helper = [DocWatchHelper watcherForPath:path];
}
```

把设备连接到 iTunes，测试这个秘诀。使用 iTunes App 选项卡界面添加和删除项目。设备的机载文件列表将会更新，以实时反映那些改变。

在使用这个秘诀时，要知道一些警告。首先，对于较大的文档，在收到了创建它们的通知之后，不应该立即阅读它们。你可能希望调查文件大小，以确定何时应该停止写入数据。第二，iTunes File Sharing 在必要时可以暂缓传输，要相应地进行编码。

秘诀 2-3　使用 kqueue 文件监测器

```
#import <fcntl.h>
#import <sys/event.h>

#define kDocumentChanged \
    @"DocumentsFolderContentsDidChangeNotification"

@interface DocWatchHelper : NSObject
{
    CFFileDescriptorRef kqref;
    CFRunLoopSourceRef rls;
}
@property (strong) NSString *path;
+ (id) watcherForPath: (NSString *) aPath;
@end

@implementation DocWatchHelper
@synthesize path;

- (void)kqueueFired
{
    int             kq;
    struct kevent   event;
    struct timespec timeout = { 0, 0 };
    int             eventCount;

    kq = CFFileDescriptorGetNativeDescriptor(self->kqref);
    assert(kq >= 0);

    eventCount = kevent(kq, NULL, 0, &event, 1, &timeout);
    assert( (eventCount >= 0) && (eventCount < 2) );

    if (eventCount == 1)
```

```objc
        [[NSNotificationCenter defaultCenter]
            postNotificationName:kDocumentChanged
            object:self];

    CFFileDescriptorEnableCallBacks(self->kqref,
        kCFFileDescriptorReadCallBack);
}

static void KQCallback(CFFileDescriptorRef kqRef,
    CFOptionFlags callBackTypes, void *info)
{
    DocWatchHelper *helper =
        (DocWatchHelper *)(__bridge id)(CFTypeRef) info;
    [helper kqueueFired];
}

- (void) beginGeneratingDocumentNotificationsInPath:
    (NSString *) docPath
{
    int                  dirFD;
    int                  kq;
    int                  retVal;
    struct kevent        eventToAdd;
    CFFileDescriptorContext context =
        { 0, (void *)(__bridge CFTypeRef) self,
            NULL, NULL, NULL };

    dirFD = open([docPath fileSystemRepresentation], O_EVTONLY);
    assert(dirFD >= 0);

    kq = kqueue();
    assert(kq >= 0);

    eventToAdd.ident = dirFD;
    eventToAdd.filter = EVFILT_VNODE;
    eventToAdd.flags = EV_ADD | EV_CLEAR;
    eventToAdd.fflags = NOTE_WRITE;
    eventToAdd.data = 0;
    eventToAdd.udata = NULL;
```

```
    retVal = kevent(kq, &eventToAdd, 1, NULL, 0, NULL);
    assert(retVal == 0);

    self->kqref = CFFileDescriptorCreate(NULL, kq,
        true, KQCallback, &context);
    rls = CFFileDescriptorCreateRunLoopSource(
        NULL, self->kqref, 0);
    assert(rls != NULL);

    CFRunLoopAddSource(CFRunLoopGetCurrent(), rls,
        kCFRunLoopDefaultMode);
    CFRelease(rls);

    CFFileDescriptorEnableCallBacks(self->kqref,
        kCFFileDescriptorReadCallBack);
}

- (void) dealloc
{
    self.path = nil;
    CFRunLoopRemoveSource(CFRunLoopGetCurrent(), rls,
        kCFRunLoopDefaultMode);
    CFFileDescriptorDisableCallBacks(self->kqref,
        kCFFileDescriptorReadCallBack);
}

+ (id) watcherForPath: (NSString *) aPath
{
    DocWatchHelper *watcher = [[self alloc] init];
    watcher.path = aPath;
    [watcher beginGeneratingDocumentNotificationsInPath:aPath];
    return watcher;
}
@end
```

获取这个秘诀的代码

要查找这个秘诀的完整示例项目，可以浏览 https://github.com/erica/iOS-6-Advanced-Cookbook，并进入第 2 章的文件夹。

2.4 秘诀：展示活动视图控制器

iOS 6 新引入的活动视图控制器可以把数据活动集成进图 2-4 所示的界面中。你只需付出最少的开发成本，这种新型控制器就允许你的用户把项目复制到粘贴板上，发布给社交媒体，以及通过电子邮件和文本共享它们等。内置的活动包括：Facebook（脸书）、Twitter（推特）、Weibo（微博）、SMS、邮件、打印、复制到粘贴板以及把数据分配给联系人。应用程序也可以定义它们自己的自定义服务，在本节后面可以了解这一点：

- **UIActivityTypePostToFacebook**
- **UIActivityTypePostToTwitter**
- **UIActivityTypePostToWeibo**
- **UIActivityTypeMessage**
- **UIActivityTypeMail**
- **UIActivityTypePrint**
- **UIActivityTypeCopyToPasteboard**
- **UIActivityTypeAssignToContact**

图 2-4　UIActivityViewController 类提供了系统和自定义服务

这个列表中遗漏了两个重要的活动，即 Open in 和 QuickLook，前者用于在应用程序之间共享文档，后者用于预览文件。本章后面将讨论这种功能，并且利用一些秘诀说明了如何独立地支持这些特性，对于 QuickLook 来说，则是如何集成进活动视图控制器中。

> **注意：**
> 无需拥有一台 AirPrint 打印机以测试打印活动。Ecamm 的 Printopia（http://www.ecamm.com/mac/printopia/，19.95 美元）可以在你的本地网络上创建一台虚拟打印机，可以从设备和模拟器中使用它。可以选择打印到任何本地打印机、打印到 Mac 上的文或者打印到 Dropbox 上的文件。对于利用新的活动视图控制器从事的任何开发工作，这都是一笔极好的投资。Netputing 制造了一款名为 handyPrint 的类似产品（http://www.netputing.com/handyprint），handyPrint 接受 PayPal 捐赠。

2.4.1 展示活动视图控制器

展示控制器的方式因设备而异，可以在 iPhone 家族成员上以模态方式以及在平板电脑上以弹出窗口（popover）显示它。UIBarButtonSystemItemAction 图标提供了用于填充链接到这个控制器的栏按钮的完美方式。

最重要的是，你自己几乎不需要做任何工作。在用户选择一种活动之后，控制器将会处理所有进一步的交互，比如展示邮件或 Twitter 创作单，向机载库中添加图片，或者把它分配给联系人。

2.4.2 活动项目源

秘诀 2-4 通过代码创建并展示了视图控制器。这种实现使它的主类采用 UIActivityItemSource 协议，并把自身添加到传递给控制器的项目数组中。在获取数据项目时，采用源协议有助于控制器理解如何使用回调。这表现了创建和展示控制器的两种方式中的第一种。

协议的两个必需的方法提供项目以进行处理（将用于活动的数据），并给该项目提供一个占位符。项目对应于适合于给定活动类型的对象。可以基于传递给回调的活动的类型，来区分返回的对象。例如，你可能发推特说："我在应用程序名称中创建了一首优秀的歌曲"，但是你可能通过电子邮件发送实际的声音文件。

用于项目的占位符通常是返回的与项目相同的数据，除非具有必须处理或创建的对象。在这种情况下，可以创建一个不带有真实数据的占位符对象。

两个回调（项目和占位符）都在主线程上运行，因此要保持数据比较小。如果需要大量的处理数据，可以考虑代之以使用提供者。

2.4.3 项目提供者

UIActivityItemProvider 类允许延迟传递数据。在共享数据前，这种操作可以给你提供操纵数据的灵活性。例如，在可以把大型视频文件上传到社交共享站点或者从较大的序列中对某个音频进行二次抽样之前，可能需要先对它们进行处理。

可以子类化提供者类并实现 item 方法。它可以代替通常用于操作的 main 方法。生成处理过的数据，在知道它将异步运行并且不会阻碍用户的交互式体验的情况下它将是安全的。

2.4.4 项目源回调

回调方法允许基于彼此之间的预期用法来区分数据。使用活动类型（比如 Facebook 或者 Add to Contacts，在本节前面列出了它们），选择你想提供的准确数据。在为不同的应用选择分辨率时，这特别重要。在打印时，要保持数据具有较高的质量。在发推特时，低分辨率的图像也可能达到想要的结果。

如果数据是不变的，也就是说，传送给 Facebook 的数据与传送给电子邮件的相同，就可以直接提供数据项的数组（通常是字符串、图像和 URL）。例如，可以像下面这样创建控制器，它使用单独一幅图像：

```
UIActivityViewController *activity = [[UIActivityViewController alloc]
    initWithActivityItems:@[imageView.image]
    applicationActivities:nil];
```

这种直接的方法要简单得多。主类不需要声明项目源协议；不需要实现额外的方法。它是一种为简单项目管理活动的快速、容易的方式。

也不仅限于传递单个项目，可以根据需要在活动项目数组中包括额外的元素。下面的控制器可能把它的两幅图像添加到电子邮件中，或者把它们都保存到系统的相机胶卷中，这依赖于用户的选择。拓宽活动以使用多个项目将允许用户更高效地使用你的应用程序：

```
UIImage *secondImage = [UIImage imageNamed:@"Default.png"];
UIActivityViewController *activity = [[UIActivityViewController alloc]
    initWithActivityItems:@[imageView.image, secondImage]
    applicationActivities:nil];
```

秘诀 2-4　活动视图控制器

```
- (void) presentViewController:
    (UIViewController *)viewControllerToPresent
{
    if (IS_IPHONE)
    {
        [self presentViewController:viewControllerToPresent
            animated:YES completion:nil];
    }
    else
    {
        popover = [[UIPopoverController alloc]
            initWithContentViewController:viewControllerToPresent];
```

```
        popover.delegate = self;
        [popover presentPopoverFromBarBarButtonItem:
                self.navigationItem.leftBarButtonItem
        permittedArrowDirections:UIPopoverArrowDirectionAny
        animated:YES];
    }
}

// Return the item to process
- (id)activityViewController:
        (UIActivityViewController *)activityViewController
    itemForActivityType:(NSString *)activityType
{
    return imageView.image;
}

// Return a thumbnail version of that item
- (id)activityViewControllerPlaceholderItem:
    (UIActivityViewController *)activityViewController
{
    return imageView.image;
}

// Create and present the view controller
- (void) action
{
    UIActivityViewController *activity =
        [[UIActivityViewController alloc]
            initWithActivityItems:@[self] applicationActivities:nil];
    [self presentViewController:activity];
}
```

获取这个秘诀的代码

要查找这个秘诀的完整示例项目,可以浏览 https://github.com/erica/iOS-6-Advanced-Cookbook,并进入第 2 章的文件夹。

2.4.5 添加服务

每个应用程序都可以通过子类化 UIActivity 类并展示一个自定义的视图控制器,来提供特定于应用程序的服务。视图控制器使用户能够以某种方式处理传递的数据。程序清单 2-1 介绍了一种最基本的活动,用于展示一个简单的文本视图。该视图列出了通过活动控制器传递给它的项目,它显示了每个项目的类和描述。

这个程序清单包括两个独特的类的详细信息。第一个类实现一个简单的文本控制器,并且打算在导航层次结构内使用。它包括一个视图和一个处理程序,前者用于展示文本,后者用于在用户点按 Done 时通过发送 activityDidFinish:来更新 UIActivity 调用实例。

添加一种方式以使活动完成是重要的,尤其是当控制器没有自然的终点时。在你的动作把数据上传到 FTP 服务器时,你就知道它何时完成。如果它发出消息,你就知道状态何时发布。在这个示例中,它取决于用户确定这个活动何时完成。确保视图控制器包含一个指回这个活动的弱属性,以便在工作结束时发送确实完成方法。

活动类包含许多必需的和可选的项目。应该实现这个程序清单中显示的所有方法。支持自定义的活动的方法包括以下一些。

- **activityType**:返回一个描述活动类型的独特字符串。在系统提供的活动中,与这个字符串对应的活动之一是 UIActivityTypePostToFacebook。使用类似的命名模式。这个字符串用于确定特定的活动类型以及它将做什么。在这个程序清单中,我返回的是 "@"CustomActivityTypeListItemsAndTypes"",它描述了活动。

- **activityTitle**:提供你想在活动控制器中显示的文本。图 2-5 中的自定义文本就是由这个方法返回的。在描述自定义的动作时可以使用活动的描述。可以遵照 Apple 的指导,例如,"Save to Camera Roll"(保存到相机胶卷)、"Print"(打印)、"Copy"(复制)。你的标题应该完成短语 "I want to..."(我想要……),例如,"I want to print"(我想要打印)、"I want to copy"(我想要复制),或者在这个示例中,"I want to list items"(我想要列出项目)。使用标题大小写形式,并且除了像 "to" 和 "and" 这样的次要单词之外,还要大写每个单词的首字母。

- **activityImage**:返回一幅图像让控制器使用。控制器将添加一个反斜杠,并把图像转换成一幅单值位图(one-value bitmap),然后把它分层放置在顶部。在透明背景上使用简单的艺术作品(在 iPhone 上是 57 像素×57 像素,在 iPad 上是 72 像素×72 像素,对于 Retina 屏幕比例则要加倍),构建图标图像的内容。你将希望在插入艺术作品时至少距离每一边 15%,以留出空间插入控制器提供的圆角矩形,给图像加上框架。

- **canPerformWithActivityItems**:扫描传递的项目,并且决定控制器是否可以处理它们。如果是,就返回 YES。

- **prepareWithActivityItems**:存储传递的项目以便以后使用(在这里,把它们分配给一个局部实例变量),并且执行任何必要的预处理。

- **activityViewController**:使用以前传递给你的活动项目,返回一个完全初始化的、像样的视图控制器。这个控制器将被自动展示给用户,她在那里执行承诺的动作之前可以自定义选项。

2.4 秘诀：展示活动视图控制器

图 2-5 添加自定义的应用程序活动

添加自定义的动作允许应用程序扩展其数据处理可能性，同时把一些特性集成进一致的系统提供的界面中。它是一种强大的 iOS 特性。最强大的活动选择将与系统服务相集成（比如复制到粘贴板，或者保存到相册），或者提供对脱离设备的 API 的连接，比如 Facebook、Twitter、Dropbox 和 FTP。

这个示例只是简单地列出项目，代表一种弱用例。没有理由不能把相同的特性提供为正常的应用程序中的屏幕。在考虑"动作"时，要尝试延伸到应用程序之外。把用户的数据与扩展到正常的 GUI 之外的共享与处理特性连接起来。

程序清单 2-1 应用程序活动

```
// All activities present a view controller. This custom controller
// provides a full-sized text view.
@interface TextViewController : UIViewController
    @property (nonatomic, readonly) UITextView *textView;
    @property (nonatomic, weak) UIActivity *activity;
@end
@implementation TextViewController

// Make sure you provide a done handler of some kind, such as this
// or an integrated button that finishes and wraps up
- (void) done
{
    [_activity activityDidFinish:YES];
}
// Just a super-basic text view controller
- (id) init
{
    if (!(self = [super init])) return nil;
    _textView = [[UITextView alloc] init];
```

```objc
        _textView.font = [UIFont fontWithName:@"Futura" size:16.0f];
        _textView.editable = NO;
        
        [self.view addSubview:_textView];
        PREPCONSTRAINTS(_textView);
        STRETCH_VIEW(self.view, _textView);
        
        // Prepare a Done button
        self.navigationItem.rightBarButtonItem =
            BARBUTTON(@"Done", @selector(done));
        
        return self;
}
@end

@interface MyActivity : UIActivity
@end
@implementation MyActivity
{
    NSArray *items;
}

// A unique type name
- (NSString *)activityType
{
    return @"CustomActivityTypeListItemsAndTypes";
}
// The title listed on the controller
- (NSString *) activityTitle
{
    return @"Cookbook";
}
// A custom image, displayed as a bitmap over a textured background
// This one says "iOS" in a rounded rect edge
- (UIImage *) activityImage
{
    CGRect rect = CGRectMake(0.0f, 0.0f, 75.0f, 75.0f);
    UIGraphicsBeginImageContext(rect.size);
    rect = CGRectInset(rect, 15.0f, 15.0f);
    UIBezierPath *path = [UIBezierPath
```

2.4 秘诀：展示活动视图控制器   61

```
        bezierPathWithRoundedRect:rect cornerRadius:4.0f];
    [path stroke];
    rect = CGRectInset(rect, 0.0f, 10.0f);
    [@"iOS" drawInRect:rect
        withFont:[UIFont fontWithName:@"Futura" size:18.0f]
        lineBreakMode:NSLineBreakByWordWrapping
        alignment:NSTextAlignmentCenter];
    UIImage *image = UIGraphicsGetImageFromCurrentImageContext();
    UIGraphicsEndImageContext();

    return image;
}

// Specify if you can respond to these items
- (BOOL)canPerformWithActivityItems:(NSArray *)activityItems
{
    return YES;
}

// Store the items locally for later use
- (void)prepareWithActivityItems:(NSArray *)activityItems
{
    items = activityItems;
}

// Return a view controller, in this case one that lists
// its items and their classes
- (UIViewController *) activityViewController
{
    TextViewController *tvc = [[TextViewController alloc] init];
    tvc.activity = self;
    UITextView *textView = tvc.textView;

    NSMutableString *string = [NSMutableString string];
    for (id item in items)
        [string appendFormat:
            @"%@: %@\n", [item class], [item description]];
    textView.text = string;
    // Make sure to provide some kind of done: handler in
    // your main controller.
```

```
    UINavigationController *nav = [[UINavigationController alloc]
        initWithRootViewController:tvc];
    return nav;
}
@end
```

2.4.6 项目和服务

为每个项目展示的服务因传递的数据种类而异。表 2-1 列出了由源数据类型提供的活动。如同在本章中所看到的，预览控制器支持扩展到了这些基础类型之外。

- iOS 的 Quick Look（快速查看）框架把活动控制器集成到它的文件预览中。Quick Look 提供的活动控制器可以打印并通过电子邮件发送许多类型的文档。一些文档类型也支持其他的活动。
- 文档交互控制器（Document Interaction Controller）提供了"open in"特性，允许在应用程序之间共享文件。它将把活动添加到它的"选项"样式表示中，并把活动与"open in"选择结合起来。

表 2-1 用于数据类型的活动类型

源	提供的活动
NSString 字符串（单个或多个）	Mail（邮件）、Message（消息）、Twitter（推特）、Facebook（脸书）、Weibo（微博）和 Copy（复制）
UIImage 图像（单个）	Mail（邮件）、Twitter（推特）、Facebook（脸书）、Weibo（微博）、Assign to Contact（分配给联系人）、Save to Camera Roll（保存到相机胶卷）、Print（打印）和 Copy（复制）
UIImage 图像（多个）	Mail（邮件）、Facebook（脸书）、Assign to Contact（分配给联系人）、Save to Camera Roll（保存到相机胶卷）、Print（打印）和 Copy（复制）
UIColor 颜色	Copy（复制）
NSURL URL	Mail（邮件）、Message（消息）、Twitter（推特）、Facebook（脸书）、Weibo（微博）和 Copy（复制）
NSDictionary 字典	字典中包含的对象的任何支持的活动。悲哀的是，对于数组则不是如此，不支持它们
不支持的项目	例如，NSData、NSArray、NSDate 或 NSNumber：什么也没有，一个空视图控制器
多种不同的项目	所有支持类型的联合体；例如，对于字符串+图像，可以获得：Mail（邮件）、Message（消息）、Twitter（推特）、Facebook（脸书）、Weibo（微博）、Assign to Contact（分配给联系人）、Save to Camera Roll（保存到相机胶卷）、Print（打印）和 Copy（复制）

2.4.7 支持 HTML 电子邮件

如果你想使用电子邮件活动发送 HTML，就要确保项目的文本字符串开始于 "@"<html>""。只要实现项目源协议，并且依据用户所选的活动返回合适的项目，就可以把基于 HTML 的电子邮件文本与普通的 Twitter 内容区分开。

2.4.8 排除活动

可以通过给 excludedActivityTypes 属性提供一份活动类型的列表，明确地排除一些活动：

```
UIActivityViewController *activity = [[UIActivityViewController alloc]
    initWithActivityItems:items applicationActivities:@[appActivity]];
activity.excludedActivityTypes = @[UIActivityTypeMail];
```

2.5 秘诀：Quick Look 预览控制器

Quick Look 预览控制器类允许用户预览许多文档类型。这个控制器支持文本、图像、PDF、RTF、iWork 文件、Microsoft Office 文档（Office 97 及更高版本，包括 doc、ppt、xls 等）和逗号分隔的值（comma-separated value，csv）文件。你提供一种受支持的文件类型，Quick Look 控制器将为用户显示它。集成的系统提供的活动视图控制器有助于共享预览的文档，如图 2-6 所示。

图 2-6　Quick Look 控制器以模态方式展示，并且显示用户点按了动作按钮之后的屏幕。Quick Look 控制器可以处理广泛的文档类型，使用户能够先查看文件内容，然后再决定要应用于它们的动作。大多数 Quick Look 类型都支持 Mail（邮件）和 Print（打印），其中许多类型支持 Copy（复制），图像文件甚至提供了更多的选项

可以推送或展示预览控制器。控制器能够适应这两种情形，与导航栈和模态表示协同工作。秘诀 2-5 演示了两种方法。

2.5.1 实现 Quick Look

Quick Look 支持需要几个简单的步骤。

(1) 在主控制器类中声明 QLPreviewControllerDataSource 协议。

(2) 实现 numberOfPreviewItemsInPreviewController:和 previewController:previewItemAtIndex:数据源方法。其中第一个方法返回要预览的项目计数；第二个方法则返回索引所引用的预览项目。

(3) 预览项目必须遵守 QLPreviewItem 协议，该协议包含两个必需的属性：预览标题和项目 URL。秘诀 2-5 创建了一个符合要求的 QuickItem 类，该类实现了一个绝对最低限度的方法，用于支持数据源。

在满足了所有这些要求之后，代码将准备好创建一个新的预览控制器，设置它的数据源，然后展示或推送它。

秘诀 2-5　Quick Look

```
@interface QuickItem : NSObject <QLPreviewItem>
@property (nonatomic, strong) NSString *path;
@property (readonly) NSString *previewItemTitle;
@property (readonly) NSURL *previewItemURL;
@end

@implementation QuickItem

// Title for preview item
- (NSString *) previewItemTitle
{
    return [_path lastPathComponent];
}

// URL for preview item
- (NSURL *) previewItemURL
{
    return [NSURL fileURLWithPath:_path];
}
@end

#define FILE_PATH [NSHomeDirectory() \
```

```
    stringByAppendingPathComponent:@"Documents/PDFSample.pdf"]

@interface TestBedViewController : UIViewController
    <QLPreviewControllerDataSource>
@end

@implementation TestBedViewController
- (NSInteger) numberOfPreviewItemsInPreviewController:
    (QLPreviewController *) controller
{
    return 1;
}
- (id <QLPreviewItem>) previewController:
       (QLPreviewController *) controller
    previewItemAtIndex: (NSInteger) index;
{
    QuickItem *item = [[QuickItem alloc] init];
    item.path = FILE_PATH;
    return item;
}
// Push onto navigation stack
- (void) push
{
    QLPreviewController *controller =
        [[QLPreviewController alloc] init];
    controller.dataSource = self;
    [self.navigationController
        pushViewController:controller animated:YES];
}


// Use modal presentation
- (void) present
{
    QLPreviewController *controller =
        [[QLPreviewController alloc] init];
    controller.dataSource = self;
    [self presentViewController:controller
        animated:YES completion:nil];
}
```

```
- (void) loadView
{
    self.view.backgroundColor = [UIColor whiteColor];

    self.navigationItem.rightBarButtonItem =
        BARBUTTON(@"Push", @selector(push));
    self.navigationItem.leftBarButtonItem =
        BARBUTTON(@"Present", @selector(present));
}
@end
```

> **获取这个秘诀的代码**
>
> 要查找这个秘诀的完整示例项目，可以浏览 https://github.com/erica/iOS-6-Advanced-Cookbook，并进入第 2 章的文件夹。

2.6 秘诀：添加 QuickLook 动作

值得注意的是，QuickLook 不存在于系统提供的活动视图控制器所展示的标准动作集中。你可以轻松地添加一个自定义的动作以提供这种特性，从而提供与秘诀 2-5 创建的相同预览。这就是秘诀 2-6 所做的工作，它把秘诀 2-5 的功能包装进一个自定义的 QLActivity 类中。

秘诀 2-6 将对符合要求的项目执行比你在本章中已经见过的更彻底的搜索。它将搜索传递给它的项目数组，直至找到一个本地文件 URL，它可以将其用于文档预览。如果它没有找到这样一个 URL，就会从 canPerformWithActivityItems:方法返回 NO，并且不会列出在活动控制器上。

秘诀 2-6 Quick Look

```
@implementation QLActivity
{
    NSArray *items;
    NSArray *qlitems;
    QLPreviewController *controller;
}

// Activity Customization
- (NSString *)activityType
{
    return @"CustomQuickLookActivity";
}
```

```objc
- (NSString *) activityTitle
{
    return @"QuickLook";
}
- (UIImage *) activityImage
{
    return [UIImage imageNamed:@"QL.png"];
}
// Items must include at least one file URL
- (BOOL)canPerformWithActivityItems:(NSArray *)activityItems
{
    for (NSObject *item in activityItems)
        if ([item isKindOfClass:[NSURL class]])
        {
            NSURL *url = (NSURL *)item;
            if (url.isFileURL) return YES;
        }
    return NO;
}
// QuickLook callbacks
- (NSInteger) numberOfPreviewItemsInPreviewController:
    (QLPreviewController *) controller
{
    return qlitems.count;
}
- (id <QLPreviewItem>) previewController: (QLPreviewController *)
    controller previewItemAtIndex: (NSInteger) index;
{
    return qlitems[index];
}
// Item preparation
- (void)prepareWithActivityItems:(NSArray *)activityItems
{
    items = activityItems;

    controller = [[QLPreviewController alloc] init];
    controller.dataSource = self;
    controller.delegate = self;
```

```objc
        NSMutableArray *finalArray = [NSMutableArray array];
        
        for (NSObject *item in items)
        {
            if ([item isKindOfClass:[NSURL class]])
            {
                NSURL *url = (NSURL *)item;
                if (url.isFileURL)
                {
                    QuickItem *item = [[QuickItem alloc] init];
                    item.path = url.path;
                    [finalArray addObject:item];
                }
            }
        }
        qlitems = finalArray;
}

- (void) previewControllerDidDismiss:
    (QLPreviewController *)controller
{
    [self activityDidFinish:YES];
}
- (UIViewController *) activityViewController
{
    return controller;
}
@end
```

获取这个秘诀的代码

要查找这个秘诀的完整示例项目,可以浏览 https://github.com/erica/iOS-6-Advanced-Cookbook,并进入第 2 章的文件夹。

2.7 秘诀:使用文档交互控制器

UIDocumentInteractionController 类允许应用程序给用户展示一个选项菜单,允许他们以各种方式使用文档文件,利用这个类,用户可以利用以下特性。

- iOS 应用程序之间的文档共享(即"在某个应用程序中打开这个文档")。

- 使用 QuickLook 进行文档预览。
- 活动控制器选项，比如打印、共享和社交网络。

你已经在本章前面的动作中见过后两种特性。文档交互类在这些特性顶部添加了应用程序间的共享，如图 2-7 所示。控制器被展示为菜单，这使用户能够指定他们想怎样与给定的文档交互。

图 2-7　UIDocumentInteractionController 显示了用于 iPhone（左图）和 iPad（右图）的选项风格。两种表示都包括两个页面的图标，如靠近显示屏底部的页面控制器所示

在 iOS 6 中，"open in"（打开在）选项的数量不再会限制它在早期的 OS 版本中的使用方式，这就是你为什么会在菜单底部看到页面指示器的原因。用户可以轻扫屏幕，查看"open in"（打开在）选项的完整补充。

控制器提供了两种基本的菜单风格。"open"（打开）风格只提供了"open in"（打开在）选择，使用菜单空间提供尽可能多的目的地选择。"options"（选项）风格（见图 2-7）提供了所有交互选项的列表，包括"open in"（打开在）、快速查看和任何支持的动作。它实质上是你将从标准的 Actions（"动作"）菜单中获得的所有良好的选项，以及"open in"（打开在）额外选项。你必须明确地添加快速查看回调，但它需要做一点工作。

2.7.1　创建文档交互控制器实例

每个文档交互控制器都特定于单个文档文件。这个文件通常存储在用户的 Documents（"文档"）文件夹中：

```
dic = [UIDocumentInteractionController
    interactionControllerWithURL:fileURL];
```

你提供一个本地文件 URL，并且使用选项变化（基本上是 Action 菜单）或者打开菜单（仅仅是"open in"(打开在)项目)。两种表示风格都来自于一个栏按钮或者屏幕上的矩形：

- `presentOptionsMenuFromRect:inView:animated:`
- `presentOptionsMenuFromBarButtonItem:animated:`
- `presentOpenInMenuFromRect:inView:animated:`
- `presentOpenInMenuFromBarButtonItem:animated:`

iPad 使用你传递的栏按钮或矩形来展示一个弹出窗口(popover)。在 iPhone 上，实现的功能将展示一个模态控制器视图。如你所期望的，更多的流水账会占据 iPad 上的空间，其中用户可能点按其他的栏按钮，可能会关闭弹出窗口，等等。

你将希望在展示每个 iPad 栏按钮项目关联的控制器之后禁用它们，以及在它们不起作用之后重新启用它们。这很重要，因为你不希望用户再次点按正在使用的栏按钮，并且需要处理需要接管不同弹出窗口的情形。基本上，如果不仔细监测哪些按钮是活动的以及哪个弹出窗口正在使用中，将有可能发生各种不愉快的情况。秘诀 2-7 预防了这些情况。

2.7.2 文档交互控制器属性

每个文档交互控制器都提供了许多属性，可以在你的委托回调中使用它们。

- URL 属性允许向控制器查询它正在服务的文件，它与你在创建控制器时传递的 URL 相同。
- UTI 属性用于确定哪些应用程序可以打开文档。它使用本章前面讨论过的系统提供的功能，基于文件名和元数据查找最佳的 UTI 匹配。你可以在代码中覆盖它，以手动设置属性。
- name 属性指定了 URL 的最后一个路径成分，它提供了一种快捷方式，指定一个用户可解释的名称。
- 使用 icons 属性获取正在使用的文件类型的图标。声明支持某些文件类型的应用程序将在它们的声明中提供图像链接(稍后将讨论如何声明文件支持)。这些图像对应于为 kUTTypeIconFileKey 键存储的值，这在本章前面讨论过。
- annotation 属性提供了一种方式，连同文件一起把自定义的数据传递给任何将打开该文件的应用程序。这个属性没有标准的用法；尽管如此，仍然必须把项目设置为某个顶级的属性列表对象，即字典、数组、数据、字符串、数字和日期。由于没有共同的标准，人们倾向于把这个属性的使用控制在最小范围内，除非开发人员在他们自己发布的应用程序套件中共享信息。

2.7.3 提供文档 Quick Look 支持

通过实现 3 个委托回调给控制器添加 Quick Look 支持。这些方法声明哪个视图控制器将用于展示预览、哪个视图将宿主它，以及用于预览尺寸的框架。你可能偶尔会有具有说服力的理由，在平板电脑上的有限屏幕空间内使用子视图控制器（比如在拆分视图中，利用仅仅一部分空间进行预览），但是对于 iPhone 家族，几乎没有任何理由不允许预览占据整个屏幕：

```
#pragma mark QuickLook
- (UIViewController *)
    documentInteractionControllerViewControllerForPreview:
        (UIDocumentInteractionController *)controller
{
    return self;
}

- (UIView *) documentInteractionControllerViewForPreview:
    (UIDocumentInteractionController *)controller
{
    return self.view;
}

- (CGRect) documentInteractionControllerRectForPreview:
    (UIDocumentInteractionController *)controller
{
    return self.view.frame;
}
```

2.7.4 检查打开菜单

在使用文档交互控制器时，Options（选项）菜单几乎总会提供有效的菜单选择，尤其是当实现 Quick Look 回调时。不过，你可能会或者可能不会有任何"open-in"（打开在）选项可以使用。这些选项依赖于你提供给控制器的文件数据以及用户在他们的设备上安装的应用程序。

当设备上没有安装支持你正在使用的文件类型的应用程序时，将会发生没有打开选项的情况。这可能是由于鲜为人知的文件类型引起的，但是更常见的是由于用户还没有购买并安装相关的应用程序。

因此总是要检查是否提供了"Open"（打开）菜单项。秘诀 2-7 执行了一个相当丑陋的测试，以查看外部程序是否把它们自身提供为给定 URL 的展示器和编辑器。它的工作方式如下：它创建一个新的、临时的控制器并尝试展示它。如果它获得成功，符合需要的文件目的地就会存在并被安装到设备上。如果没有成功，就没有这样的应用程序，并且应该禁用任何"Open In"（打开在）按钮。

在 iPad 上，必须在 viewDidAppear:中或者以后（也就是说在建立了窗口之后）运行这项检查。方法在展示控制器之后将立即使之消失。最终用户应该不会注意到它，并且没有任何调用使用动画。

显然，这是一个相当糟糕的实现，但它具有在布置界面或者开始处理一个新文件时进行测试的优点。我鼓励你在 bugreporter.apple.com 上提出一个增强请求。

进一步警告：尽管这种测试可以在主视图上工作（如同这个秘诀中一样），但它可能在 iPad 上的弹出窗口中的非标准表示中引发问题。

> **注意：**
> 通常极少在同一个应用程序中同时使用"选项"和"打开"项目。秘诀 2-7 为 Options（选项）菜单使用系统提供的 Action（动作）项目。你可能想为只使用"打开"风格的应用程序使用它来代替 "Open in..."（打开在）文本。

秘诀 2-7　文档交互控制器

```
@implementation TestBedViewController
{
    NSURL *fileURL;
    UIDocumentInteractionController *dic;
    BOOL canOpen;
}

#pragma mark QuickLook
- (UIViewController *)
    documentInteractionControllerViewControllerForPreview:
        (UIDocumentInteractionController *)controller
{
    return self;
}

- (UIView *) documentInteractionControllerViewForPreview:
    (UIDocumentInteractionController *)controller
{
    return self.view;
}

- (CGRect) documentInteractionControllerRectForPreview:
    (UIDocumentInteractionController *)controller
{
    return self.view.frame;
}

#pragma mark Options / Open in Menu
```

```objectivec
// Clean up after dismissing options menu
- (void) documentInteractionControllerDidDismissOptionsMenu:
    (UIDocumentInteractionController *) controller
{
    self.navigationItem.leftBarButtonItem.enabled = YES;
    dic = nil;
}
// Clean up after dismissing open menu
- (void) documentInteractionControllerDidDismissOpenInMenu:
    (UIDocumentInteractionController *) controller
{
    self.navigationItem.rightBarButtonItem.enabled = canOpen;
    dic = nil;
}

// Before presenting a controller, check to see if there's an
// existing one that needs dismissing
- (void) dismissIfNeeded
{
    if (dic)
    {
        [dic dismissMenuAnimated:YES];
        self.navigationItem.rightBarButtonItem.enabled = canOpen;
        self.navigationItem.leftBarButtonItem.enabled = YES;
    }
}
// Present the options menu
- (void) action: (UIBarButtonItem *) bbi
{
    [self dismissIfNeeded];
    dic = [UIDocumentInteractionController interactionControllerWithURL:fileURL];
    dic.delegate = self;
    self.navigationItem.leftBarButtonItem.enabled = NO;
    [dic presentOptionsMenuFromBarButtonItem:bbi animated:YES];
}
// Present the open-in menu
- (void) open: (UIBarButtonItem *) bbi
{
    [self dismissIfNeeded];
```

```objc
        dic = [UIDocumentInteractionController interactionControllerWithURL:fileURL];
        dic.delegate = self;
        self.navigationItem.rightBarButtonItem.enabled = NO;
        [dic presentOpenInMenuFromBarButtonItem:bbi animated:YES];
}


#pragma mark Test for Open-ability
-(BOOL)canOpen: (NSURL *) aFileURL
{
        UIDocumentInteractionController *tmp =
            [UIDocumentInteractionController
                interactionControllerWithURL:aFileURL];
        BOOL success = [tmp presentOpenInMenuFromRect:CGRectMake(0,0,1,1)
            inView:self.view animated:NO];
        [tmp dismissMenuAnimated:NO];
        return success;
}
- (void) viewDidAppear:(BOOL)animated
{
        // Only enable right button if the file can be opened
        canOpen = [self canOpen:fileURL];
        self.navigationItem.rightBarButtonItem.enabled = canOpen;
}


#pragma mark View management
- (void) loadView
{
        self.view.backgroundColor = [UIColor whiteColor];
        self.navigationItem.rightBarButtonItem =
            BARBUTTON(@"Open in...", @selector(open:));
        self.navigationItem.leftBarButtonItem =
            SYSBARBUTTON(UIBarButtonSystemItemAction,
                @selector(action:));

        NSString *filePath = [NSHomeDirectory()
            stringByAppendingPathComponent:@"Documents/DICImage.jpg"];
        fileURL = [NSURL fileURLWithPath:filePath];
}
@end
```

> **获取这个秘诀的代码**
> 要查找这个秘诀的完整示例项目，可以浏览 https://github.com/erica/iOS-6-Advanced-Cookbook，并进入第 2 章的文件夹。

2.8 秘诀：声明文档支持

应用程序文档并不仅限于它们创建或者从 Internet 下载的文件。如你在前一个秘诀中所发现的，应用程序可能处理某些文件类型。它们可能打开从其他应用程序传递过来的项目。你已经从发送方的角度见过了文档共享，它使用"open in"（打开在）控制器把文件导出到其他应用程序。现在应该从接收方的角度探讨它。

应用程序在它们的 Info.plist 属性列表中声明它们对某些文件类型的支持。Launch Services（启动服务）系统将读取该数据，并创建被文档交互控制器使用的文件-应用程序之间的关联。

尽管可以直接编辑属性列表，但是 Xcode 4 提供了一种简单的形式，作为 Project > Target > Info screen 的一部分。定位 Document Types 区域，你将发现它位于 Custom iOS Target Properties 下面。打开这个区域，并单击"+"，添加受支持的新文档类型。图 2-8 显示了它对于接受 JPEG 图像文档的应用程序来说看起来像是什么样子的。

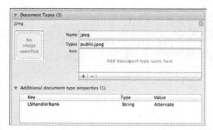

图 2-8 在 Xcode 的 Target > Info 窗格中声明受支持的文档类型

这个声明包含 3 个最低限度的细节：名称、一个或多个 UTI 以及一个处理程序等级，在这里是替代者。

- 名称是必需的和任意的。它应该能够描述正在使用的文档的种类，但它在 iOS 上也有点像是一种事后追加的东西。当在 Macintosh 上使用时这个字段将更有意义（它是 Finder 使用的"种类"字符串），但它不是可选的。
- 在输入时指定一个或多个 UTI。这个示例只指定了 public.jpeg。在列出多个项目时，可以在项目之间添加逗号。例如，你可能具有一种打开 public.jpeg、public.tiff 和 public.png 的"图像"文档类型。当需要限制文件支持时，可以枚举特定的类型。尽管声明 public.image 将包含全部 3 种类型，它也可能允许打开不受支持的图像样式。

- 启动服务处理程序等级描述了应用程序如何在处理这种文件类型的竞争中看待它自身。"所有者"说这是创建这类文件的原始应用程序,"替代者"(如图2-8所示)则提供了辅助的查看方式。可以在额外的文档类型属性中手动添加 LSHandlerRank 键。

可以选择指定图标文件。它们在 OS X 中被用作文档图标,与 iOS 世界具有最少的重叠。在唯一一种情况下,我可以认为你可能把这些图标看作 iTunes 中的 Apps 选项卡,此时将使用 File Sharing 区域添加和移除项目。图标通常具有两种尺寸:320像素×320像素(UTType Size320IconFile)和64像素×64像素(UTTypeSize64IconFile),通常限制于应用程序创建并为其定义一种自定义类型的文件。

Xcode 在底下使用这种交互形式在应用程序的 Info.plist 中构建一个 CFBundleDocumentTypes 数组。下面的代码段以其 Info.plist 形式显示了图2-8中的信息。

```
<key>CFBundleDocumentTypes</key>
<array>
    <dict>
        <key>CFBundleTypeIconFiles</key>
        <array/>
        <key>CFBundleTypeName</key>
        <string>jpg</string>
        <key>LSHandlerRank</key>
        <string>Alternate</string>
        <key>LSItemContentTypes</key>
        <array>
            <string>public.jpeg</string>
        </array>
    </dict>
</array>
```

2.8.1 创建自定义的文档类型

当应用程序构建新的文档类型时,应该在 Target > Info 编辑器的 Exported UTIs 区域中声明它们,如图2-9所示。这将注册系统支持这种文件类型,并把你标识为该类型的所有者。

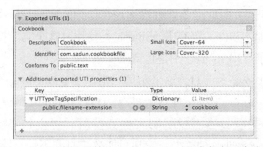

图2-9 在 Target > Info 编辑器的 Exported UTIs 区域中声明自定义的文件类型

要定义新的类型，可以提供一个自定义的 UTI（在这里是 com.sadun.cookbookfile）、文档艺术作品（大小为 64 像素和 320 像素），并且指定标识文件类型的文件扩展名。与声明文档支持一样，Xcode 将把一个导出的声明数组构建到项目的 Info.plist 文件中。下面显示了图 2-9 中所示的声明看起来可能像是什么样子的。

```
<key>UTExportedTypeDeclarations</key>
<array>
    <dict>
        <key>UTTypeConformsTo</key>
        <array>
            <string>public.text</string>
        </array>
        <key>UTTypeDescription</key>
        <string>Cookbook</string>
        <key>UTTypeIdentifier</key>
        <string>com.sadun.cookbookfile</string>
        <key>UTTypeSize320IconFile</key>
        <string>Cover-320</string>
        <key>UTTypeSize64IconFile</key>
        <string>Cover-64</string>
        <key>UTTypeTagSpecification</key>
        <dict>
            <key>public.filename-extension</key>
            <string>cookbook</string>
        </dict>
    </dict>
</array>
```

如果像这样添加到项目中，应用程序应该会使用 com.sadun.cookbookfile UTI，打开具有 cookbook 扩展名的任何文件。

2.8.2　实现文档支持

当应用程序提供了文档支持时，每当 "Inbox"（收件箱）文件夹变成活动状态时都应该检查它。确切地讲，查看 Inbox 文件夹是否出现在 Documents 文件夹中。如果是，就应该把那个收件箱中的元素移到属于它们的位置，通常是在主 Documents 目录中。在清空了收件箱后，就删除它。这提供了最佳的用户体验，尤其对于通过 iTunes 进行任何文件共享则更是如此，其中的 Inbox 及其作用可能使用户混淆：

```
- (void)applicationDidBecomeActive:(UIApplication *)application
{
```

```
    // perform inbox test here
}
```

当把项目移到 Documents 中时，要检查名称冲突，并且使用替代路径名称（通常是通过追加一个连字符，其后接着一个数字）来避免重写任何现有的文件。下面的方法有助于查找目标路径的替代名称。在经过 1000 次尝试后，它会放弃，但是严重的是，任何用户都不应该宿主许多重复的文档名称。如果他们这样做，总体应用程序设计就会存在严重的错误。

秘诀 2-8 展示了扫描 Inbox 并把文件移到合适位置的细节。它将在清空 Inbox 之后移除它。可以看到，任何像这样的方法都是文件管理器密集型的。它主要涉及处理可能在整个任务过程中弹出的各种可能的错误组合，这对于小文件支持应该运行得很快。如果必须处理大文件，比如视频或音频，就要确保在它自己的操作队列上执行这种处理。

如果计划支持 public.data 文件（也就是说，将打开任何内容），则可能希望使用 UIWebView 实例显示那些文件。参阅 Technical Q&A QA1630（http://developer.apple.com/library/ios/#qa/qa1630），可了解关于 iOS 在那些视图中能够以及不能显示哪些文档类型的详细信息。Web 视图可以展示大多数音频和视频资源，以及 Excel、Keynote、Numbers、Pages、PDF、PowerPoint 和 Word 资源，还包括简单的 HTML。

秘诀 2-8　处理传入的文档

```
#define DOCUMENTS_PATH [NSHomeDirectory() \
    stringByAppendingPathComponent:@"Documents"]
#define INBOX_PATH     [DOCUMENTS_PATH \
    stringByAppendingPathComponent:@"Inbox"]

@implementation InboxHelper
+ (NSString *) findAlternativeNameForPath: (NSString *) path
{
    NSString *ext = path.pathExtension;
    NSString *base = [path stringByDeletingPathExtension];

    for (int i = 1; i < 999; i++)
    {
        NSString *dest =
            [NSString stringWithFormat:@"%@-%d.%@", base, i, ext];

        // if the file does not yet exist, use this destination path
        if (![[NSFileManager defaultManager]
            fileExistsAtPath:dest])
            return dest;
    }
```

```objc
        NSLog(@"Exhausted possible names for file %@. Bailing.",
            path.lastPathComponent);
        return nil;
}
- (void) checkAndProcessInbox
{
    // Does the Inbox exist? If not, we're done
    BOOL isDir;
    if (![[NSFileManager defaultManager]
        fileExistsAtPath:INBOX_PATH isDirectory:&isDir])
            return;

    NSError *error;
    BOOL success;

    // If the Inbox is not a folder, remove it.
    if (!isDir)
    {
        success = [[NSFileManager defaultManager]
            removeItemAtPath:INBOX_PATH error:&error];
        if (!success)
        {
            NSLog(@"Error deleting Inbox file (not directory): %@",
                error.localizedFailureReason);
            return;
        }
    }

    // Retrieve a list of files in the Inbox
    NSArray *fileArray = [[NSFileManager defaultManager]
        contentsOfDirectoryAtPath:INBOX_PATH error:&error];
    if (!fileArray)
    {
        NSLog(@"Error reading contents of Inbox: %@",
            error.localizedFailureReason);
        return;
    }

    // Remember the number of items
    NSUInteger initialCount = fileArray.count;
```

```objc
// Iterate through each file, moving it to Documents
for (NSString *filename in fileArray)
{
    NSString *source =
        [INBOX_PATH stringByAppendingPathComponent:filename];
    NSString *dest = [DOCUMENTS_PATH
        stringByAppendingPathComponent:filename];

    // Is the file already there?
    BOOL exists =
        [[NSFileManager defaultManager] fileExistsAtPath:dest];
    if (exists) dest = [self findAlternativeNameForPath:dest];
    if (!dest)
    {
        NSLog(@"Error. File name conflict not resolved");
        continue;
    }

    // Move file into place
    success = [[NSFileManager defaultManager]
        moveItemAtPath:source toPath:dest error:&error];
    if (!success)
    {
        NSLog(@"Error moving file from Inbox: %@",
            error.localizedFailureReason);
        continue;
    }
}

// Inbox should now be empty
fileArray = [[NSFileManager defaultManager]
    contentsOfDirectoryAtPath:INBOX_PATH error:&error];
if (!fileArray)
{
    NSLog(@"Error reading contents of Inbox: %@",
        error.localizedFailureReason);
    return;
}
```

```
    if (fileArray.count)
    {
        NSLog(@"Error clearing Inbox. %d items remain",
            fileArray.count);
        return;
    }

    // Remove the inbox
    success = [[NSFileManager defaultManager]
        removeItemAtPath:INBOX_PATH error:&error];
    if (!success)
    {
        NSLog(@"Error removing inbox: %@",
            error.localizedFailureReason);
        return;
    }
    NSLog(@"Moved %d items from the Inbox", initialCount);
}
@end
```

获取这个秘诀的代码

要查找这个秘诀的完整示例项目，可以浏览 https://github.com/erica/iOS-6-Advanced-Cookbook，并进入第 2 章的文件夹。

2.9 秘诀：创建基于 URL 的服务

Apple 的内置应用程序提供了多种可以通过 URL 调用访问的服务。可以要求 Safari 打开 Web 页面，让 Maps 显示一幅地图，或者使用 mailto:风格的 URL 开始在 Mail 中撰写一封信件。URL 模式指出现在冒号之前的 URL 的第一部分，比如 http 或 ftp。

这些服务可以工作，因为 iOS 知道如何将 URL 模式匹配到应用程序。以 http:开头的 URL 将在 Mobile Safari 中打开。mailto:URL 总会链接到 Mail。你可能不知道的是：你可以定义自己的 URL 模式，并在应用程序中实现它们。并非所有标准的模式都在 iOS 上受支持，FTP 模式就不能使用。

无论何时 Mobile Safari 或另一个应用程序打开那种类型的 URL，自定义的模式都将允许应用程序启动。例如，如果应用程序注册 xyz，那么任何 xyz:链接都会直接到达应用程序以进行处理，其中将把它们传递给应用程序委托的 URL 打开方法。你不必在那里添加任何特殊的编码。如果你只想运行应用程序，添加模式和打开 URL 就支持跨应用程序的启动。

处理程序扩展了启动，以允许应用程序利用传递给它的 URL 做某件事情。它们可能打开特定的数据文件，获取特定的名称，显示某一幅图像，或者处理调用中包括的信息。

2.9.1 声明模式

要声明 URL 模式，可以编辑 Target > Info 编辑器的 URL Types 区域（参见图 2-10），并列出你将使用的 URL 模式。由该声明创建的 Info.plist 区域看起来将如下所示：

```
<key>CFBundleURLTypes</key>
<array>
    <dict>
        <key>CFBundleURLName</key>
        <string>com.sadun.urlSchemeDemonstration</string>
        <key>CFBundleURLSchemes</key>
        <array>
            <string>xyz</string>
        </array>
    </dict>
</array>
```

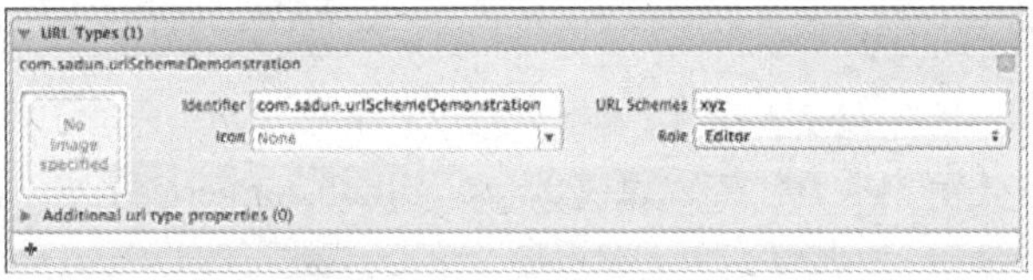

图 2-10　在 Target > Info 编辑器的 URL Types 区域中添加自定义的 URL 模式

CFBundleURLTypes 条目包括一个字典数组，描述了应用程序可以打开和处理的 URL 类型。每个字典都相当简单，它们包含两个键：**CFBundleURLName**（定义任意的标识符）和 **CFBundleURLSchemes** 的数组。

模式数组提供了一个前缀列表，它们属于抽象的名称。可以添加一种或多种模式，下面的示例只描述了其中一种。你可能想利用"x"给名称加前缀（例如，x-sadun-services）。尽管 iOS 家族不是任何标准化组织的一部分，"x"前缀还是指示这是一个未注册的名称。x-callback-url 的草案规范正在开发过程中，参见 http://x-callback-url.com。

以前的 iOS 开发人员（以及当前的 Apple 雇员）Emanuele Vulcano 在 CocoaDev Web 站点（http://cocoadev.com/index.pl?ChooseYourOwniPhoneURLScheme）上开启了正式的注册。iOS 开发人员可以在核心名单中共享他们的模式，以便你可以发现想要使用的服务，并且宣传你自己提供的服务。注册表将列出服务和它们的 URL 模式，并将描述这些服务可以怎样被其他开发人员使用。其他的注册表包括 http://handleopenurl.com、http://wiki.akosma.com/IPhone_URL_Schemes 和 http://applookup.com/Home。

2.9.2 测试 URL

可以测试一种 URL 服务是否可用。如果 UIApplication 的 canOpenURL:方法返回 YES，就保证 openURL:可以启动另一个应用程序打开那个 URL。但是不保证该 URL 是有效的，而只能保证它的模式被正确地注册到现有的应用程序：

```
if ([[UIApplication sharedApplication] canOpenURL:aURL])
    [[UIApplication sharedApplication] openURL:aURL];
```

2.9.3 添加处理程序方法

要处理 URL 请求，可以实现特定于 URL 的应用程序委托方法，如秘诀 2-9 所示。不幸的是，仅当应用程序已经在运行时，才可以保证将会触发该方法。如果它没有触发，并且 URL 请求启动了应用程序，控制首先会转到启动方法（将要完成和确实完成）。

你想确保正常的 application:didFinishLaunchingWithOptions:返回 YES。这允许控制传递给 application:openURL:sourceApplication:annotation:，以便可以处理传入的 URL。

秘诀 2-9 提供 URL 模式支持

```
// Called if the app is open or if the launch returns YES
- (BOOL)application:(UIApplication *)application
    openURL:(NSURL *)url
    sourceApplication:(NSString *)sourceApplication
    annotation:(id)annotation
{
    NSString *logString = [NSString stringWithFormat:
        @"DID OPEN: URL[%@] App[%@] Annotation[%@]\n",
        url, sourceApplication, annotation];
    tbvc.textView.text =
        [logString stringByAppendingString:tbvc.textView.text];
    return YES;
}

// Make sure to return YES
- (BOOL)application:(UIApplication *)application
    didFinishLaunchingWithOptions:(NSDictionary *)launchOptions
{
    window = [[UIWindow alloc]
        initWithFrame:[[UIScreen mainScreen] bounds]];
    tbvc = [[TestBedViewController alloc] init];
```

```
UINavigationController *nav = [[UINavigationController alloc]
    initWithRootViewController:tbvc];
window.rootViewController = nav;
[window makeKeyAndVisible];
return YES;
}
```

获取这个秘诀的代码

要查找这个秘诀的完整示例项目，可以浏览 https://github.com/erica/iOS-6-Advanced-Cookbook，并进入第 2 章的文件夹。

2.10 小结

你希望跨应用程序共享数据并且利用系统提供的动作吗？本章说明了如何实现你的愿望。你了解了 UTI 以及如何将它们用于跨应用程序指定数据角色，看到了粘贴板的工作方式以及如何利用 iTunes 共享文件，还学习了监测文件夹并且发现了如何实现自定义的 URL。你深入研究了文档交互控制器，并且看到了如何添加对各类操作的支持，从打印到复制再到预览。在结束本章的学习之前，要思考以下几点。

- 你从未受限于 Apple 提供的内置 UTI，但是在决定添加你自己的 UTI 时应该遵循它的指导。确保使用自定义的预留域命名，并在导出的定义中添加尽可能多的详细信息（公共 URL 定义页面、典型的图标和文件扩展名），一定要精确。
- 顺应性数组有助于确定你正在处理哪一类事情。知道它是一幅图像还是一个文本文件或电影有助于更好地处理与任何文件关联的数据。
- 常规的粘贴板提供了一种极佳的方式来处理共享数据，但是，如果具有特定于应用程序的交叉通信的需求，就没有理由不使用自定义的粘贴板来共享信息。只要知道粘贴板上的数据在重新启动后将不会持续存在即可。
- Documents 文件夹属于用户，而不属于你。要记住这一点，并且要殷勤地管理该目录。
- 文档交互控制器取代了许多开发人员使用自定义的 URL 模式的许多理由。使用控制器可提供用户需要的应用程序间的交互，不要害怕引入注释支持，它有助于使应用程序之间的转换变得容易。
- 不要提供 "open In" 菜单项，除非有机载应用程序准备好支持那个按钮。你在本章中学到的解决方案是不太成熟的，但它好于通过客户支持与生气、沮丧或困惑的用户打交道。考虑提供一个由这个方法提供支持的警报，当没有其他的应用程序可用时给出解释。

第 3 章
Core Text

从个人计算机出现以来，细致入微的文本布局就是计算体验的一个重要部分。甚至最早的消费者操作系统都包括了字处理与属性化的文本特性。数十年前，交互式文档准备工具提供了字体选择、文本样式设置（比如加粗、加下划线和倾斜）、缩进等。

这些相同的元素构成了 iOS 体验一个重大部分。开发人员使用属性化的字符串和 Core Text 框架利用最少量的开发工作提供高雅时髦的文本元素。本章介绍了属性化文本处理，并且探索了如何将文本特性构建到应用程序中。你将学到把属性化字符串添加到公共 UIKit 元素中，如何创建 Core Text 强化的视图，以及如何为自由的文本排版突破更大的界限。在学完本章后，你将发现 Core Text 带给 iOS 的强大能力。

3.1 Core Text 和 iOS

Core Text 是一种 Mac 和 iOS 技术，它以编程方式支持排版任务，包括属性规范和布局。它的 API 允许定义字体、颜色、对齐方式、单词换行以及其他移动字符串的特性，这些字符串不仅仅是字符的简单集合。最初作为一种 Mac OS X 10.5 框架，Core Text 在过去几个版本的 iOS 平台上进展缓慢。你将发现，甚至在 iOS 6 中，迁移的过程仍在继续。

3.1.1 属性

大量的文字排版操作都包含属性，尤其是属性化的字符串。属性是一组应用于某个范围内的文本的特性，比如字体选择或文本颜色。顾名思义，属性化的字符串将给所选的子串添加一些特征。属性化的字符串同时包含文本信息和应用于该字符串的特定于范围的属性。

为了理解属性的工作方式以及怎样结合它们，可以考虑下面的字符串：

This **is a sample** *string* that demonstrates how attributes can combine.

在这个非 iOS 的示例中，受图书出版的现实情况所限，上面的字符串为它的第 6~16 个字符使用了一种加粗属性，并为第 11~23 个字符使用了倾斜属性。

在 iOS 上，应用于字符串的属性种类不同于编写图书时使用的那些属性。例如，不会添加强调或加粗效果。它们是通过改变字体来设置的，例如，从 Courier 改为 Courier-Bold 或者 Courier-BoldOblique。你可以指定该字体应该应用的范围，就像可以指定范围来设置给定的笔画颜色或者文本对齐方式一样。

其他属性以类似的方式工作。可以更新某个范围内的字符的颜色、添加阴影或者改变字体大小。

3.1.2 C 语言与 Objective C

第一批基于 C 语言的 Core Text 属性首次出现在 iOS 3.2 中。它们使用 Core Foundation 模式，同时采用了少许 Objective C 友好的方式。下面给出了一个示例，显示了原始的 Core Text 看起来像什么样子。接着是一个来自 iOS 5 Cookbook 的代码段，它演示了如何创建一个 CTParagraphStyleRef，用于定义自定义的对齐方式和换行模式：

```
uint8_t theAlignment = [self ctAlignment];
CTParagraphStyleSetting alignSetting = {
    kCTParagraphStyleSpecifierAlignment, sizeof(uint8_t),
    &theAlignment};


uint8_t theLineBreak = [self ctBreakMode];
CTParagraphStyleSetting wordBreakSetting = {
    kCTParagraphStyleSpecifierLineBreakMode,
    sizeof(uint8_t),&theLineBreak};

CTParagraphStyleSetting settings[2] = {alignSetting, wordBreakSetting};
CTParagraphStyleRef paraStyle = CTParagraphStyleCreate(settings, 2);
```

下面的 iOS 6 代码与之形成了鲜明的对比，它们执行相同的任务。下面这个代码段创建一个 NSMutableParagraphStyle 实例，并利用上一个示例中使用的相同对齐方式和换行模式值自定义其属性。这段代码要简单得多，因为它利用了新的 Objective-C 类：

```
NSMutableParagraphStyle *paraStyle =
    [[NSMutableParagraphStyle alloc] init];
paraStyle.alignment = [self ctAlignment];
paraStyle.lineBreakMode = [self ctBreakMode];
```

iOS 6 为几乎所有的 Core Text 引入了 Objective-C 实现，允许忽略库的 Core Foundation 调用。还有几个值得注意的例外情况，本章后面将探讨它们。也就是说，iOS 现在进入了自定义文本的黄金时代，可以给 UIKit 中几乎所有的内容添加字体、颜色、间距等。

3.1.3 UIKit

在 iOS 6 中对 UIKit 进行了更新，使之能够更得心应手地处理文本。许多 UIKit 类（包括文本框、文本视图、标签和按钮）允许把属性化（Core Text 风格）的字符串分配给它们的 attributedText 属性，就像把普通的 NSString 分配给它们的 text 属性一样：

textField.attributedText = myAttributedString;

此外，iOS 现在还提供了一个新的（并且比较小的）特定于 UIKit 的文本属性词汇表，比如字体、颜色和阴影。它们可以用于导航栏、分段控件和栏项目（即栏式元素）。可以通过调用 setTitleTextAttributes:（导航栏）和 setTitleTextAttributes:forState:（分段控件和栏项目）来设置属性。可以使用以下字典键和值传递属性字典。

- **UITextAttributeFont**：提供一个 UIFont 实例。
- **UITextAttributeTextColor**：提供一个 UIColor 实例。
- **UITextAttributeTextShadowColor**：提供一个 UIColor 实例。
- **UITextAttributeTextShadowOffset**：提供一个包装 UIOffset 结构的 NSValue 实例。偏移量包括两个浮点值，分别在水平方向和垂直方向。使用 UIOffsetMake()通过一对浮点值构建结构。

例如，下面这个代码段把分段控件的文本颜色设置为以淡灰色显示其选中状态。无论何时选择这个控件，文本颜色都会从白色变为灰色：

```
NSDictionary *attributeDictionary =
    @{UITextAttributeTextColor : [UIColor lightGrayColor]};
[segmentedControl setTitleTextAttributes:attributeDictionary
    forState:UIControlStateSelected];
```

当 Apple 继续把排版特性融入其控件和视图中时，这个 UIKit 词汇表将随着时间的推移而变得更大。

3.2 属性化字符串

可以使用 NSAttributedString 类或其可变版本 NSMutableAttributedString 类的成员来定义特征。可变版本提供了大得多的灵活性，使你能够单独对属性进行分层处理，而不是同时添加所有的一切。

要创建属性化字符串，可以分配它，并利用文本和属性字典对其进行初始化。下面的代码段演示了如何构建一个"Hello World"字符串，并利用较大的灰色字体显示它。图 3-1 显示了这里创建的属性化字符串。

```
NSMutableDictionary *attributes = [NSMutableDictionary dictionary];
attributes[NSFontAttributeName] =
    [UIFont fontWithName:@"Futura" size:36.0f];
attributes[NSForegroundColorAttributeName] =
    [UIColor grayColor];

attributedString = [[NSAttributedString alloc]
    initWithString:@"Hello World" attributes: attributes];
textView.attributedText = attributedString;
```

排版关注的是创建属性字典。可以使用以下键填充字典，并把它们传递给属性化字符串。对于可变的实例，可以指定一个应用它们的字符范围。

图 3-1　属性化字符串存储诸如字体选择和颜色之类的信息

- **NSFontAttributeName**：一个 UIFont 对象，用于设置文本颜色。
- **NSParagraphStyleAttributeName**：一个 NSParagraphStyle，用于指定许多段落设置，包括对齐方式、换行模式、缩进等。
- **NSForegroundColorAttributeName** 和 **NSBackgroundColorAttributeName**：UIColor 对象，用于设置文本的颜色和文本后面显示的颜色。在当前 iOS 版本中，NSStrokeColorAttributeName 等同于前景色。
- **NSStrokeWidthAttributeName**：一个 NSNumber，存储一个浮点值，将笔画宽度定义为字体大小的百分比。负数将同时勾画和填充文本；正数则创建一种"空心"表示，并且勾画每个字符字形的边缘，如图 3-2 所示。

图 3-2　正笔画值将描绘字符字形的轮廓，但是不会填充其内部

```
attributes[NSStrokeWidthAttributeName] = @(3.0);
```

- **NSStrikethroughStyleAttributeName** 和 **NSUnderlineStyleAttributeName**：这些键指定一个项目是否使用删除线或下划线。在实际中，它们实质上是 Boolean（布尔）型的 NSNumber 实例，存储 0 或 1。从理论上讲，它们可能在将来的 iOS 版本中扩展到额外的值，因此 Apple 把属性值定义为 NSUnderlineStyleNoneI（0）或 NSUnderlineStyleSingle（1）。
- **NSShadowAttributeName**：一个 NSShadow 对象，用于设置阴影的颜色、偏移量和模糊半径，如图 3-3 所示。下面这个代码段对每个 Apple 的文档使用一个 CGSize 偏移量，但是当然这很快就会更新为 UIOffset 支持。

图 3-3　在 iOS 上使用新的 NSShadow 类添加阴影

```
NSShadow *shadow = [[NSShadow alloc] init];
shadow.shadowBlurRadius = 3.0f;
shadow.shadowOffset = CGSizeMake(4.0f, 4.0f);
attributes[NSShadowAttributeName] = shadow;
```

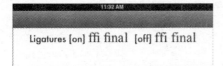

图 3-4 连写把某些字母组合（比如"f"和"i"）结合进单个字符字形中。当禁用连写时，注意"i"上单独的圆点

- **NSLigatureAttributeName**：连写指可以把单个字形（字符图片）绑定在一起的方式，比如"f"和"I"，如图 3-4 所示。这个键引用一个 NSNumber，可以选择"不使用连写"（0）、"使用默认连写"（1）或者"使用所有连写"（2）。iOS 6 不支持值 2。

  要知道的是，UIKit 视图可能不支持连写。我不得不使用直接的 Core Text 绘图创建图 3-4。我最初分配属性化字符串的 UITextView 不能正确地呈现"fi"连写。

- **NSBaselineOffsetAttributeName**：一个 NSNumber，这个浮点数指示相对于正常文本基线（即字母停留的位置）的位移。一些字母（比如"j"和"q"）将在布局期间落在基线下方。负数很可能把文本向基线上方偏移，正数则将其向下偏移（在编写本书时，违反了这个特性）。

- **NSKernAttributeName**：一个 NSNumber，在 iOS 6 中没有使用它，它最终指示是启用（1）还是禁用（0）字距调整。字距调整允许排字机调整字母之间的间距，使它们自然重叠，比如当把字母 A 放在字母 V 旁边时，例如，AV。

- **NSVerticalGlyphFormAttributeName**：一个 NSNumber，在 iOS 6 中没有使用它，它最终将支持水平（0）和垂直（1）文本布局。

3.2.1 段落样式

段落样式存储在它们自己的对象中，它们是 NSParagraphStyle 类的成员。使用 NSMutableParagraphStyle 类的可变版本设置样式的具体细节。把值分配给实例的属性，确保匹配必需的数据类型。下面的代码段创建图 3-5 中所示的表示效果，在段落之间使用超大的间距，并在第一行使用了慷慨的缩进效果。

```
NSMutableParagraphStyle *paragraphStyle = [[NSMutableParagraphStyle alloc] init];
paragraphStyle.alignment = NSTextAlignmentLeft;
paragraphStyle.lineBreakMode = NSLineBreakByWordWrapping;
paragraphStyle.firstLineHeadIndent = 36.0f;
paragraphStyle.lineSpacing = 8.0f;
paragraphStyle.paragraphSpacing = 24.0f;
attributes[NSParagraphStyleAttributeName] = paragraphStyle;
```

其中大多数值都与磅有关，比如行或段落之间的间距以及缩进。如果你认真仔细，就可以逐个段落地控制这些特性，为每个段落分配新的属性字典。

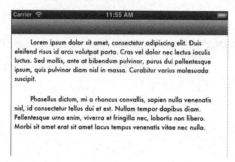

图 3-5　段落样式属性包括缩进、段落之间的间距、对齐方式等

> **注意：**
> 确保为每次属性更新都创建段落样式对象的副本。如果不这样做，就可能创建一种属性化的结果，其中所有的段落样式都指向相同的对象。这样，更新一个段落将会更新所有的段落。

3.3　秘诀：基本的属性化字符串

秘诀 3-1 演示了通过创建用户控制的段落样式和字体颜色，来构建属性化字符串。为了支持图 3-6 中所示的界面，用户从顶部的导航栏中选择一种对齐方式（左对齐、居中、右对齐和两端对齐）和一种颜色（黑色、红色、绿色和蓝色）。当用户选择这些选项时，秘诀 3-1 中的 **setupText** 方法将应用它们，以便可以实时更新显示。它构建了一个属性字典，并使用该字典创建属性化字符串，然后把它加载进中心文本视图中。

由于这个秘诀创建了一个简单的 **NSAttributedString** 实例，将在整个字符串中应用属性。在下一节中可以看到，使用可变实例有助于构建更复杂的表示。

秘诀 3-1　结合使用基本的属性化字符串与文本视图

```
- (void) setupText
{
    // Establish a new dictionary
    NSMutableDictionary *attributes = [NSMutableDictionary dictionary];

    // Create the paragraph style
    NSMutableParagraphStyle *paragraphStyle =
        [[NSMutableParagraphStyle alloc] init];
```

```
    paragraphStyle.alignment = alignment;
    paragraphStyle.paragraphSpacing = 12.0f;
    attributes[NSParagraphStyleAttributeName] = paragraphStyle;

    // Load up the attributes dictionary
    attributes[NSFontAttributeName] =
        [UIFont fontWithName:@"Futura" size:14.0f];
    attributes[NSForegroundColorAttributeName] = color;

    // Build the attributed string
    attributedString = [[NSAttributedString alloc]
        initWithString:lorem attributes: attributes];

    // Display the string
    textView.attributedText = attributedString;
}
```

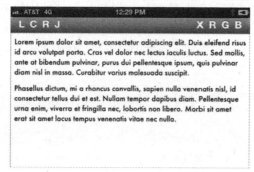

图 3-6 秘诀 3-1 构建了一个基本的界面，使用户能够设置字体颜色和文本对齐方式

> **获取这个秘诀的代码**
> 要查找这个秘诀的完整示例项目，可以浏览 https://github.com/erica/iOS-6-Advanced-Cookbook，并进入第 3 章的文件夹。

3.4 秘诀：可变的属性化字符串

顾名思义，NSMutableAttributedString 类提供了一些方式，用于改变已经创建好的属性化字符串。在这个类提供的许多方法当中，有一些方法用于替换、删除、插入和追加内容。

- **使用 replaceCharactersInRange:withString**：用新字符替换原来的字符。新字符将继承所替换的范围内的第一个字符处的任何活动属性。与之相比，replaceCharacters-

InRange:withAttributedString:则会引入传递的字符串中使用的任何属性和字符。
- 要删除字符，可以应用 deleteCharactersInRange:。
- **insertAttributedString:atIndex:** 和 **appendAttributedString:** 所做的事情正好就像它们的名称所暗示的那样。新添加的内容提供它自己的字符和属性。

这个类还允许调整属性，这是通过对某个范围应用一些特征来实现的，它通常是用于文本视图实例的所选范围。
- **setAttributes:range:** 将完全替换属于指定范围的任何字符的属性字典。
- **addAttributes:range:** 在给定范围内的任何现有属性之上分层放置传递的字典中的属性。
- **addAttribute:value:range:** 向范围中添加单个属性。
- **removeAttribute:range:** 从范围内的字符中删除单个属性。

设置一个属性字典不是没有风险；它完全消除了任何现有的属性，利用传递的字典中包括的任何属性彻底替换它们。任何未指定的属性都可能恢复 iOS 的默认值。例如，你可能尝试更改所选范围的颜色，并且注意到字体意想不到地快速恢复为 12 磅的 Helvetica。

对于修改，可代之以使用 addAttribute:方法。第二个和第三个方法将把新属性分层放置在现有的属性上，并且允许利用新值替换活动属性中的项目。利用这些方法，可以添加尚未指定的属性，或者修改现有的属性。

最后一个方法允许删除单个属性，比如颜色、阴影或字体选择，而不会影响当前生效的任何其他的属性。

图 3-7 应用特征并跨越范围界限

在实际应用中，用户创建的文本选择通常会跨越属性界限。图 3-7 提供了一个示例。单词 "eleifend"、"risus" 和 "id" 以不同的字体大小显示，分别是 36 磅、24 磅和 18 磅。想象一下用户选择 "fend ris"，并且请求改变色彩。要应用这个更新，addAttribute:value:range:方法必须单独处理每一组统一的特征。在这种情况下，选择将分解较大字体的 "fend" 和较小字体的 "ris"。

秘诀 3-2 再现了 addAttribute:value:range:功能。之所以包括进这个秘诀，是因为它在遍历属性范围方面极其有用，这就是一个非常好的示例。这段代码展示了如何分解属性范围，以及如何从应用程序中查询它们。该秘诀将遍历文本视图的所选范围，它将为在某个位置发现的属性查询其属性化字符串，并获取该属性字典的完整的有效范围。

在视图的初始状态中，这个范围将包含存储到视图中的所有文本，它通常比用户做出的任何选择都要大得多。为了处理有效的属性范围大于用户选择的情形，秘诀 3-2 进行了横切，返回受限于用户选择的范围。

它把当前位置的现有属性复制到一个可变字典中，并设置传递给方法的新属性。然后把这些属性复制回属性化字符串中，并把当前位置前移——朝着当前有效范围的末尾，或者朝着选择的末尾。

在图 3-7 中，"fend ris"选择包括两个基于字体大小的属性范围。当在这个示例中运行时，秘诀 3-2 首先会把较大的"fend"涂成红色，然后进行迭代，以对剩下的"ris"进行着色。在实际应用中，选择可以证明甚至更复杂，包含许多要更新的有效范围。

可以从这个秘诀中获得的要点如下。

- 属性化字符串同时包含字符和属性。
- 每个字符范围都与一个属性集相关联。可以通过查询任意字符索引处的属性，来获得完整的范围。这些范围不会重叠。
- 可变的属性化字符串允许把新属性分层放置在字符串某些部分中的属性之上，或者删除这些属性。这样做通常会创建新的范围，以便使每个范围保持受限于一个属性集。
- 可以替换字符串中任意位置的字符，它们将继承那个范围内的任何活动的属性。当插入或追加属性化字符串时，添加的资料将保留定义它时的属性。

秘诀 3-2　通过迭代范围分层放置属性

```
- (void) applyAttribute: (id) attributeValue withName: (NSString *) keyName
{
    // Replicating this approach through custom code
    // [attributedString addAttribute:keyName
    //      value:attributeValue range:range];

    NSRange range = textView.selectedRange;
    if (range.location == NSNotFound) return;

    // Keep track of attribute range effects
    CGFloat fullExtent = range.location + range.length;
    CGFloat currentLocation = range.location;

    // Iterate through each effective range within the selection
    while (currentLocation < fullExtent)
    {
        // Extract the current effective range
        NSRange effectiveRange;
        NSDictionary *currentAttributes =
```

```
            [attributedString attributesAtIndex:currentLocation
                effectiveRange:&effectiveRange];

        // Limit that range to the current selection
        NSRange intersection = NSIntersectionRange(range, effectiveRange);

        // Extract and modify the current attributes
        NSMutableDictionary *attributes = [NSMutableDictionary
            dictionaryWithDictionary:currentAttributes];
        attributes[keyName] = attributeValue;

        // Apply those attributes back. This uses "set attributes"
        // instead of "apply attributes" for demonstration
        [attributedString setAttributes:attributes range:intersection];

        // Update the current location to move past the effective range
        currentLocation = effectiveRange.location + effectiveRange.length;
    }
    [self setAttrStringFrom:textView.attributedText to:attributedString];
}
```

获取这个秘诀的代码

要查找这个秘诀的完整示例项目，可以浏览 https://github.com/erica/iOS-6-Advanced-Cookbook，并进入第 3 章的文件夹。

3.5 响应者样式的神秘性

iOS 6 中引入了 3 个响应者方法，允许对象对所选文本应用样式。这些方法是在 NSObject 中定义的，但是打算用于 UIResponder。这 3 个方法是 toggleBoldface:、toggleItalics: 和 toggleUnderline:，可以看到，它们有一点粗糙。它们将对视图中所选的任何文本应用属性。

这些方法似乎受限于文本视图和文本框。可以通过把 allowsEditingTextAttributes 属性设置为 YES 来启用它们。标准的 iOS 6 上下文文本菜单进行了更新，以包括新的 BIU 选项，如图 3-8 所示。视图将处理其余的实现细节，而不需要以编程方式做任何进一步的工作。可以通过文本视图或文本框的 attributedText 属性访问结果。

这个新特性在许多层面上都令人费解。开始时，UIFont 实例不允许直接访问下面 3 个属性：加粗、倾斜或下划线。

下划线属性绑定到一个相当复杂的 Core Text 系统中，其中只有单线（NSUnderlineStyleSingle）目前在字符串属性（NSUnderlineStyleAttributeName）中是公开的。其他原始的 Core Text 下划线样式包括双线、勾号线和多种模式，比如图 3-8 中使用的模式，用于着重指出可能的错误拼写。

图 3-8　iOS 的上下文菜单进行了更新，以便在文本视图和文本框中启用 allowsEditingTextAttributes 之后包括进 BIU 选项

加粗和倾斜特征代表字体系列的变化。管理它们需要了解基于 C 语言的 API，如程序清单 3-1 所示。该程序清单详细说明了如何从字体系列名称（比如 Trebuchet MS）和特征字典中获取字体描述符。可以使用该描述符提取特定于特征的字体名称（比如 TrebuchetMS-Italic 或 Trebuchet-BoldItalic），然后利用该名称构建一个 UIFont 实例。

当用户应用这些选项开关时，文本视图委托不会接收到对其内容的更新，这增加了神秘性。不会发送常规的委托回调，比如 textViewDidChange:。如果你想监测改变，以及更新 Undo/Redo（撤销/重做）能力（如图 3-8 中所示），可以添加一个计时器，轮询撤销管理器的栈上的项目。

最后一点神秘性是：文本视图不会存储可变的属性化字符串实例。因此，Apple 利用属性更新来更新视图内容的任何机制都会改变，并且替换文本。

这种神秘性可能会增加基本的重新设计工作量，在 iOS 6 完整版初次露面时还没有及时完成它。不过，它的承诺令人兴奋。目前，iOS 6 中的任何特性都是合乎标准的。在很大程度上，它是 Core Foundation 文本调用方面的巨大改进，如程序清单 3-1 所示。

程序清单 3-1　从其特征中返回一种字体

```
// Handle bold and italics traits
- (UIFont *) fontWithTraits
{
    // Core font elements
    NSString *familyName = self.font.familyName;
    CGFloat newFontSize = self.font.pointSize;

    // Return core font
    if (!self.bold && !self.italic)
        return [UIFont fontWithName:familyName size:newFontSize];
```

```objc
    // Create traits value
    NSUInteger appliedTraits = 0;
    if (self.bold) appliedTraits = kCTFontBoldTrait;
    if (self.italic) appliedTraits =
        appliedTraits | kCTFontItalicTrait;
    NSNumber *traitsValue = @(appliedTraits);

    // Build dictionary from family name and traits
    NSDictionary *traitDictionary =
        @{(NSString *)kCTFontSymbolicTrait:traitsValue};
    NSDictionary *dict =
    @{
        (NSString *)kCTFontFamilyNameAttribute:familyName,
        (NSString *)kCTFontTraitsAttribute:traitDictionary,
    };

    // Extract font descriptor
    CFDictionaryRef dictRef = CFBridgingRetain(dict);
        CTFontDescriptorRef desc =
            CTFontDescriptorCreateWithAttributes(dictRef);
    CFRelease(dictRef);

    // If this failed, return core font
    if (!desc)
        return [UIFont fontWithName:familyName size:newFontSize];

    // Otherwise, extract the new font name e.g. whatever-bold
    CTFontRef ctFont = CTFontCreateWithFontDescriptor(
        desc, self.font.pointSize, NULL);
    NSString *newFontName = CFBridgingRelease(
        CTFontCopyName(ctFont, kCTFontPostScriptNameKey));

    // Create font with trait-name
    return [UIFont fontWithName:newFontName size:newFontSize];
}
```

3.6 秘诀：属性栈

在编写这本 Cookbook 的 iOS 5 版本期间，我开发了一个字符串辅助类。这个类把基于 C 语言的 Core Foundation 隐藏在属性化字符串创建背后，并添加了一个 Objective C 包装器。它还添加了一些属性（property），用于为追加到字符串上的新文本定义属性（attribute）。例如，你可能设置一种新的前景色，并且添加的文本将采用该颜色。

快速前进到 iOS 6。在 iOS 6 问世时，它引入了 Objective C 类和常量，以支持属性化字符串。使用自定义类的大约一半的原因现在已经消失了。但是，在我抛弃秘诀的过程中可能发生别的事情：读者和早期的测试员对存储状态特性反响强烈。他们喜欢自己可以设置属性、打包文本、更新属性，以及打包更多的文本。

从此，属性栈的概念有所扩展，它可以提供保存的状态特性，类似于 CGContextSaveGState()。它将允许推送压入一组属性，比如新的颜色或字体大小，然后在添加了新文本之后，弹回以前的状态。

秘诀 3-3 改进了原始的包装器，引入了栈支持。像它带给人们的启发一样，它展示了可以转变为属性（attribute）字典的属性（property）。不过，它还引入了背道而驰的功能。这种实现从存储的字典中获取属性字符串，以便对象可以移回其保存的栈，并且它的属性将会更新以与之匹配。

栈本身是一个简单的数组，压入状态（通过 saveContext）将复制最上面的字典，并把它存储在栈上。恢复状态（通过 popContext）将弹出栈，并把它的属性（attribute）导出给对象属性（property）。

秘诀 3-3 还提供了瞬态属性。performTransientAttributeBlock:方法将自动化保存、执行、弹出序列，因此任何更改都不会永久影响存储的属性栈。下面显示了这种方法的工作方式。在嵌入式瞬态块内，一些属性（property）将设置新添加的文本的属性（attribute）。

```
string.font = [UIFont fontWithName:@"Arial" size:24.0f];
[string performTransientAttributeBlock:^(){
    string.foregroundColor = [UIColor redColor];
    string.bold = YES;
    [string appendFormat:@"This is red and bold. "];
    [string performTransientAttributeBlock:^(){
        string.font = [UIFont fontWithName:@"Georgia" size:24.0f];
        string.bold = NO;
        string.underline = YES;
        string.strokeColor = [UIColor greenColor];
        string.strokeWidth = 2.0f;
        [string appendFormat:
            @"This is Green Georgia Outline, Underlined Not Bolded. "];
    }];
```

```
            string.bold = NO;
            string.foregroundColor = COOKBOOK_PURPLE_COLOR;
            [string appendFormat:@"This is not green *or* bolded.\n\n"];
        }];
        [string appendFormat:@"Back to the original attributes."];
```

当外部块结束时,字符串将恢复到基本的 24 磅 Arial 状态。其他的属性更新将不会在块的生存期之外继续存在。在同步执行每个块时,在块完全完成之后,将会执行放在块后面的任何命令。图 3-9 显示了这段代码的实际应用。

图 3-9　瞬态属性块的实际应用

秘诀 3-3　利用 Objective-C 包装器构建属性化字符串

```
{
    NSMutableArray *stack;
    NSMutableDictionary *top;
}

// Initialize with a new string, a stack, and a current dictionary (top)
- (id) init
{
    if (!(self = [super init])) return self;

    _string = [[NSMutableAttributedString alloc] init];
    stack = [NSMutableArray array];
    top = [NSMutableDictionary dictionary];

    graphStyle = [[NSMutableParagraphStyle alloc] init];
    _font = [UIFont fontWithName:@"Helvetica" size:12.0f];

    return self;
}

// Execute a block with temporary traits
- (void) performTransientAttributeBlock: (AttributesBlock) block
{
    [self saveContext];
    block();
    [self popContext];
}

// Transform object properties into an attributes dictionary
```

```objc
- (NSDictionary *) attributes
{
    NSMutableDictionary *attributes = [NSMutableDictionary dictionary];

    // Font and Para Style
    self.font = [self fontWithTraits];
    [attributes setObject:self.font forKey:NSFontAttributeName];
    [attributes setObject:self.paragraphStyle
        forKey:NSParagraphStyleAttributeName];

    // Colors
    if (self.foregroundColor)
        [attributes setObject:self.foregroundColor
            forKey:NSForegroundColorAttributeName];
    if (self.backgroundColor)
        [attributes setObject:self.backgroundColor
            forKey:NSBackgroundColorAttributeName];
    if (self.strokeColor)
        [attributes setObject:self.strokeColor
            forKey:NSStrokeColorAttributeName];

    // Other Styles
    [attributes setObject:@(self.strokeWidth)
        forKey:NSStrokeWidthAttributeName];
    [attributes setObject:@(self.underline)
        forKey:NSUnderlineStyleAttributeName];
    [attributes setObject:@(self.strikethrough)
        forKey:NSStrikethroughStyleAttributeName];
    if (self.shadow)
        [attributes setObject:self.shadow
            forKey:NSShadowAttributeName];
    [attributes setObject:@(self.useLigatures)
        forKey:NSLigatureAttributeName];

    return attributes;
}

// Expose the current attribute set through properties
- (void) setAttributesFromDictionary: (NSDictionary *) dictionary
{
```

```objc
    // Update active properties
    graphStyle = dictionary[NSParagraphStyleAttributeName];

    // Establish font
    _font = dictionary[NSFontAttributeName];
    CTFontSymbolicTraits traits =
        CTFontGetSymbolicTraits((__bridge CTFontRef)(self.font));
    self.bold = (traits & kCTFontBoldTrait) != 0;
    self.italic = (traits & kCTFontItalicTrait) != 0;

    // Colors
    _foregroundColor = dictionary[NSForegroundColorAttributeName];
    _backgroundColor = dictionary[NSBackgroundColorAttributeName];
    _strokeColor = dictionary[NSStrokeColorAttributeName];

    // Other
    _strokeWidth = ((NSNumber *)dictionary[
        NSStrokeWidthAttributeName]).floatValue;
    _underline = ((NSNumber *)dictionary[
        NSUnderlineStyleAttributeName]).boolValue;
    _strikethrough = ((NSNumber *)dictionary[
        NSStrikethroughStyleAttributeName]).boolValue;
    _useLigatures = ((NSNumber *)dictionary[
        NSLigatureAttributeName]).boolValue;
    _shadow = dictionary[NSShadowAttributeName];
}

#pragma mark - Stack Operations -

// Push the top dictionary onto the stack
- (void)saveContext
{
    [stack addObject:self.attributes];
    top = [NSMutableDictionary dictionaryWithDictionary:top];

    // Create a copy of the style, to point to a distinct object
    graphStyle = [graphStyle mutableCopy];
    top[NSParagraphStyleAttributeName] = graphStyle;
}
```

```
// Pop the top dictionary off the stack, and set its attributes
- (BOOL) popContext
{
    if (!stack.count) return NO;

    // Pop it
    top = [stack lastObject];
    [stack removeLastObject];
    [self setAttributesFromDictionary:top];

    return (stack.count > 0);
}
```

获取这个秘诀的代码

要查找这个秘诀的完整示例项目，可以浏览 https://github.com/erica/iOS-6-Advanced-Cookbook，并进入第 3 章的文件夹。

3.7　秘诀：使用伪 HTML 创建属性化文本

给字符串逐个添加属性，甚至在得到辅助类帮助的情况下，这个过程也会变得单调、乏味。这就是为什么使用 HTML 的原因，它是为文本标记构建的一个熟悉的标准，可以转换和简化布局任务。制作基本的 HTML 内容比以迭代方式实现创建类似表示的 Core Text 调用更容易。考虑下面的字符串：

```
<h1>Using Core Text with Markup</h1>
<p>This is an example <i>of some text that uses <b>HTML styling</b></i>.</p>
```

用于这个简单字符串的符合 ISO 标准的 HTML 布局将需要相当多的 Core Text 编码，以匹配所要求的标记，但是为什么要付出这么多的努力？如果你愿意灵活一些，可以自动化一种类似于 HTML 的解决方案，为你创建 Core Text 编码。

UIWebView 类天生支持通过 HTML 进行排版，并且以一种严格支持标准的方式达此目的。不过，可以利用少得多的代码制作 Core Text 视图，并且不会有内存开销。例如，你可能不希望把一打 Web 视图添加到 UITableView 单元中，但是 Core Text 支持的替代方案只需少量的内存开销，就可以工作得更好。

挑战在于把 HTML（或者类似于 HTML）注释的字符串转变成一个属性化字符串。秘诀 3-4 解决了这个问题，提供一个增强的 HTML 子集，将其内容解析为一个属性化字符串。

注意：
要了解出众的 HTML 解决方案，可以检查 Oliver Drobnik 的优秀的属性化字符串扩展库。在 https://github.com/Cocoanetics/DTCoreText 上可以找到他的库。

图 3-10 伪标记及其属性化字符串表示

这个类比标准 HTML 提供了更多的特性。它的简单的扫描和匹配方法意味着只需几分钟就能添加自己设计的标签。我添加了对两个便利标签的支持，它们不依赖于标准的 HTML，这两个标签是在靠近 stringFromMarkup: 方法末尾发现的 color 和 size 标签。图 3-10 并排显示了运行的应用程序以及驱动其表示的伪标记文件。

这个类通过扫描封闭在尖括号中的标签来工作。标签使用秘诀 3-3 中讨论过的基于相同栈的字符串类来设置文本属性。代码使用当前的字符串属性集来追加每个文本序列。

像这样的实用程序最终会存在，以服务于单独的开发需要，它们不一定支持像 HTML 这样的标准。如果可以利用少量工作添加新标签以便自己使用，那么像这个秘诀这样的自定义解决方案可能在应用程序开发过程中节省许多工作。不要觉得必须受标准制约。

在构建了（当然还要进行严格的调试）像秘诀 3-4 那样的解析器之后，它可以在将来的项目中节省许多时间。它允许重用 Core Text 例程，而不必因为引入了许多 Core Text 开销而烦恼。

秘诀 3-4　伪 HTML 标记

```
+ (NSAttributedString *) stringFromMarkup: (NSString *) aString
{
    // Core Fonts
    UIFont *baseFont = [UIFont fontWithName:@"Palatino" size:14.0f];
    UIFont *headerFont = [UIFont fontWithName:@"Palatino" size:14.0f];

    // Prepare to scan
    NSScanner *scanner = [NSScanner scannerWithString:aString];
    [scanner setCharactersToBeSkipped:[
        NSCharacterSet newlineCharacterSet]];
    // Initialize a string helper
    FancyString *string = [FancyString string];
    string.font = baseFont;
```

```objc
// Iterate through the string
while (scanner.scanLocation < aString.length)
{
    NSString *contentText = nil;
    
    // Scan until the next tag
    [scanner scanUpToString:@"<" intoString:&contentText];
    
    // Process entities and append the text
    contentText = [contentText
        stringByReplacingOccurrencesOfString:@"&lt;"
        withString:@"<"];
    if (contentText)
        [string appendFormat:@"%@", contentText];
    
    
    // Scan through the tag
    NSString *tagText = nil;
    [scanner scanUpToString:@">" intoString:&tagText];
    if (scanner.scanLocation < aString.length)
        scanner.scanLocation += 1;
    tagText = [tagText stringByAppendingString:@">"];
    
    // -- PROCESS TAGS --
    
    // Header Tags
    if (STRMATCH(tagText, @"</h")) // finish any headline
    {
        [string popContext];
        [string appendFormat:@"\n"];
        continue;
    }
    
    if (STRMATCH(tagText, @"<h"))
    {
        int hlevel = 0;
        if (STRMATCH(tagText, @"<h1>")) hlevel = 1;
        else if (STRMATCH(tagText, @"<h2>")) hlevel = 2;
        else if (STRMATCH(tagText, @"<h3>")) hlevel = 3;
```

```objc
        [string performTransientAttributeBlock:^(){
            // add a wee spacer
            string.font = [UIFont boldSystemFontOfSize:8.0f];
            [string appendFormat:@"\n"];
        }];
        
        [string saveContext];
        string.bold = YES;
        string.font = [UIFont fontWithName:headerFont.fontName
            size:20.0f + MAX(0, (4 - hlevel)) * 4.0f];
    }
    
    // Bold and Italics
    if (STRMATCH(tagText, @"<b>")) string.bold = YES;
    if (STRMATCH(tagText, @"</b>")) string.bold = NO;
    if (STRMATCH(tagText, @"<i>")) string.italic = YES;
    if (STRMATCH(tagText, @"</i>")) string.italic = NO;
    
    // Paragraph and line break tags
    if (STRMATCH(tagText, @"<br>")) [string appendFormat:@"\n"];
    if (STRMATCH(tagText, @"</p>")) [string appendFormat:@"\n\n"];
    
    // Color
    if (STRMATCH(tagText, @"<color"))
    {
        if STRMATCH(tagText, @"blue")
            string.foregroundColor = [UIColor blueColor];
        if STRMATCH(tagText, @"red")
            string.foregroundColor = [UIColor redColor];
        if STRMATCH(tagText, @"green")
            string.foregroundColor = [UIColor greenColor];
    }
    if (STRMATCH(tagText, @"</color>"))
        string.foregroundColor = nil;
    
    // Size
    if (STRMATCH(tagText, @"<size"))
    {
        // Scan the value for the new font size
        NSScanner *newScanner = [NSScanner scannerWithString:tagText];
```

```
            NSCharacterSet *cs =
                [[NSCharacterSet decimalDigitCharacterSet] invertedSet];
            [newScanner setCharactersToBeSkipped:cs];

            CGFloat fontSize;
            [newScanner scanFloat:&fontSize];
            [string saveContext];
            string.font = [UIFont fontWithName:string.font.fontName
                size:fontSize];
        }

        if (STRMATCH(tagText, @"</size>"))
            [string popContext];
    }

    return string.string;
}
```

获取这个秘诀的代码

要查找这个秘诀的完整示例项目,可以浏览 https://github.com/erica/iOS-6-Advanced-Cookbook,并进入第 3 章的文件夹。

3.8 利用 Core Text 绘图

属性化字符串扩展到了那些展示它们的 UIKit 类之外,比如文本视图、标签、刷新控件等。UIKit 提供了字符串绘图扩展,允许创建支持属性化字符串的自定义的 UIView 类。在 iOS 6 中,几乎不用做任何工作就能实现这一点。程序清单 3-2 详细描述了一个自定义的视图,它存储一个属性化字符串,并通过 drawRect:方法绘制其内容。

程序清单 3-2　属性化字符串视图

```
@interface ASView : UIView
@property (nonatomic, strong) NSAttributedString *attributedString;
@end

@implementation ASView
- (id) initWithFrame:(CGRect)frame
{
    if (self = [super initWithFrame:frame])
```

```
        self.backgroundColor = [UIColor clearColor];
    return self;
}

- (void) drawRect:(CGRect)rect
{
    [super drawRect: rect];
    [_attributedString drawInRect:self.bounds];
}
@end
```

比较程序清单 3-2 与程序清单 3-3，其中后者显示了在 iOS 6 之前使用 Core Text 调用构建的相同的自定义视图类。新的 drawRect:方法非常简单，演示了多少属性化文本特性把它们自身集成进 UIKit 中，并且利用一个健壮、有用的 Objective-C API 隐藏了 Core Text 的基于 C 语言的 API。

程序清单 3–3　属性化字符串的 Core Text 视图

```
@interface CTView : UIView
@property (nonatomic, strong) NSAttributedString *attributedString;
@end

@implementation CTView
- (id) initWithFrame:(CGRect)frame
{
    if (self = [super initWithFrame:frame])
        self.backgroundColor = [UIColor clearColor];
    return self;
}

- (void) drawRect:(CGRect)rect
{
    [super drawRect: rect];
    CGContextRef context = UIGraphicsGetCurrentContext();

    // Flip the context
    CGContextSetTextMatrix(context, CGAffineTransformIdentity);
    CGContextTranslateCTM(context, 0, self.bounds.size.height);
    CGContextScaleCTM(context, 1.0, -1.0);
```

```objc
    // Slightly inset from the edges of the view
    CGMutablePathRef path = CGPathCreateMutable();
    CGRect insetRect = CGRectInset(self.frame, 20.0f, 20.0f);
    CGPathAddRect(path, NULL, insetRect);

    // Build the framesetter
    CTFramesetterRef framesetter =
        CTFramesetterCreateWithAttributedString(
            (__bridge CFAttributedStringRef)_attributedString);

    // Draw the text
    CTFrameRef destFrame = CTFramesetterCreateFrame(
        framesetter, CFRangeMake(0, _attributedString.length),
        path, NULL);
    CTFrameDraw(destFrame, context);

    // Clean up
    CFRelease(framesetter);
    CFRelease(path);
    CFRelease(destFrame);
}
@end
```

程序清单 3-3 中的 drawRect:方法使用原始的 Core Text 执行被程序清单 3-2 中的 UIKit 字符串绘图调用隐藏的所有任务。它首先将翻转它的环境，这允许文本从左上角开始绘图，这在 UIKit 中是标准行为，但是在原始的 Core Text 中则不然。然后，它将使用视图界限作为起点并从那里稍微往里移动一点，创建一个内嵌（inset）矩形。最后，它将建立一个框架设置器对象，该对象负责管理分解为各个字符的元素。

这些元素称为线（line），每根线都包含许多字符顺串（run）。每个顺串又包含一系列文本字形（glyph），它们是字符图像，比如 "a" 和 "b" 的图片。一个顺串中的每个字形共享公共的属性，比如字体大小和颜色。最后，框架设置器使用所有这些信息构建一个框架（frame），它然后可以把文本绘制到图形环境中。

图 3-11　Core Text 自动处理从矩形路径到椭圆形路径的改变

通过更改路径，可以相当容易地实现眼睛捕获的视觉效果。在本章后面的秘诀中将会看到其中一些效果。对于这个示例，在调整内嵌矩形的大小时，可以发现文本会自动适应并调整它的包装，以匹配它的新界限。改变形状的其他方式包括向路径中添加椭圆和圆弧。Core Text能够处理非矩形路径。图 3-11 显示了在程序清单 3-3 中把使用矩形框架（CGPathAddRect()）改为使用椭圆形框架（CGPathAddEllipse- InRect()）的结果。

3.9 创建图像图案

向 Core Text 布局中添加图像涉及两个关键的步骤。首先，在翻转环境坐标系统之前绘制那些图像。UIImage 绘图例程假定左上角有一个原点。然后，修改文本路径，使得框架设置器在设想的图像位置绘制文本。

程序清单 3-4 创建了一个手工布局，其中将插入两幅图像。在图 3-12 中可以看到由这段代码创建的排版效果。文本环绕在图像周围，遵循在 drawRect:方法中构建的路径图案。这个示例使用一种右边参差不齐、左边对齐的段落样式，它解释了每幅图像左边为什么会出现不同的间隙。

图 3-12　Core Graphics 路径的图案区域为图片提供了空间。Core Text 排字机可以容纳这些间隙

程序清单 3–4　向 Core Text 流中添加图像

```
// Flip a rectangle within an outer coordinate system
CGRect CGRectFlipVertical(CGRect innerRect, CGRect outerRect)
{
    CGRect rect = innerRect;
    rect.origin.y = outerRect.origin.y + outerRect.size.height -
        (rect.origin.y + rect.size.height);
    return rect;
}

- (void) drawRect: (CGRect) rect
{
    NSAttributedString *string = [self string];
```

```
[super drawRect: rect];
CGContextRef context = UIGraphicsGetCurrentContext();
CGRect aRect = CGRectInset(self.bounds, 10.0f, 10.0f);

// Draw the background
[[UIColor whiteColor] set];
CGContextFillRect(context, self.bounds);

// Perform the image drawing in the normal context geometry
CGRect imageRect1 =
    CGRectMake(150.0f, 500.0f, 200.0f, 300.0f);
[[UIImage imageNamed:@"Default.png"]
    drawInRect:CGRectInset(imageRect2, 10.0f, 10.0f)];

// Draw the Bear (public domain)
CGRect imageRect2 =
    CGRectMake(500.0f, 100.0f, 187.5f, 150.0f);
[[UIImage imageNamed:@"Bear.jpg"]
    drawInRect:CGRectInset(imageRect1, 10.0f, 10.0f)];

// Flip the context for the Core Text layout
CGContextSetTextMatrix(context, CGAffineTransformIdentity);
CGContextTranslateCTM(context, 0, self.frame.size.height);
CGContextScaleCTM(context, 1.0, -1.0);

// Start the path
CGMutablePathRef path = CGPathCreateMutable();
CGPathAddRect(path, NULL, aRect);

// Cut out the two image areas
CGPathAddRect(path, NULL,
    CGRectFlipVertical(imageRect1, self.bounds));
CGPathAddRect(path, NULL,
    CGRectFlipVertical(imageRect2, self.bounds));

// Create framesetter
CTFramesetterRef framesetter =
    CTFramesetterCreateWithAttributedString(
        (__bridge CFAttributedStringRef)string);
// Draw the text
```

```
    CTFrameRef theFrame = CTFramesetterCreateFrame(framesetter,
        CFRangeMake(0, string.length), path, NULL);
    CTFrameDraw(theFrame, context);

    // Clean up
    CFRelease(path);
    CFRelease(theFrame);
    CFRelease(framesetter);
}
```

3.10 秘诀：在滚动视图上绘制 Core Text

尽管 UIKit 对字符串绘图的支持简化了文本布局，但是知道 Core Text 仍然很重要，可将其用于更复杂但并非不同寻常的情况。程序清单 3-4 演示了如何在图像周围布置文本。秘诀 3-5 显示了如何构建 Core Text 强化的滚动文本视图。

如果认真工作，可以将自定义的 Core Text 呈现扩展到滚动视图。秘诀 3-5 修改了程序清单 3-3 的 Core Text 绘图，将属性化文本绘制到自定义的滚动视图上。更新中包括一个方法，每当视图的几何形状改变（通常来自于重新定位 iPhone）时它都会计算滚动视图的内容尺寸；还包括数学运算，用于在绘图时考虑进较大的内容尺寸。

updateContentSize 方法在工作时将请求建议的框架大小，该方法将传递源文本、视图宽度和虚拟高度（即最大浮点数）。这允许框架设置器计算文本在宽度固定的情况下将占据多大的垂直高度，允许它忽略任何高度限制。该方法使用返回的框架来设置滚动视图的内容尺寸。

在 drawRect:方法中，秘诀 3-5 的数学运算必须补偿绘图发生在滚动视图的大内容视图内（而不是其框架内）的事实。环境翻转和目标矩形都会使用内容尺寸。

最后，在 initWithFrame:方法中，将滚动视图的内容模式设置为重绘。无论何时视图的界限改变了，这都会强制它进行刷新。

秘诀 3-5　Core Text 和滚动视图

```
@interface CTView : UIScrollView <UIScrollViewDelegate>
@property (nonatomic, strong) NSAttributedString *attributedString;
@end

@implementation CTView
- (id) initWithFrame:(CGRect)frame
{
```

```objc
    if (self = [super initWithFrame:frame])
    {
        self.backgroundColor = [UIColor clearColor];
        self.contentMode = UIViewContentModeRedraw;
        self.delegate = self;
    }
    return self;
}

// Calculates the content size
- (void) updateContentSize
{
    CTFramesetterRef framesetter =
        CTFramesetterCreateWithAttributedString(
            (__bridge CFAttributedStringRef)_attributedString);
    CFRange destRange = CFRangeMake(0, 0);
    CFRange sourceRange = CFRangeMake(0, _attributedString.length);
    CGSize frameSize = CTFramesetterSuggestFrameSizeWithConstraints(
        framesetter, sourceRange, NULL,
        CGSizeMake(self.frame.size.width, CGFLOAT_MAX), &destRange);
    self.contentSize = CGSizeMake(self.bounds.size.width, frameSize.height);
    CFRelease(framesetter);
}

// This is a scroll-view specific drawRect
- (void) drawRect:(CGRect)rect
{
    [super drawRect: rect];
    CGContextRef context = UIGraphicsGetCurrentContext();

    // Flip the context
    CGContextSetTextMatrix(context, CGAffineTransformIdentity);
    CGContextTranslateCTM(context, 0, self.contentSize.height);
    CGContextScaleCTM(context, 1.0, -1.0);

    CGMutablePathRef path = CGPathCreateMutable();
    CGRect destRect = (CGRect){.size = self.contentSize};
    CGPathAddRect(path, NULL, destRect);

    // Create framesetter
```

```
    CTFramesetterRef framesetter =
        CTFramesetterCreateWithAttributedString(
            (__bridge CFAttributedStringRef)_attributedString);

    // Draw the text
    CTFrameRef theFrame = CTFramesetterCreateFrame(framesetter,
        CFRangeMake(0, _attributedString.length), path, NULL);
    CTFrameDraw(theFrame, context);
    // Clean up
    CFRelease(path);
    CFRelease(theFrame);
    CFRelease(framesetter);
}
@end
```

获取这个秘诀的代码

要查找这个秘诀的完整示例项目，可以浏览 https://github.com/erica/iOS-6-Advanced-Cookbook，并进入第 3 章的文件夹。

3.11 秘诀：探讨字体

现今，iOS 带有一个相当健壮的字体集合，可以轻松地表现你的核心内容。字体的范围可能因设备和操作系统而异，因此检查设备看看哪些字体可用总是最好的做法。

秘诀 3-6 遍历 UIFont 的字体系列及其成员的集合，以构建一份报告，它嵌入在文本视图中（参见图 3-13）。对于使用秘诀 3-3 中的助手类创建的示例，秘诀 3-6 通过暂时对字符串的属性应用每种字体而扩展了它们。

图 3-13　属性化字符串有助于查看机载字体

使用这个调查应用程序,可以发现新字体名称,以及预览它们看起来像是什么样子的。

> **注意:**
> 这个秘诀禁用了对助手类中的 fontWithTraits 的调用。这确保每种字体都会准确地使用,而不必遵循助手类的 bold 和 italics 属性。

秘诀 3-6 创建字体列表

```
string = [FancyString string];
string.ignoreTraits = YES;

UIFont *headerFont = [UIFont fontWithName:@"Futura" size:24.0f];
UIFont *familyFont = [UIFont fontWithName:@"Futura" size:18.0f];

for (NSString *familyName in [UIFont familyNames])
{
    string.font = headerFont;
    string.foregroundColor = [UIColor redColor];
    
    // \u25BC is a downward-pointing triangle
    [string appendFormat:@"\u25BC %@\n", familyName];
    
    string.foregroundColor = nil;
    string.font = familyFont;
    
    for (NSString *fontName in [UIFont fontNamesForFamilyName:familyName])
    {
        // \u25A0 is a square
        [string appendFormat:@"\t\u25A0 %@: ", fontName];
        [string performTransientAttributeBlock:^(){
            string.font = [UIFont fontWithName:fontName size:18.0f];
            string.foregroundColor = [UIColor darkGrayColor];
            [string appendFormat:quickBrownDogString];
        }];
    }
    [string appendFormat:@"\n"];
}
textView.attributedText = string.string;
```

获取这个秘诀的代码

要查找这个秘诀的完整示例项目,可以浏览 https://github.com/erica/iOS-6-Advanced-Cookbook,并进入第 3 章的文件夹。

3.12 向应用程序中添加自定义的字体

既然 iOS 支持把自定义的字体集成进应用程序中,那么你永远也不会受限于设备的机载字体库。这个秘诀演示了如何把自己的字体添加到应用程序中以便使用它们。图 3-14 显示了一幅截屏图,它使用了 Pirulen 字体以及一些基本的 Lorem Ipsum 文本。

向 Xcode 项目中添加 TrueType 字体。然后编辑应用程序的 Info.plist 文件,声明 UIAppFonts(即"由应用程序提供的字体")。这是你添加的文件名的数组。图 3-15 定义了单个字体条目,即 pirulen.ttf。

图 3-14 向应用程序中添加自定义的字体允许利用独特的文本润色效果设置应用程序的风格

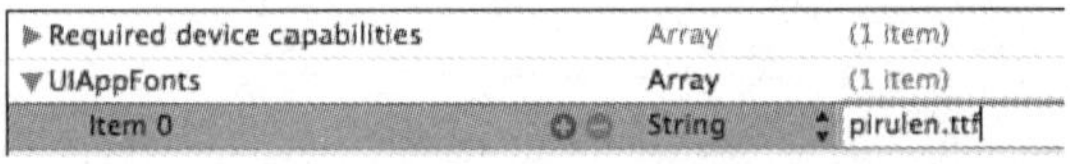

图 3-15 编辑 Info.plist 文件,添加自定义的字体

要使用新字体,可以像通常所做的那样创建字体。确保正确地拼写字体名称,并且大小写也是正确的。

```
// This call to UIFont uses a custom font name
textView.font = [UIFont fontWithName:@"pirulen"
        size:IS_IPAD ? 28.0f : 12.0f];
```

3.13 秘诀:把 Core Text 进行分页处理

在秘诀 3-4 中可以看到,通过标记产生属性化字符串有助于把表示与实现分隔开。它是一个灵活的方法，可让你在不影响代码基（code base）的情况下编辑源材料。但是，当文本的范围超出一个页面时，问题就随之而来了。对于这些情况，你将希望逐个页面地把文本分成一些区域。

秘诀 3-7 显示了如何执行该任务。它使用一个 Core Text 框架设置器，基于给定的页面大小（基于磅）返回建议的分页数组。可以把这种方法与伪 HTML 标记和页面视图控制器紧密结合起来，制作图 3-16 所示的图书。这本图书读入文本，把它从标记源转换为属性化字符串，然后对它进行分页处理，以便在分页式控制器中显示。

秘诀 3-7 的分页方法强化了图 3-16 所示的界面。该方法使用 Core Text 遍历属性化字符串，逐个页面地收集范围信息。Core Text 的 CTFramesetterSuggestFrameSizeWithConstraints 方法接受一个起始范围和一个目标大小，产生适合其属性化字符串的目标范围。尽管这个秘诀没有使用任何控制属性，你还是可能想为自己的实现探索它们。Apple 的技术性 Q&A QA1698（http://developer.apple.com/library/ios/#qa/qa1698）提供了一些指示。

图 3-16　Core Text 框架设置器允许把属性化字符串分成适合给定页面大小的区域

秘诀 3–7　多页式 Core Text

```
- (NSArray *) findPageSplitsForString: (NSAttributedString *)theString
    withPageSize: (CGSize) pageSize
{
    NSInteger stringLength = theString.length;
    NSMutableArray *pages = [NSMutableArray array];

    CTFramesetterRef frameSetter =
        CTFramesetterCreateWithAttributedString(
            (__bridge CFAttributedStringRef) theString);

    CFRange baseRange = {0,0};
    CFRange targetRange = {0,0};
    do {
        CTFramesetterSuggestFrameSizeWithConstraints(
            frameSetter, baseRange, NULL, pageSize, &targetRange);
        NSRange destRange = {baseRange.location, targetRange.length};
        [pages addObject:[NSValue valueWithRange:destRange]];
        baseRange.location += targetRange.length;
```

```
        } while(baseRange.location < stringLength);
        CFRelease(frameSetter);
        return pages;
}
```

获取这个秘诀的代码

要查找这个秘诀的完整示例项目，可以浏览 https://github.com/erica/iOS-6-Advanced-Cookbook，并进入第 3 章的文件夹。

3.14 秘诀：把属性化文本绘制到 PDF 中

你迄今为止看到的 Core Text 和属性化清单都是把文本绘制到视图中，并在 drawRect 更新中呈现它们。秘诀 3-8 把绘图方式从视图转换为 PDF。它将构建一个新的 PDF 文件，逐个页面地添加内容，并且使用秘诀 3-7 中介绍的相同的页面布局方法。

绘制到 PDF 环境中类似于绘制到图像环境中，它们之间的区别出现在开始一个新 PDF 页面的调用中。秘诀 3-8 构建了它的多页式 PDF 文档，并把它存储到在方法调用中传递的路径中。

秘诀 3-8 绘制到 PDF

```
- (void) dumpToPDFFile: (NSString *) pdfPath
{
    // This is an arbitrary page size. Adjust as desired.
    CGRect theBounds = CGRectMake(0.0f, 0.0f, 480.0f, 640.0f);
    CGRect insetRect = CGRectInset(theBounds, 0.0f, 10.0f);

    NSArray *pageSplits = [self findPageSplitsForString:string
         withPageSize:insetRect.size];
    int offset = 0;

    UIGraphicsBeginPDFContextToFile(pdfPath, theBounds, nil);

    for (NSValue *pageStart in pageSplits)
    {
        UIGraphicsBeginPDFPage();
        NSRange offsetRange = {offset, pageStart.rangeValue.length};
        NSAttributedString *subString =
            [string attributedSubstringFromRange:offsetRange];
        offset += offsetRange.length;
```

```
        [subString drawInRect:insetRect];
    }
    UIGraphicsEndPDFContext();
}

- (void) createPDF
{
    NSString *path = [[NSBundle mainBundle]
        pathForResource:@"data" ofType:@"txt"];
    NSString *markup = [NSString stringWithContentsOfFile:path
        encoding:NSUTF8StringEncoding error:nil];
    string = [MarkupHelper stringFromMarkup:markup];

    NSString *destPath = [NSHomeDirectory()
        stringByAppendingPathComponent:@"Documents/results.pdf"];
    [self dumpToPDFFile:destPath];

    // Display the PDF in QuickLook
    QLPreviewController *controller =
        [[QLPreviewController alloc] init];
    controller.dataSource = self;
    [self.navigationController pushViewController:controller animated:YES];
}
```

获取这个秘诀的代码

要查找这个秘诀的完整示例项目,可以浏览 https://github.com/erica/iOS-6-Advanced-Cookbook,并进入第 3 章的文件夹。

3.15 秘诀:大电话文本

在 Macintosh 上,Address Book(地址簿)的大文本显示是我最喜爱的桌面特性之一。它可以放大文本,提供短信息的易于阅读的显示效果,你甚至可能在整个房间内都能阅读这些信息。那么,为什么不在 iPhone、iPad 或 iPod Touch 上使用相同的使之变大和易读的指导思想呢,如图 3-17 所示。具有高对比度显示效果并且不受设备旋转影响的大文本可能是许多应用程序的真正资产。

图 3-17 高对比度的大文本显示效果有助于用户以一种高度可读的方式显示较短的信息片断

这是由于大文本并不仅仅是电话号码。你可能显示名字、分级解锁码或电子邮件地址，以及那些电话号码，使得更容易可视化地把信息传送给使用另一部 iOS 设备、真实的记事本、游戏控制台、计算机或者甚至是第三方智能手机的人。这里的诱因是文本的长度比较短、尺寸比较大并且易于阅读。

具有一种算法工具用于显示难以阅读的文本是一个优秀的软件特性，我非常自豪地从 Address Book（地址簿）中偷学到了这个特性，它启发我创建了本章最后一个秘诀中的可重用的 Objective-C 类。秘诀 3-9 的 BigTextView 类打算覆盖关键的设备窗口，并且不会响应设备方向的改变，以确保当用户重新定位设备时文本不会开始移动，尤其是在显示信息给别人看时。

当前的实现利用两次点按来关闭视图，以防止在典型的处理期间过早地关闭它。

秘诀 3-9 大文本，确实很大的文本

```
// Center a rectangle inside a parent
CGRect rectCenteredInRect(CGRect rect, CGRect mainRect)
{
    return CGRectOffset(rect,
                        CGRectGetMidX(mainRect)-CGRectGetMidX(rect),
                        CGRectGetMidY(mainRect)-CGRectGetMidY(rect));
}
// Big Text View
@implementation BigTextView
{
    NSString *baseString;
}

- (id) initWithString: (NSString *) theString
{
    if (self = [super initWithFrame:CGRectZero])
    {
        baseString = theString;

        // Require a double-tap to dismiss
        UITapGestureRecognizer *tapRecognizer =
            [[UITapGestureRecognizer alloc] initWithTarget:self
                action:@selector(dismiss)];
        tapRecognizer.numberOfTapsRequired = 2;
        [self addGestureRecognizer:tapRecognizer];
    }
    return self;
```

```objc
}

- (void) dismiss
{
    [self removeFromSuperview];
}
+ (void) bigTextWithString:(NSString *)theString
{
    BigTextView *theView =
        [[BigTextView alloc] initWithString:theString];

    // Create a dark, translucent backdrop
    theView.backgroundColor =
        [[UIColor darkGrayColor] colorWithAlphaComponent:0.5f];

    // Constrain the view to stretch to its parent size
    UIWindow* window = [[UIApplication sharedApplication] keyWindow];
        [window addSubview:theView];
    PREPCONSTRAINTS(theView);
        STRETCH_VIEW(window, theView);

    return;
}


// Draw out the message
- (void) drawRect:(CGRect)rect
{
    [super drawRect:rect];
    CGContextRef context = UIGraphicsGetCurrentContext();

    // Create a geometry with width greater than height
    CGRect orientedRect = self.bounds;
    if (orientedRect.size.height > orientedRect.size.width)
        orientedRect.size = CGSizeMake(
            orientedRect.size.height, orientedRect.size.width);

    // Rotate 90 deg to write text horizontally
    // along window's vertical axis
    CGContextRotateCTM(context, -M_PI_2);
    CGContextTranslateCTM(context, -self.frame.size.height, 0.0f);
```

```objc
// Draw a light gray rounded-corner backsplash
[[[UIColor darkGrayColor] colorWithAlphaComponent:0.75f] set];
CGRect insetRect = CGRectInset(orientedRect,
    orientedRect.size.width * 0.05f,
    orientedRect.size.height * 0.35f);
[[UIBezierPath bezierPathWithRoundedRect:insetRect
    cornerRadius:32.0f] fill];
CGContextFillPath(context);

// Inset again for the text
insetRect = CGRectInset(insetRect,
    insetRect.size.width * 0.05f,
    insetRect.size.height * 0.05f);

// Iterate until finding a set of font
// traits that fits this rectangle
UIFont *textFont;
NSString *fontFace = @"HelveticaNeue-Bold";
CGSize fullSize = CGSizeMake(CGFLOAT_MAX, CGFLOAT_MAX);
for (CGFloat fontSize = 18; fontSize < 300; fontSize++ )
{
    // Search until the font size is too big
    textFont = [UIFont fontWithName:fontFace size: fontSize];
    CGSize textSize = [baseString sizeWithFont:textFont
        constrainedToSize:fullSize];
    if (textSize.width > insetRect.size.width)
    {
        // Ease back on font size to prior level
        textFont = [UIFont fontWithName:fontFace
            size: fontSize - 1];
        break;
    }
}

// Establish a frame that just encloses the text at the maximum size
CGSize textSize = [baseString sizeWithFont:textFont
    constrainedToSize:fullSize];
CGRect textFrame = (CGRect){.size = textSize};
CGRect centerRect = rectCenteredInRect(textFrame, insetRect);
```

```
    // Draw the string in white
    [[UIColor whiteColor] set];
    [baseString drawInRect:centerRect withFont:textFont];
}
@end
```

获取这个秘诀的代码

要查找这个秘诀的完整示例项目，可以浏览 https://github.com/erica/iOS-6-Advanced-Cookbook，并进入第 3 章的文件夹。

3.16 小结

本章介绍了许多方式，用于在 iOS 应用程序中创造性地使用文本。在本章中，你学习了属性化字符串，以及使用 Core Text 构建解决方案。在学习下一章之前，要思考以下几点。

- 借助合适的实用类，可以非常简单地使用属性化字符串。它们通常提供了一种容易的方式，利用少量的编码即可创建充满美感并且令人愉悦的表示。既然 UIKit 为如此多的类支持属性化字符串，现在就应该从 NSString 潮流中脱身而出，并进入属性的强大世界中。
- 一些排版特性似乎忽略了 iOS 6 的界限。尽管 iOS 6 非常适合于处理文本，但是 iOS 7 应该会令人惊异。本章详细说明了在许多方面（比如给字体添加特征）iOS 6 并没有完全脱离 Core Text 进行开发。诸如加粗和倾斜之类的响应者特性在 iOS 擅长的领域提供了诱人的品味。
- 严肃的工作需要严肃的排版。不要害怕使用 Core Text 来规划混合有文本与图像的复杂页面。你已经见过了一些示例，演示了如何创建灵活的路径。可以使用它们构建各种令人惊叹的布局。
- 如果计划结合使用自定义的排版与页面视图控制器，可以安排时间设计数学运算和布局——尤其是当打算让用户调整字体大小以及从纵向布局转向横向布局时。Apple 组建了一个完整的团队来处理 iBooks，有一个很好的理由让他们这样做。
- iOS 带有一个出众、健壮的字体集，但是并不限制你只能使用那个字体集。可以自由地许可和捆绑所需要的字体，只要它们与 iOS 兼容即可。
- 你在寻找沿着贝塞尔路径布置文本的秘诀吗？这是下一章要介绍的知识。第 4 章 "几何学" 创造了与底层问题的更好匹配，它更多地涉及数学知识，而不是 Core Text，在下一章中就会学到这些。

第 4 章
几何学

尽管与（比如说）Core Animation 或 Open GL 相比，UIKit 不太需要应用数学，但是在处理贝塞尔路径和视图变换时，几何学仍然扮演着重要的作用。为什么需要几何学呢？它有助于以非标准的方式操纵视图，包括沿着自定义的路径布置文本，以及沿着路径运行动画。如果你在一提到贝塞尔曲线、凸包和样条时就目光呆滞，本章可以帮助澄清这些术语，使你能够向工具箱中添加一些功能强大的自定义选项。

4.1 秘诀：获取贝塞尔路径中的点

在数学中，贝塞尔曲线指参数化的平滑曲线，它是由应用于线段上的控制点创建的。在 UIKit 中，贝塞尔路径（path）定义了通过直线和曲线线段构建的形状，它可能包括贝塞尔曲线和圆弧。UIBezierPath 类的成员存储线段序列，它们可能是开放的或封闭的、连续的或不连续的。路径可能包括以下成分。

- 直线，通过 moveToPoint:和 addLineToPoint:调用创建。
- 三次贝塞尔曲线段，通过调用 addCurveToPoint:controlPoint1:controlPoint2:创建。
- 二次贝塞尔曲线段，通过 addQuadCurveToPoint:controlPoint:调用创建。
- 圆弧，通过调用 addArcToCenter:radius:startAngle:endAngle:clockwise:添加。

类方法允许构建矩形、椭圆、圆角矩形和圆弧，允许通过单独一个调用来访问公共的路径样式。在处理自由形式的绘图时，通常以交互方式创建路径，以响应用户在此期间的触摸。

尽管贝塞尔路径为处理绘制的路径提供了许多方便，但是它们没有直接提供一种方式，用于获取它们的点的数组。秘诀 4-1 通过收集源点的数组来执行该任务。它使用 CGPathApply 函数迭代贝塞尔路径的元素，并且提供了一个自定义的函数（getPointsFromBezier()），在每个元素上调用它。

这个函数获取元素类型（比如移动到点、添加直接到点、添加曲线到点、添加二次曲线到点或者封闭子路径），并使用该信息取出组成路径的点。这种移入和移出路径的能力以及点表示法为你利用两种方法提供了一种方式。

路径对象提供了对绘画特性（比如填充和笔画）的核心级访问。可以设置路径的线条宽度和虚线样式。这使得很容易把路径纳入到视图的 drawRect:实现中。简单调用[myPath stroke]将执行显示路径的轮廓所需的所有绘图工作。

从路径转到点上允许对自定义的绘图应用数学函数。可以平滑路径、找到它的边界框、应用稀释点的函数等。点提供了对组成路径的数据的具体访问。

在这一整章中，可以找到涉及两个方向的例程：从路径对象到点数组以及相反的方向。秘诀4-1 通过提供一个属性（points）和一个类方法（pathwithPoints:）使之成为可能，其中前者返回从贝塞尔路径中提取的点的数组，后者则用于通过点的数组建立贝塞尔路径。

秘诀 4-1　提取贝塞尔路径中的点

```
#define POINT(_INDEX_) \
    [(NSValue *)[points objectAtIndex:_INDEX_] CGPointValue]
#define VALUE(_INDEX_) \
    [NSValue valueWithCGPoint:points[_INDEX_]]

void getPointsFromBezier(void *info, const CGPathElement *element)
{
    NSMutableArray *bezierPoints = (__bridge NSMutableArray *)info;
    CGPathElementType type = element->type;
    CGPoint *points = element->points;
    if (type != kCGPathElementCloseSubpath)
    {
        if ((type == kCGPathElementAddLineToPoint) ||
            (type == kCGPathElementMoveToPoint))
            [bezierPoints addObject:VALUE(0)];
        else if (type == kCGPathElementAddQuadCurveToPoint)
            [bezierPoints addObject:VALUE(1)];
        else if (type == kCGPathElementAddCurveToPoint)
            [bezierPoints addObject:VALUE(2)];
    }
}

- (NSArray *)points
{
    NSMutableArray *points = [NSMutableArray array];
    CGPathApply(self.CGPath, (__bridge void *)points, getPointsFromBezier);
    return points;
```

```
+ (UIBezierPath *) pathWithPoints: (NSArray *) points
{
    UIBezierPath *path = [UIBezierPath bezierPath];
    if (points.count == 0) return path;
    [path moveToPoint:POINT(0)];
    for (int i = 1; i < points.count; i++)
        [path addLineToPoint:POINT(i)];
    return path;
}
```

> **获取这个秘诀的代码**
> 要查找这个秘诀的完整示例项目，可以浏览 https://github.com/erica/iOS-6-Advanced-Cookbook，并进入第 4 章的文件夹。

4.2 稀释点

在秘诀 4-1 中创建的点是通过触摸事件构建的。每次用户移动他们的手指时，就会有一个新的点加入到底层的贝塞尔路径中。这样，路径很快就会变得混乱不堪，如图 4-1（左图）所示。这幅截屏图显示了在用户交互期间捕获的触摸点。每个小圆圈都代表当用户轻划屏幕时的触摸事件。

其中许多点是共线的，代表冗余的元素。图 4-1（右图）显示了相同路径的稀释表示，其中删除了众多项目。这个版本删除了不能改变线条的

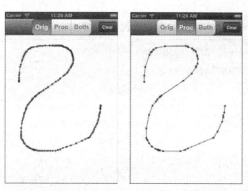

图 4-1 稀释允许删除共线的点，以减少用于构建曲线的项目数量。如果在描绘笔画或者支持撤销/重做，这就很重要。它允许在减少应用程序的内存开销的同时，提高其性能和响应速度

点，因此你看到的是几条较长的直线段，而没有任何圆形。创建这种流线型版本的数学运算测试了对齐方式，并且删除了在定义形状时不起作用的点。

考虑图 4-2。它显示了 3 个点：A、P 和 B。线段 AP 与 PB 之间形成的角度接近于扁平的，几乎但不等于 180°。在这个序列中删除点 P 将产生一条线段（AB），但接近于原始线段（APB），但是要少一个点。这利用最低的精度损失创建了更简单的表示。在图 4-1 中可以看到，尽管数据

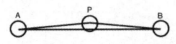

图 4-2　线段 AP 与 PB 之间形成的角度几乎是扁平的

点的数量减少了 4~5 倍，但是前后两个版本几乎完全相同。

秘诀 4-2 详细说明了稀释算法。它遍历路径上的点，并使用秘诀 4-1 中的方法获取点。它将检查每个点（图 4-2 中的点 P），回溯一个点（到点 A）并前进一个点（到点 B）。它创建了两个向量：AP 和 BP，并计算它们的点积。

点积是把两个向量的元素相乘的代数运算。结果直接与两个向量之间的角度的余弦相关。秘诀 4-2 把余弦值与测试值做比较，它的范围在 -1（π 或 180° 的余弦值）~-0.865（150° 的余弦值）之间。这对应于一个公差，可以把它提供作为一个函数参数。其范围在 0（无公差，角度必须正好是 180°）~1（最高的公新式，允许最大 150° 的角度）。

当公差逐渐增大接近于 1 时，算法允许更大地偏离严格的共线性。使用这项检查，函数将删除共线和接近共线的点，返回极大地稀释过的贝塞尔路径。

在节省内存与描述的精度之间有一个折衷，这类似于 JPEG 图像压缩。较小的压缩节省的内存也较少，但是对图像质量的影响几乎察觉不到。如果压缩率太大，节省的空间将导致图像变得模糊，无法识别。作为开发人员，你将决定在不使用户体验降级的情况下可以使用哪种程度的精度。使用较小的值，可以获得更精确的表示；使用较大的值，可以获得精度较低但是内存更有限的表示。

秘诀 4-2　稀释贝塞尔路径中的点

```
#define POINT(_INDEX_) \
    [(NSValue *)[points objectAtIndex:_INDEX_] CGPointValue]

// Return dot product of two vectors normalized
static float dotproduct (CGPoint v1, CGPoint v2)
{
    float dot = (v1.x * v2.x) + (v1.y * v2.y);
    float a = ABS(sqrt(v1.x * v1.x + v1.y * v1.y)); // magnitude a
    float b = ABS(sqrt(v2.x * v2.x + v2.y * v2.y)); // magnitude b
    dot /= (a * b);
    
    return dot;
}

// Pass a tolerance within 0 to 1.
// 0 tolerance uses the tightest checking for colinearity
// As the values loosen, colinearity will be allowed for angles
// further from 180 degrees, up to 150 degrees at a tolerance of 1
```

```objc
UIBezierPath *thinPath(UIBezierPath *path, CGFloat tolerance)
{
    // Retrieve the points
    NSArray *points = path.points;
    if (points.count < 3) return path;
    
    // Create a new output path
    UIBezierPath *newPath = [UIBezierPath bezierPath];
    CGPoint p1 = POINT(0);
    [newPath moveToPoint:p1];
    
    CGPoint mostRecent = p1;
    int count = 1;
    
    // -1 = 180 degrees, -0.985 = 170 degrees,
    // -0.865 = 150 degrees
    CGFloat checkValue = -1.0f + .135 * tolerance;
    
    // Add only those points that are inflections
    for (int i = 1; i < (points.count - 1); i++)
    {
        CGPoint p2 = POINT(i);
        CGPoint p3 = POINT(i+1);
        
        // Cast vectors around p2 origin
        CGPoint v1 = CGPointMake(p1.x - p2.x, p1.y - p2.y);
        CGPoint v2 = CGPointMake(p3.x - p2.x, p3.y - p2.y);
        float dot = dotproduct(v1, v2);
        
        // Colinear items need to be as close as possible to 180 degrees
        // That means as close to -1 as possible
        
        if (dot < checkValue) continue;
        p1 = p2;
        
        mostRecent = POINT(i);
        [newPath addLineToPoint:mostRecent];
        count++;
    }
```

```
    // Add final point
    CGPoint finalPoint = POINT(points.count - 1);
    if (!CGPointEqualToPoint(finalPoint, mostRecent))
        [newPath addLineToPoint:finalPoint];
    return newPath;
}
```

> **获取这个秘诀的代码**
>
> 要查找这个秘诀的完整示例项目，可以浏览 https://github.com/erica/iOS-6-Advanced-Cookbook，并进入第 4 章的文件夹。

4.3 秘诀：平滑绘图

依赖于使用的设备和涉及的同时处理的数量，捕获用户姿势可能产生比想要的更粗糙的结果。图 4-3 显示了源于粒状输入的棱角种类。触摸事件通常受限于 CPU 需求，并且受制于任何系统事件。使用实时的平滑算法，通过使用基本的样条算法在点之间进行插补操作，可以抵消这些限制。

Catmull-Rom 样条提供了最简单的方法之一，用于在关键点之间创建连续的曲线。这种算法确保提供的每个初始点都将保持为最终曲线的一部分，因此得到的路径也将保留原始路径的形状。可以选择在每个参考点

图 4-3 可以实时应用 Catmull-Rom 平滑，以改进触摸事件之间的弧度。这里显示的图像基于完全相同的姿势输入，分别显示了应用和未应用平滑的效果

之间要插入的点数量。要在处理能力与更大的平滑度之间进行折衷。添加的点越多，消耗的 CPU 资源也越多。

在使用与本章配套的示例代码时可以看到，少量的平滑也要经历一段较长的过程。即便如此，更新的 iOS 设备的响应速度是如此之快，以至于很难从一开始就绘制出锯齿状特别鲜明的线条。

Catmull-Rom 算法将插入点以创建平滑的表示。秘诀 4-3 实现了这种算法，它一次使用 4 个点，计算每个序列中的第二个和第三个点之间的中间值。它使用指定的粒度，在这些点之间创建额外的项目。要产生图 4-3 中所示的结果，示例代码使用了粒度 4。秘诀 4-3 提供了仅仅一种实时几何处理的示例，可以把它添加到自己的应用程序中。

秘诀 4-3　Catmull-Rom 样条算法

```
#define POINT(_INDEX_) \
    [(NSValue *)[points objectAtIndex:_INDEX_] CGPointValue]

// Return a smoothed path using the supplied granularity
UIBezierPath *smoothedPath(UIBezierPath *path, NSInteger granularity)
{
    NSMutableArray *points = [path.points mutableCopy];
    if (points.count < 4) return [path copy];
    [points insertObject:[points objectAtIndex:0] atIndex:0];

    [points addObject:[points lastObject]];

    UIBezierPath *smoothedPath = [UIBezierPath bezierPath];

    // Copy traits
    smoothedPath.lineWidth = path.lineWidth;

    // Draw out the first 3 points (0..2)
    [smoothedPath moveToPoint:POINT(0)];

    for (int index = 1; index < 3; index++)
        [smoothedPath addLineToPoint:POINT(index)];

    for (int index = 4; index < points.count; index++)
    {
        CGPoint p0 = POINT(index - 3);
        CGPoint p1 = POINT(index - 2);
        CGPoint p2 = POINT(index - 1);
        CGPoint p3 = POINT(index);

        // now add n points starting at p1 + dx/dy up
        // until p2 using Catmull-Rom splines
        for (int i = 1; i < granularity; i++)
        {
            float t = (float) i * (1.0f / (float) granularity);
            float tt = t * t;
            float ttt = tt * t;
```

```
            CGPoint pi; // intermediate point
            pi.x = 0.5 * (2*p1.x+(p2.x-p0.x)*t +
                (2*p0.x-5*p1.x+4*p2.x-p3.x)*tt +
                (3*p1.x-p0.x-3*p2.x+p3.x)*ttt);
            pi.y = 0.5 * (2*p1.y+(p2.y-p0.y)*t +
                (2*p0.y-5*p1.y+4*p2.y-p3.y)*tt +
                (3*p1.y-p0.y-3*p2.y+p3.y)*ttt);
            [smoothedPath addLineToPoint:pi];
        }

        // Now add p2
        [smoothedPath addLineToPoint:p2];
    }

    // finish by adding the last point
    [smoothedPath addLineToPoint:POINT(points.count - 1)];

    return smoothedPath;
}
```

获取这个秘诀的代码

要查找这个秘诀的完整示例项目，可以浏览 https://github.com/erica/iOS-6-Advanced-Cookbook，并进入第 4 章的文件夹。

4.4 秘诀：基于速度的笔画

基于速度的粗细度通过模仿墨水流动给交互式绘图添加一种现实感。钢笔移动得越快，它可以放出的墨水就越少。更缓慢的移动可以放出更多的墨水，从而创建更粗的区域。图 4-4 显示了在进行和不进行基于速度的粗细度调整的情况下相同的描摹分别看起来像是什么样子的。在捕获签名或创建"钢笔和墨水"风格的界面时，把线条粗细度与用户速度关联起来是最常用的方法。

秘诀 4-4 使用在绘图期间捕获的点速度。

图 4-4 用户创建的描摹，图中分别显示了进行（右图）和不进行（左图）基于速度的粗细度调整的效果

一个名为 FlowPath 的自定义类存储日期以及点的位置。把时间偏移量和经过的距离结合起来确定要应用哪种笔画宽度。

为了把改变减弱一点，这个方法（受到整个 Internet 上的许多出众的解决方案启发）使用了一个加权因子。这限制了可能影响当前速度的加速运动，在计算新的笔画宽度之前，可以利用当前和以前的速度。

你将希望对这个秘诀进行调整，以产生尽可能好的视觉设计。它提供了你开始工作所需的所有基础，但是把"完美笔画"作为一个练习留给读者完成。要知道的是，这个例程为每种路径长度使用统一的笔画宽度。更详尽的实现可能使用样条平滑路径，并沿着每条插入的线段调整宽度。

秘诀 4-4　基于速度的笔画

```
#define POINT(_INDEX_) \
    [(NSValue *)[points objectAtIndex:_INDEX_] CGPointValue]

// Calculate the velocity between two point events
- (CGFloat) velocityFrom:(int) j to:(int) i
{
    CGFloat dPos = distance(POINT(j), POINT(i));
    CGFloat dTime = (DATE(j) - DATE(i));
    return dPos / dTime;
}

// Return a stroke width based on velocity
// Feel free to tweak this all you like
- (CGFloat) strokeWidth: (CGFloat) velocity
{
    CGFloat multiplier = 2.0f;
    CGFloat base = 5.0f;
    CGFloat adjusted = base - (log2f(velocity) / multiplier);
    adjusted = MIN(MAX(adjusted, 0.4), base);
    return multiplier * adjusted * _lineWidth;
}

// Create a Bezier path from p0 to p1 for drawing
UIBezierPath *bPath(CGPoint p0, CGPoint p1)
{
    UIBezierPath *path = [UIBezierPath bezierPath];
    [path moveToPoint:p0];
    [path addLineToPoint:p1];
```

```
        return path;
    }

    // Stroke the custom FlowPath
    - (void) stroke
    {
        if (points.count < 2) return;

        // Store the most recent velocity
        CGFloat lastVelocity = [self velocityFrom:1 to:0];
        // Adjustable weighting for slight filtering
        CGFloat weight = 0.5f;

        UIBezierPath *path;
        for (int i = 1; i < points.count; i++)
        {
            // Adjust the velocity so it doesn't change too much
            // at any given time
            CGFloat velocity = [self velocityFrom:i to:i-1];
            velocity = weight*velocity + (1.0f - weight)*lastVelocity;
            lastVelocity = velocity;
            CGFloat strokeWidth = [self strokeWidth:velocity];

            // Stroke each segment
            path = bPath(POINT(i - 1), POINT(i));
            path.lineWidth = strokeWidth;
            [path stroke];
        }
    }
```

获取这个秘诀的代码

要查找这个秘诀的完整示例项目，可以浏览 https://github.com/erica/iOS-6-Advanced-Cookbook，并进入第 4 章的文件夹。

4.5 秘诀：限制贝塞尔路径

贝塞尔路径产生不规则的形状。因此，可能需要获取路径的边界框（封闭路径的最小矩形）或凸包（封闭路径的最小凸面形状）以执行相交测试，或者提供可视化背景。图 4-5 显示了用于脸部素描的边界框和凸包。

贝塞尔路径的内置 bounds 属性对应于曲线的边界框。秘诀 4-5 显示了如何创建路径的凸包。它在工作时，将沿着 *x* 轴然后沿着 *y* 轴用几何方法对点进行排序。该秘诀将计算底下的凸包，然后计算顶部的凸包，并测试每个点，确定它是在对象内（位于下凸包的左边或者上凸包的右边）还是对象外（位于下凸包的右边或者上凸包的左边）。如果点位于对象外，方法就会扩展凸包以容纳它。

在计算了边界点之后，凸包方法将构建一个新的 UIBezierPath 实例以存储它们，并返回那个对象。

图 4-5　人脸的 UIBezierPath 绘图。外部的淡色矩形标记路径的边界框；内部的深色轮廓是路径的凸包

秘诀 4-5　边界框和凸包

```
#define POINT(_INDEX_) \
    [(NSValue *)[points objectAtIndex:_INDEX_] CGPointValue]

@implementation UIBezierPath (Bounding)

// Create a zero-sized rectangle at a point
static CGRect pointRect(CGPoint point)
{
    return (CGRect){.origin=point};
}

// Return an array of sorted points along X and then Y
- (NSArray *) sortedPoints
{
    NSArray *sorted = [self.points sortedArrayUsingComparator:
        ^NSComparisonResult(id item1, id item2)
    {
        NSValue *v1 = (NSValue *) item1;
        NSValue *v2 = (NSValue *) item2;
        CGPoint p1 = v1.CGPointValue;
        CGPoint p2 = v2.CGPointValue;

        if (p1.x == p2.x)
            return [@(p1.y) compare:@(p2.y)];
        else
```

```objc
            return [@(p1.x) compare:@(p2.x)];
    }];
    return sorted;
}

// Test a point's half-plane
static float halfPlane(CGPoint p1, CGPoint p2, CGPoint testPoint)
{
    return (p2.x-p1.x)*(testPoint.y-p1.y) - (testPoint.x-p1.x)*(p2.y-p1.y);
}
// Return a path's convex hull
- (UIBezierPath *) convexHull
{
    /*
    minmin = top left, min x, min y
    minmax = bottom left, min x, max y
    maxmin = top right, max x, min y
    maxmax = bottom right, max x, max y
    */
    
    NSMutableArray *output = [NSMutableArray array];
    NSInteger bottom = 0;
    NSInteger top = -1;
    NSInteger i;
    
    // Pre-sort the points
    NSArray *points = self.sortedPoints;
    NSInteger lastIndex = points.count - 1;
    
    // Location of top-left corner
    NSInteger minmin = 0;
    CGFloat xmin = POINT(0).x;
    
    // Locate minmax, bottom left
    for (i = 1; i <= lastIndex; i++)
        if (POINT(i).x != xmin)
            break;
    NSInteger minmax = i - 1;
// If the bottom left is the final item
// check whether to add both minmin & minmax
```

```
if (minmax == lastIndex)
{
    output[++top] = points[minmin];
    if (POINT(minmax).y != POINT(minmin).y)
    {
        // add the second point, and close the path
        output[++top] = points[minmax];
        output[++top] = points[minmin];
    }

    for (int i = top + 1; i < output.count; i++)
        [output removeObjectAtIndex:i];

    return [UIBezierPath pathWithPoints:output];
}

// Search for top right, max x, min y by moving
// back from max x, max y at final index
NSInteger maxmin = lastIndex;
CGFloat xmax = POINT(lastIndex).x;
for (i = lastIndex - 1; i >= 0; i--)
    if (POINT(i).x != xmax)
        break;
maxmin = i + 1;

// Compute Lower Hull
output[++top] = points[minmin]; // top left
i = minmax; // bottom left

while (++i < maxmin) // top right
{
    // Test against TopLeft-TopRight
    if ((halfPlane(POINT(minmin),
        POINT(maxmin), POINT(i)) >= 0) &&
        (i < maxmin))
        continue;

    while (top > 0)
    {
        // Find points that extend the hull and add them
```

```
        if (halfPlane([output[top - 1] CGPointValue],
            [output[top] CGPointValue], POINT(i)) > 0)
            break;
        else
            top--;
    }
    output[++top] = points[i];
}

// Ensure the hull is continuous when going from lower to upper
NSInteger maxmax = lastIndex;
if (maxmax != maxmin)
    output[++top] = points[maxmax];

// Compute Upper Hull
bottom = top;
i = maxmin;
while (--i >= minmax)
{
    if ((halfPlane(POINT(maxmax),
        POINT(minmax), POINT(i)) >= 0) &&
        (i > minmax))
        continue;

    while (top > bottom)
    {
        // Add points that extend the hull
        if (halfPlane([output[top - 1] CGPointValue],
            [output[top] CGPointValue], POINT(i)) > 0)
            break;
        else
            top--;
    }
    output[++top] = points[i];
}

// Again ensure continuity at the end
if (minmax != minmin)
    output[++top] = points[minmin];
```

```
    NSMutableArray *results = [NSMutableArray array];
    for (int i = 0; i <= top; i++)
        [results addObject:output[i]];

    return [UIBezierPath pathWithPoints:results];
}
@end
```

获取这个秘诀的代码

要查找这个秘诀的完整示例项目，可以浏览 https://github.com/erica/iOS-6-Advanced-Cookbook，并进入第 4 章的文件夹。

4.6 秘诀：放入路径

UIKit 贝塞尔路径提供了一种容易的方式，在应用程序中使用向量图形。它们的几何表示允许在不考虑像素的情况下缩放和放置艺术作品。秘诀 4-6 详细说明了在自定义的矩形中绘制路径所涉及的工作，如图 4-6 所示。这幅图片的自定义线条画在绘制时将实时投射到较小的内嵌方框中。

这个秘诀在工作时，将会计算路径的边界框，并使该尺寸能够放入作为方法参数提供的目标矩形中。这只是检查水平和垂直缩放因子的简单事情。较小的比例胜出了，并且会偏置结果，以在目标矩形内居中显示缩放后的路径。

这个秘诀将使用简单的几何学把原始矩形中的每个点都投射到目标矩形中。它将从初始原点计算点的向量，对其进行缩放，然后从目标原点强制转换该向量。这种解决方案对于基于点的 UIBezierPath 实例工作得很好，但是在面对支持圆弧和线条的路径对象时则表现不佳。继续阅读下面的内容，以找到一种可以同时处理基于点的路径和基于曲线的路径的更一般的解决方案。

图 4-6　用户绘制的较大的贝塞尔路径将投射到较小的深色方框中

秘诀 4-6　把路径放入自定义的矩形中

```
// Determine the scale that allows a size to fit into
// a destination rectangle
CGFloat AspectScaleFit(CGSize sourceSize, CGRect destRect)
{
    CGSize destSize = destRect.size;
    CGFloat scaleW = destSize.width / sourceSize.width;
    CGFloat scaleH = destSize.height / sourceSize.height;
    return MIN(scaleW, scaleH);
}
```

```objc
// Create a rectangle that will fit an item while preserving
// its original aspect
CGRect AspectFitRect(CGSize sourceSize, CGRect destRect)
{
    CGSize destSize = destRect.size;
    CGFloat destScale = AspectScaleFit(sourceSize, destRect);

    CGFloat newWidth = sourceSize.width * destScale;
    CGFloat newHeight = sourceSize.height * destScale;

    float dWidth = ((destSize.width - newWidth) / 2.0f);
    float dHeight = ((destSize.height - newHeight) / 2.0f);

    CGRect rect = CGRectMake(destRect.origin.x + dWidth,
        destRect.origin.y + dHeight, newWidth, newHeight);
    return rect;
}

// Add two points
CGPoint PointAddPoint(CGPoint p1, CGPoint p2)
{
    return CGPointMake(p1.x + p2.x, p1.y + p2.y);
}

// Subtract a point from a point
CGPoint PointSubtractPoint(CGPoint p1, CGPoint p2)
{
    return CGPointMake(p1.x - p2.x, p1.y - p2.y);
}

// Project a point from a native rectangle into a destination
// rectangle
NSValue *adjustPoint(CGPoint p, CGRect native, CGRect dest)
{
    CGFloat scaleX = dest.size.width / native.size.width;
    CGFloat scaleY = dest.size.height / native.size.height;

    CGPoint point = PointSubtractPoint(p, native.origin);
    point.x *= scaleX;
    point.y *= scaleY;
```

```
    CGPoint destPoint = PointAddPoint(point, dest.origin);

    return [NSValue valueWithCGPoint:destPoint];
}

// Fit a path into a rectangle
- (UIBezierPath *) fitInRect: (CGRect) destRect
{
    // Calculate an aspect-preserving destination rectangle
    NSArray *points = self.points;
    CGRect bounding = self.bounds;
    CGRect fitRect = AspectFitRect(bounding.size, destRect);

    // Project each point from the original to the
    // destination rectangle
    NSMutableArray *adjustedPoints = [NSMutableArray array];
    for (int i = 0; i < points.count; i++)
        [adjustedPoints addObject:adjustPoint(
            POINT(i), bounding, fitRect)];
    return [UIBezierPath pathWithPoints:adjustedPoints];
}
```

获取这个秘诀的代码

要查找这个秘诀的完整示例项目,可以浏览 https://github.com/erica/iOS-6-Advanced-Cookbook,并进入第 4 章的文件夹。

4.7 处理曲线

UIBezierPath 不仅仅是你迄今为止看到的点和直线。可以通过复杂的元素构造实例,包括三次曲线和二次曲线。每个元素都具有一种类型和 0~3 个参数,其中前者描述了元素在路径中扮演的角色,后者则和于构造元素。路径封闭元素不使用任何参数。三次贝塞尔曲线元素有 3 个:一个端点和两个控制点。

程序清单 4-1 扩展了秘诀 4-1 中的概念,用于获取路径元素而不是点。每个元素都被表示为一个数组,其中包含元素类型,其后接着它的参数。bezierElements 方法返回该数组。

也很容易走到另一个方向。程序清单的 pathWithElements:类方法接受这个元素数组,返回通过这些元素构造的原始路径。与秘诀 4-1 一样,这一对方法允许像点一样获取路径元素,评估并调整它们,然后利用更新的项目重新构造正确的路径。

程序清单 4-1　贝塞尔曲线元素

```objc
// Construct an array of Bezier Elements
void getBezierElements(void *info, const CGPathElement *element)
{
    NSMutableArray *bezierElements = (__bridge NSMutableArray *)info;
    CGPathElementType type = element->type;
    CGPoint *points = element->points;

    switch (type)
    {
        case kCGPathElementCloseSubpath:
            [bezierElements addObject:@[@(type)]];
            break;
        case kCGPathElementMoveToPoint:
        case kCGPathElementAddLineToPoint:
            [bezierElements addObject:@[@(type), VALUE(0)]];
            break;
        case kCGPathElementAddQuadCurveToPoint:
            [bezierElements addObject:
                @[@(type), VALUE(0), VALUE(1)]];
            break;
        case kCGPathElementAddCurveToPoint:
            [bezierElements addObject:
                @[@(type), VALUE(0), VALUE(1), VALUE(2)]];
            break;
    }
}
// Retrieve the element array
- (NSArray *) bezierElements
{
    NSMutableArray *elements = [NSMutableArray array];
    CGPathApply(self.CGPath,
        (__bridge void *)elements, getBezierElements);
    return elements;
}
// Construct a path from its elements
+ (UIBezierPath *) pathWithElements: (NSArray *) elements
{
    UIBezierPath *path = [UIBezierPath bezierPath];
```

```objc
    if (elements.count == 0) return path;
    
    for (NSArray *points in elements)
    {
        if (!points.count) continue;
        CGPathElementType elementType = [points[0] integerValue];
        switch (elementType)
        {
            case kCGPathElementCloseSubpath:
                [path closePath];
                break;
            case kCGPathElementMoveToPoint:
                if (points.count == 2)
                    [path moveToPoint:POINT(1)];
                break;
            case kCGPathElementAddLineToPoint:
                if (points.count == 2)
                    [path addLineToPoint:POINT(1)];
                break;
            case kCGPathElementAddQuadCurveToPoint:
                if (points.count == 3)
                    [path addQuadCurveToPoint:POINT(2)
                        controlPoint:POINT(1)];
                break;
            case kCGPathElementAddCurveToPoint:
                if (points.count == 4)
                    [path addCurveToPoint:POINT(3)
                        controlPoint1:POINT(1) controlPoint2:POINT(2)];
                break;
        }
    }
    return path;
}
```

4.7.1 放入元素

在可以访问元素之后，就可以调整系统提供的路径，比如圆角矩形、椭圆等，就像基于点的路径那样容易。程序清单 4-2 更新了秘诀 4-2 中的概念，以允许针对一般的贝塞尔路径情况投射点和控制点。图 4-7 显示了怎样移动通过二次和三次贝塞尔路径构造的路径，并将其调整为目标矩形。

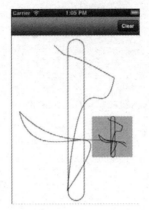

图 4-7　可以通过二次和三次曲线和直线构造 UIBezierPath。该路径由一个圆角矩形和几条随机的三次贝塞尔曲线组成，并使用从原始路径中恢复的元素投射到一个自定义的矩形中

程序清单 4-2 在工作时将遍历元素数组。它将从其原始几何形状中调整每个点（目标点和控制点），把它投射到目标矩形中。然后把调整过的元素传递给程序清单 4-1 中的重新构造类方法，返回更新过的贝塞尔路径。

程序清单 4-2　放入基于元素的贝塞尔路径

```
// Project a Bezier Path into a custom rectangle
- (UIBezierPath *) fitElementsInRect: (CGRect) destRect
{
    CGRect bounding = self.bounds;
    CGRect fitRect = AspectFitRect(bounding.size, destRect);

    NSArray *elements = self.bezierElements;
    NSMutableArray *adjustedElements = [NSMutableArray array];
    for (NSArray *points in elements)
    {
        if (!points.count) continue;
        NSMutableArray *outArray = [NSMutableArray array];
        [outArray addObject:points[0]]; // NSNumber, type
        for (int i = 1; i < points.count; i++)
            [outArray addObject:adjustPoint(
                POINT(i), bounding, fitRect)];
        [adjustedElements addObject:outArray];
    }
    return [UIBezierPath pathWithElements:adjustedElements];
}
```

4.8 秘诀：沿着贝塞尔路径移动项目

动画代表贝塞尔路径的常见应用。例如，你可能沿着用户绘制的自定义路径移动视图。图 4-8 中的矩形沿着其路径移动，把它自己定位到所行进的路径上。秘诀 4-7 通过在给出一个偏称量时返回一个点和一个斜率，来支持这种行为。

偏移量可能在 0~1 之间变化，代表沿着路径行进的百分比。0%的进程将返回起点；100%则返回终点。为了确定它们之间的点，这个秘诀将计算沿着路径的总距离（length），并通过它的成分点构建一个数组，在 pointPercentArray 方法中预先计算它们的进程。

一个简单的搜索和插补操作将允许 pointAtPercent:slope:方法返回一个点，例如，沿着路径正好 67.25%的位置。该方法将确定位于那个项目之前和之后的项目。这允许它返回通过这些点定义的斜率（dy/dx），为计算动画式对象的旋转角度提供一个基础。为了计算角度，可以给反正切函数提供一个斜率。使用仿射变换应用旋转。

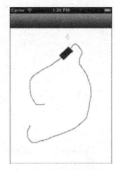

图 4-8 用户的描摹提供了矩形移动的路径。它将调整其旋转角度，以匹配路径的旋转角度

秘诀 4-7　从贝塞尔路径中获取点和斜率

```
// Return distance between two points
static float distance (CGPoint p1, CGPoint p2)
{
    float dx = p2.x - p1.x;
    float dy = p2.y - p1.y;

    return sqrt(dx*dx + dy*dy);
}

// Return the length of a Bezier path
- (CGFloat) length
{
    NSArray *points = self.points;
    float totalPointLength = 0.0f;
    for (int i = 1; i < points.count; i++)
        totalPointLength += distance(POINT(i), POINT(i-1));
    return totalPointLength;
}
```

144 第 4 章 几何学

```objc
- (NSArray *) pointPercentArray
{
    // Use total length to calculate the percent of path
    // consumed at each control point
    NSArray *points = self.points;
    int pointCount = points.count;
    
    float totalPointLength = self.length;
    float distanceTravelled = 0.0f;
    
    NSMutableArray *pointPercentArray = [NSMutableArray array];
    [pointPercentArray addObject:@(0.0)];
    
    for (int i = 1; i < pointCount; i++)
    {
        distanceTravelled += distance(POINT(i), POINT(i-1));
        [pointPercentArray addObject:
            @(distanceTravelled / totalPointLength)];
    }
    // Add a final item just to stop with. Probably not needed.
    [pointPercentArray addObject:[NSNumber numberWithFloat:1.1f]]; // 110%
    
    return pointPercentArray;
}

// Return a point and its slope at a given offset
- (CGPoint) pointAtPercent: (CGFloat) percent withSlope: (CGPoint *) slope
{
    NSArray *points = self.points;
    NSArray *percentArray = self.pointPercentArray;
    CFIndex lastPointIndex = points.count - 1;
    
    if (!points.count)
        return CGPointZero;
    
    // Check for 0% and 100%
    if (percent <= 0.0f) return POINT(0);
    if (percent >= 1.0f) return POINT(lastPointIndex);
    
    // Find a corresponding pair of points in the path
```

```
    CFIndex index = 1;
    while ((index < percentArray.count) &&
           (percent >
            ((NSNumber *)percentArray[index]).floatValue))
        index++;

    // Calculate the intermediate distance between the two points
    CGPoint point1 = POINT(index -1);
    CGPoint point2 = POINT(index);

    float percent1 =
        [[percentArray objectAtIndex:index - 1] floatValue];
    float percent2 =
        [[percentArray objectAtIndex:index] floatValue];
    float percentOffset =
        (percent - percent1) / (percent2 - percent1);

    float dx = point2.x - point1.x;
    float dy = point2.y - point1.y;

    // Store dy, dx for retrieving arctan
    if (slope) *slope = CGPointMake(dx, dy);

    // Calculate new point
    CGFloat newX = point1.x + (percentOffset * dx);
    CGFloat newY = point1.y + (percentOffset * dy);
    CGPoint targetPoint = CGPointMake(newX, newY);

    return targetPoint;
}
```

> **获取这个秘诀的代码**
> 要查找这个秘诀的完整示例项目，可以浏览 https://github.com/erica/iOS-6-Advanced-Cookbook，并进入第 4 章的文件夹。

4.9 秘诀：沿着贝塞尔路径绘制属性化文本

由秘诀 4-7 返回的插值和斜率数学运算并不仅限于动画。这些方法可以沿着贝塞尔路径设置文本（见图 4-9）。秘诀 4-8 的文本布局例程遵守字符串属性，因此可以混合与匹配字体、颜色和大小，同时遵循底层的路径几何学。

该秘诀计算属性化字符串中的每个字符所呈现的尺寸，确定边界高度和宽度。知道这个尺寸使秘诀能够确定每个字符（或字形）将消耗多少路径：

```
- (CGSize) renderedSize
{
    CGRect bounding = [self boundingRectWithSize:CGSizeMake(
        CGFLOAT_MAX, CGFLOAT_MAX) options:0 context:nil];
    return bounding.size;
}
```

如果布置在直线上并且使用该实例作为路径长度的百分比，它将计算字形中心将出现在什么位置。秘诀 4-7 的点的斜率函数可以返回这种放置方式的位置和角度。

该方法通过平移和旋转环境来执行这种放置。这允许使用 NSAttributedString 的 drawAtPoint:方法来呈现字符串。在绘制每个字形之后，图形栈将把环境弹回其原始状态。

在消耗了整个路径之后，例程将停止添加文本，并从视图中裁剪任何剩余的字符。如果你想确保显示整个字符串，可以使用边界矩形例程（boundingRectWithSize:和 size-WithFont:）来计算哪些字体适应给定的尺寸。

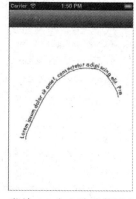

图 4-9　秘诀 4-8 小心地沿着用户绘制的路径排版它的路径，它一直都摇摆不定

秘诀 4-8　沿着贝塞尔路径布置文本

```
- (void) drawAttributedString: (NSAttributedString *) string
    withOptions: (StringRenderingOptions) renderingOptions
{
    if (!string) return;

    NSArray *points = self.points;
    int pointCount = points.count;
    if (pointCount < 2) return;

    // Please do not send over anything with a new line
    NSAttributedString *baseString = string.versionWithoutNewLines;

    // Keep a running tab of how far the glyphs have travelled to
    // be able to calculate the percent along the point path
```

```
float glyphDistance = 0.0f;

// Should the renderer squeeze/stretch the text to fit?
BOOL fitText = (renderingOptions & RenderStringToFit) != 0;
float lineLength = fitText ? baseString.renderedWidth : self.length;

// Optionally force close path
BOOL closePath = (renderingOptions & RenderStringClosePath) != 0;
if (closePath) [self addLineToPoint:POINT(0)];

// Establish the context
CGContextRef context = UIGraphicsGetCurrentContext();
CGContextSaveGState(context);

// Set the initial positions -- skip?
CGPoint textPosition = CGPointMake(0.0f, 0.0f);
CGContextSetTextPosition(context, textPosition.x, textPosition.y);

for (int loc = 0; loc < baseString.length; loc++)
{
    // Retrieve item
    NSRange range = NSMakeRange(loc, 1);
    NSAttributedString *item = [baseString attributedSubstringFromRange:range];

    // Calculate the percent travel
    CGFloat glyphWidth = item.renderedWidth;
    glyphDistance += glyphWidth;
    CGFloat percentConsumed = glyphDistance / lineLength;
    if (percentConsumed > 1.0f) break; // stop when all consumed

    // Find a corresponding pair of points in the path
    CGPoint slope;
    CGPoint targetPoint = [self pointAtPercent:percentConsumed withSlope:&slope];

    // Set the x and y offset
    CGContextTranslateCTM(context, targetPoint.x, targetPoint.y);
    CGPoint positionForThisGlyph = CGPointMake(textPosition.x, textPosition.y);

    // Rotate
    float angle = atan(slope.y / slope.x);
```

```
        if (slope.x < 0) angle += M_PI; // going left, update the angle
        CGContextRotateCTM(context, angle);
        // Place the glyph
        positionForThisGlyph.x -= glyphWidth;
        if ((renderingOptions & RenderStringOutsidePath) != 0)
        {
            positionForThisGlyph.y -= item.renderedHeight;
        }
        else if ((renderingOptions & RenderStringInsidePath) != 0)
        {
            // no op
        }
        else // over path or default
        {
            positionForThisGlyph.y -= item.renderedHeight / 2.0f;
        }

        // Draw the glyph
        [item drawAtPoint:positionForThisGlyph]; // was textPosition

        // Reset context transforms
        CGContextRotateCTM(context, -angle);
        CGContextTranslateCTM(context, -targetPoint.x, -targetPoint.y);
    }

    CGContextRestoreGState(context);
}
```

> **获取这个秘诀的代码**
> 要查找这个秘诀的完整示例项目，可以浏览 https://github.com/erica/iOS-6-Advanced-Cookbook，并进入第 4 章的文件夹。

4.10 秘诀：视图变换

仿射变换代表 UIKit 中最常用、最可怕的特性之一。依赖于直接交互，在处理姿势识别器、动画以及任何类型的视图缩放和旋转时通常会遇到它们。变换的挫折因子（frustration factor）很大程度上与底层结构的不透明性和缺少人类容易适应的方法有关。执行简单的调整就可以把 CGAffineTransform 结构转换为更友好的基于 Objective-C 的属性和方法。

4.10.1 基本变换

仿射变换允许在应用程序中缩放、旋转和平移 UIView 对象。一般会创建一种变换，并使用以下模式之一把它应用于视图。可以使用 make 函数之一创建和应用新的变换。

```
float angle = theta * (PI / 100);
CGAffineTransform transform = CGAffineTransformMakeRotation(angle);
myView.transform = transform;
```

或者可以 action 函数之一把新的改变分层放置在现在的变换上：旋转、缩放或平移。

```
CGAffineTransform transform = CGAffineTransformRotate(myView.transform, angle);
myView.transform = transform;
```

创建新变换将重置已经应用于视图的任何改变。例如，如果已经通过变换放大了视图，那么这两个示例中的第一个示例将覆盖该缩放，并用旋转代替它。图 4-10 显示了应用这种方案之前和之后的效果。外部较大的、缩放过的视图将被未缩放、较小、旋转过的视图所代替（红圈标记视图的右上角）。

在第二个示例中（参见图 4-11），将对变换进行分层处理，使之与缩放叠加，而不是代替它。在这种情况下，缩放过的视图将会旋转，但是缩放将不会受影响。

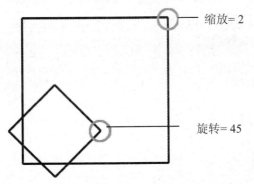

图 4-10　设置一种变换代替以前设置的任何视图变换

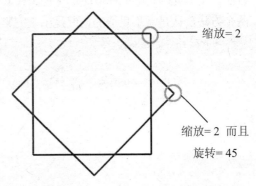

图 4-11　追加变换将保留以前的设置

4.10.2 揭密

每种变换都由底层的变换矩阵表示，它被建立如下。
这个矩阵对应于一个简单的 C 语言结构：

```
struct CGAffineTransform {
    CGFloat a;
    CGFloat b;
    CGFloat c;
```

```
    CGFloat d;
    CGFloat tx;
    CGFloat ty;
};
typedef struct CGAffineTransform CGAffineTransform;
```

UIKit 框架定义了多种特定于图形和绘图操作的辅助函数，其中包括几个特定于仿射的实用程序。可以通过 UIKit 的 NSStringFromCGAffineTransform() 函数打印出视图的变换，CGAffineTransformFromString() 函数则执行与其相反的操作。下面显示了以前讨论过的缩放（通过 1.5 的因子）和旋转（通过 $\pi/4$ 的因子）视图的变换值：

```
2012-08-31 09:43:20.837 HelloWorld[41450:c07]
    [1.06066, 1.06066, -1.06066, 1.06066, 0, 0]
```

这些原始数字不是特别有用。确切地讲，这种表示不会准确指示视图缩放和旋转了多少。幸运的是，有一种方式可用于解决这个问题。

4.10.3 获取变换值

可以通过仿射结构的 a、b、c、d、tx 和 ty 条目轻松地计算特定的变换值。下面给出了 5 个方法，用于返回视图的缩放比例（在 x 和 y 方向）、旋转角度和变换值（在 x 和 y 方向）。其中后两个方法公认是微不足道的，但是出于完整性考虑包括了它们：

```
- (CGFloat) xscale
{
    CGAffineTransform t = self.transform;
    return sqrt(t.a * t.a + t.c * t.c);
}


- (CGFloat) yscale
{
    CGAffineTransform t = self.transform;
    return sqrt(t.b * t.b + t.d * t.d);
}


- (CGFloat) rotation
{
    CGAffineTransform t = self.transform;
    return atan2f(t.b, t.a);
}


- (CGFloat) tx
```

```
{
    CGAffineTransform t = self.transform;
    return t.tx;
}

- (CGFloat) ty
{
    CGAffineTransform t = self.transform;
    return t.ty;
}
```

使用这些方法，可以精确确定要把视图旋转和缩放多少。当使用姿势识别器交互式地利用组合变换拉伸、收缩和旋转视图时这特别有用。在设置缩放界限时，这些方法也可能是有用的。例如，你可能想限制把视图刚好放大到它的正常大小的两倍，或者使之收缩到原始大小的一半以下。检查当前缩放比例，就可以让你执行这类操作。

4.10.4 设置变换值

借助正确的数学运算，很容易设置变换值（比如旋转和 x 方向的平移），就像获取它们一样。可以通过其成分计算出每种变换：

```
CGAffineTransform makeTransform(CGFloat xScale, CGFloat yScale,
    CGFloat theta, CGFloat tx, CGFloat ty)
{
    CGAffineTransform transform = CGAffineTransformIdentity;

    transform.a = xScale * cos(theta);
    transform.b = yScale * sin(theta);
    transform.c = xScale * -sin(theta);
    transform.d = yScale * cos(theta);
    transform.tx = tx;
    transform.ty = ty;

    return transform;
}
```

假如你想独立地设置视图的 y 方向上的缩放。下面说明了如何使用本书前面定义的视图属性来执行该任务：

```
- (void) setYscale: (CGFloat) yScale
{
```

```
self.transform = makeTransform(self.xscale, yScale,
    self.rotation, self.tx, self.ty);
}
```

记住：旋转只在一个方向上缩放过的视图可能会产生扭曲。图 4-12 显示了两幅图像，第一幅图像表示具有自然的 1:1.5 的宽高比的视图（该视图的宽度是 100 磅，高度是 150 磅），将其向右旋转 45° 左右。第二幅图像是具有自然的 1:1 的宽高比的视图（100×100 磅）。在 Y 方向上把它放大 1.5 倍，然后旋转相同的 45° 左右。注意扭曲。从上到下的 y 方向的缩放保持 1.5 倍（沿轴左上角到右下角的轴线），使视图歪曲以能够容纳它。

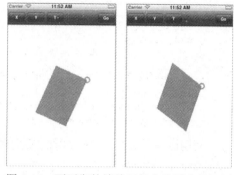

图 4-12　不平衡的缩放可能产生歪曲的视图

4.10.5　获取视图点的位置

除了询问"视图的当前旋转角度是什么"和"把它缩放多少倍"这样的问题之外，开发人员还要执行与视图变换后的几何形状相关的数学运算。为此，需要指定画面元素出现在屏幕上的什么位置。

在没有意外事件发生的情况下，视图的中心在从变换前到变换后的过渡期间将保持有意义。这个值可能会改变，尤其是在进行缩放后，但是无论应用了什么变换，这个属性都是有效的。这个中心属性总是指视图画面在父坐标系统内的几何中心。

画面不是如此有弹性。在旋转后，视图的原点可能完全脱离视图。在图 4-13 中，它在其原始画面（最小的轮廓线）的顶部显示旋转过的视图和更新过的画面（最大的灰色轮廓线）。圆圈指示视图在旋转前后的右上角。

在应用变换后，画面将更新到封闭视图的最小边界框。它的新原点（外部方框的左上角）实质上与更新过的视图原点（内部方框的左上角）毫无关系。iOS 没有提供一种方式来获取那个更新过的点。

秘诀 4-9 定义了为你执行这种数学运算的视图方法。它们返回变换后的视图的各个角：左上角、右上角、左下角和右下角。这些坐标定义在父视图中；因此，如果你想在顶部圆圈的上方添加一个新视图，可以把它的中心放在 theView.transformedTopRight。

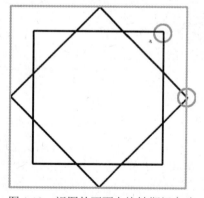

图 4-13　视图的画面在旋转期间会改变，可能不再是有意义的。外部的灰色矩形对应于旋转过的视图画面

秘诀 4-9　变换后的视图访问

```
// Coordinate utilities
- (CGPoint) offsetPointToParentCoordinates: (CGPoint) aPoint
{
    return CGPointMake(aPoint.x + self.center.x,
        aPoint.y + self.center.y);
}
- (CGPoint) pointInViewCenterTerms: (CGPoint) aPoint
{
    return CGPointMake(aPoint.x - self.center.x,
        aPoint.y - self.center.y);
}
- (CGPoint) pointInTransformedView: (CGPoint) aPoint
{
    CGPoint offsetItem = [self pointInViewCenterTerms:aPoint];
    CGPoint updatedItem = CGPointApplyAffineTransform(
        offsetItem, self.transform);
    CGPoint finalItem =
        [self offsetPointToParentCoordinates:updatedItem];
    return finalItem;
}
- (CGRect) originalFrame
{
    CGAffineTransform currentTransform = self.transform;
    self.transform = CGAffineTransformIdentity;
    CGRect originalFrame = self.frame;
    self.transform = currentTransform;

    return originalFrame;
}

    // These four methods return the positions of view elements
    // with respect to the current transform

- (CGPoint) transformedTopLeft
{
    CGRect frame = self.originalFrame;
    CGPoint point = frame.origin;
    return [self pointInTransformedView:point];
```

```objc
}

- (CGPoint) transformedTopRight
{
    CGRect frame = self.originalFrame;
    CGPoint point = frame.origin;
    point.x += frame.size.width;
    return [self pointInTransformedView:point];
}

- (CGPoint) transformedBottomRight
{
    CGRect frame = self.originalFrame;
    CGPoint point = frame.origin;
    point.x += frame.size.width;
    point.y += frame.size.height;
    return [self pointInTransformedView:point];
}

- (CGPoint) transformedBottomLeft
{
    CGRect frame = self.originalFrame;
    CGPoint point = frame.origin;
    point.y += frame.size.height;
    return [self pointInTransformedView:point];
}
```

获取这个秘诀的代码

要查找这个秘诀的完整示例项目，可以浏览 https://github.com/erica/iOS-6-Advanced-Cookbook，并进入第 4 章的文件夹。

4.11 秘诀：测试视图相交

秘诀 4-10 检查两个变换后的视图是否相交。其中的代码还可以处理未经变换的视图，以便可以把它用于任何两个视图；尽管如此，这样做有点无意义（可以为未变换过的简单画面使用 CGRectIntersectsRect() 函数）。这个自定义的相交方法最适合于其画面不代表它们的底层几何形状的视图，如图 4-13 所示。

intersectsView:方法为凸多边形应用一种轴分隔算法。对于每个视图的每条边,它将测试一个视图中的所有点是否都落在某条边的一侧,并且另一个视图中的所有点是否都落在另一侧。这种测试基于半平面(half plane)函数,它返回一个值,指示一个点是位于某条边的左侧还是右侧。

一旦它找到某条边满足这种条件,intersectsView:方法将返回 NO。如果有一根线条把一个对象中的所有点与另一个对象中的所有点分隔开,那么视图就不能在几何上相交。

如果全部 8 个测试都失败了(第一个视图上的 4 条边和第二个视图上的 4 条边),该方法就会推断出两个视图确实相交,并且返回 YES。

秘诀 4-10　获取变换值

```
// The following three methods move points into and out of the
// transform coordinate system whose origin is at the view center

- (CGPoint) offsetPointToParentCoordinates: (CGPoint) aPoint
{
    return CGPointMake(aPoint.x + self.center.x,
        aPoint.y + self.center.y);
}

- (CGPoint) pointInViewCenterTerms: (CGPoint) aPoint
{
    return CGPointMake(aPoint.x - self.center.x, aPoint.y - self.center.y);
}

- (CGPoint) pointInTransformedView: (CGPoint) aPoint
{
    CGPoint offsetItem = [self pointInViewCenterTerms:aPoint];
    CGPoint updatedItem = CGPointApplyAffineTransform(
        offsetItem, self.transform);
    CGPoint finalItem =
        [self offsetPointToParentCoordinates:updatedItem];
    return finalItem;
}

// Return the original frame without transform
- (CGRect) originalFrame
{
    CGAffineTransform currentTransform = self.transform;
    self.transform = CGAffineTransformIdentity;
```

```objc
        CGRect originalFrame = self.frame;
        self.transform = currentTransform;

        return originalFrame;
}

// These four methods return the positions of view elements
// with respect to the current transform

- (CGPoint) transformedTopLeft
{
    CGRect frame = self.originalFrame;
    CGPoint point = frame.origin;
    return [self pointInTransformedView:point];
}
- (CGPoint) transformedTopRight
{
    CGRect frame = self.originalFrame;
    CGPoint point = frame.origin;
    point.x += frame.size.width;
    return [self pointInTransformedView:point];
}

- (CGPoint) transformedBottomRight
{
    CGRect frame = self.originalFrame;
    CGPoint point = frame.origin;
    point.x += frame.size.width;
    point.y += frame.size.height;
    return [self pointInTransformedView:point];
}

- (CGPoint) transformedBottomLeft
{
    CGRect frame = self.originalFrame;
    CGPoint point = frame.origin;
    point.y += frame.size.height;
    return [self pointInTransformedView:point];
}
// Determine if two views intersect, with respect to any
```

```objc
// active transforms

// After extending a line, determine which side of the half
// plane defined by that line, a point will appear
BOOL halfPlane(CGPoint p1, CGPoint p2, CGPoint testPoint)
{
    CGPoint base = CGPointMake(p2.x - p1.x, p2.y - p1.y);
    CGPoint orthog = CGPointMake(-base.y, base.x);
    return (((orthog.x * (testPoint.x - p1.x)) +
        (orthog.y * (testPoint.y - p1.y))) >= 0);
}

// Utility test for testing view points against a proposed line
BOOL intersectionTest(CGPoint p1, CGPoint p2, UIView *aView)
{
    BOOL tlTest = halfPlane(p1, p2, aView.transformedTopLeft);
    BOOL trTest = halfPlane(p1, p2, aView.transformedTopRight);
    if (tlTest != trTest) return YES;
    BOOL brTest = halfPlane(p1, p2, aView.transformedBottomRight);
    if (tlTest != brTest) return YES;

    BOOL blTest = halfPlane(p1, p2, aView.transformedBottomLeft);
    if (tlTest != blTest) return YES;

    return NO;
}

// Determine whether the view intersects a second view
// with respect to their transforms
- (BOOL) intersectsView: (UIView *) aView
{
    if (!CGRectIntersectsRect(self.frame, aView.frame)) return NO;

    CGPoint A = self.transformedTopLeft;
    CGPoint B = self.transformedTopRight;
    CGPoint C = self.transformedBottomRight;
    CGPoint D = self.transformedBottomLeft;

    if (!intersectionTest(A, B, aView))
    {
```

```
        BOOL test = halfPlane(A, B, aView.transformedTopLeft);
        BOOL t1 = halfPlane(A, B, C);
        BOOL t2 = halfPlane(A, B, D);
        if ((t1 != test) && (t2 != test)) return NO;
}
if (!intersectionTest(B, C, aView))
{
        BOOL test = halfPlane(B, C, aView.transformedTopLeft);
        BOOL t1 = halfPlane(B, C, A);
        BOOL t2 = halfPlane(B, C, D);
        if ((t1 != test) && (t2 != test)) return NO;
}
if (!intersectionTest(C, D, aView))
{
        BOOL test = halfPlane(C, D, aView.transformedTopLeft);
        BOOL t1 = halfPlane(C, D, A);
        BOOL t2 = halfPlane(C, D, B);
        if ((t1 != test) && (t2 != test)) return NO;
}
if (!intersectionTest(D, A, aView))
{
        BOOL test = halfPlane(D, A, aView.transformedTopLeft);
        BOOL t1 = halfPlane(D, A, B);
        BOOL t2 = halfPlane(D, A, C);
        if ((t1 != test) && (t2 != test)) return NO;
}
A = aView.transformedTopLeft;
B = aView.transformedTopRight;
C = aView.transformedBottomRight;
D = aView.transformedBottomLeft;

if (!intersectionTest(A, B, self))
{
        BOOL test = halfPlane(A, B, self.transformedTopLeft);
        BOOL t1 = halfPlane(A, B, C);
        BOOL t2 = halfPlane(A, B, D);
        if ((t1 != test) && (t2 != test)) return NO;
}
if (!intersectionTest(B, C, self))
{
```

```
        BOOL test = halfPlane(B, C, self.transformedTopLeft);
        BOOL t1 = halfPlane(B, C, A);
        BOOL t2 = halfPlane(B, C, D);
        if ((t1 != test) && (t2 != test)) return NO;
    }
    if (!intersectionTest(C, D, self))
    {
        BOOL test = halfPlane(C, D, self.transformedTopLeft);
        BOOL t1 = halfPlane(C, D, A);
        BOOL t2 = halfPlane(C, D, B);
        if ((t1 != test) && (t2 != test)) return NO;
    }
    if (!intersectionTest(D, A, self))
    {
        BOOL test = halfPlane(D, A, self.transformedTopLeft);
        BOOL t1 = halfPlane(D, A, B);
        BOOL t2 = halfPlane(D, A, C);
        if ((t1 != test) && (t2 != test)) return NO;
    }
    return YES;
}
```

获取这个秘诀的代码

要查找这个秘诀的完整示例项目，可以浏览 https://github.com/erica/iOS-6-Advanced-Cookbook，并进入第 4 章的文件夹。

4.12 小结

本章考查了多种与路径和视图相关的实用的几何秘诀。无可否认，在那些解决方案中融入了许多数学知识。在继续学习下一章之前，要思考以下几点。

- 本章中的大量内容关注的是展示通常被 UIKit API 隐藏的属性。仅仅由于类以对其开发人员有意义的方式表达它自身（变换、A、B、C、D、tx、ty 等）并不意味着它在语义上也最适合于其用户（旋转、缩放、平移和点）。如果可以展示底层的数学运算，就可以使用更适合于人们使用的术语，自由地通过类别表达那些类。
- 本章中使用的许多贝塞尔路径示例都基于屏幕上的用户描摹。这些秘诀并不仅限于此。可以非常轻松地把这些例程用于以编程方式创建的路径，或者使用第三方工具（比如 PixelCut 的 PaintCode（99 美元，http://paintcodeapp.com））绘制的路径。

第 5 章
联网

作为一种连接 Internet 的设备，iPhone 及其他的 iOS 家族成员特别适合于获取远程数据和访问基于 Web 的服务。Apple 以各种网络计算及其支持技术给平台打下了坚实的基础。Core Cookbook 一书中有关联网的章节介绍了网络状态检查、同步和异步下载、JSON 和 XML 解析。本书中的这一章继续讨论这个主题，并且介绍了更高级的技术。其中包括身份验证质询、使用系统密钥链（keychain）、处理 OAuth 等。这里介绍了一些方便的方法，应该可以帮助你从事开发工作。

5.1 秘诀：安全凭证存储

在深入研究实际的联网之前，考虑一下用户凭证的作用。它们构成了任何现代网络访问方案的基础部分。使用凭证登录社交服务，上传到文件共享站点，或者从用户的账户获取信息。当今，几乎没有不涉及某种用户名和密码的网络开发任务。因此，本章中的前两个秘诀介绍了安全凭证存储。

iOS 的凭证系统提供了一种安全机制，用于存储和获取用户名和密码。凭证特定于服务器或"领域"，例如，在 Facebook 上使用的凭证将不同于在 Twitter 或 imgur 或者其他 Web 站点上使用的凭证。因此，每个服务器都有它自己的保护"空间"，只能访问特定于那台主机的凭证。

可以指定凭证的范围，确定凭证将持续多长时间，以及允许系统提供的存储系统管理该信息。对于凭证存储，可以使用 3 种持久性。

- NSURLCredentialPersistenceNone 意指凭证必须立即使用，不能把它存储起来以便将来使用。
- NSURLCredentialPersistenceForSession 只允许在这个应用程序会话期间存储凭证。
- NSURLCredentialPersistencePermanent 允许把凭证存储到用户的密钥链，并与其他应用程序共享。

下面说明了如何创建凭证。可以创建一个键/值对（以用户名和密码的形式），并指定持久性：
NSURLCredential *credential = [NSURLCredential credentialWithUser:key

```
password:value persistence: NSURLCredentialPersistencePermanent];
```

凭证还允许存储与签名身份关联的证书，以及确定服务器信任身份验证。

把凭证添加到存储系统中的方法是：把它们分配给保护空间。保护空间存储单台主机的身份验证详细信息。每个保护空间都会定义一个凭证有效的范围。例如，可能创建一个特定于 ericasadun.com 的保护空间，如下所示：

```
[[NSURLProtectionSpace alloc] initWithHost:@"ericasadun.com" port:0
    protocol:@"http" realm:nil authenticationMethod:nil]
```

把端口设置为 0 意味着 Use the Default Port（使用默认端口）。对于 http，这个默认值是端口 80。如果需要，可以传递一个不同的值。协议定义了用于访问资源的模式：通过 http、https、ftp 等。

可以把凭证保存到中心凭证存储系统中，它是一个系统提供的单例（singleton）。下面显示了如何为特定的保护空间存储凭证，其中"设置"方法将为你保存那个凭证：

```
[[NSURLCredentialStorage sharedCredentialStorage]
    setCredential:credential forProtectionSpace:self.protectionSpace];
```

秘诀 5-1 构建了一个辅助类，用于存储和获取安全信息。给每个实例都提供一台特定的主机，并将它的保护空间绑定到那台主机。利用辅助类，可以添加新的凭证，为现有凭证设置新的值，删除凭证，以及实质上把凭证存储系统视作一个字典。

之所以提供这个字典支持，是因为辅助类实现了键控下标（keyed subscript）方法。这意味着可以像下面这样使用类，并且利用 Objective-C 的快捷方式。

```
CredentialHelper *helper =
    [CredentialHelper helperWithHost:@"ericasadun.com"];
helper[@"John Doe"] = @"FooBarGaz";
NSLog(@"%@", helper[@"John Doe"]);
```

保护空间的详细信息封闭在辅助类中，使之成为一个简单的解决方案，可以满足大多数安全解决方案的要求。

秘诀 5-1　凭证的助手类

```
@implementation CredentialHelper

// Create a helper specific to a supplied host
+ (id) helperWithHost: (NSString *) host
{
    CredentialHelper *helper = [[CredentialHelper alloc] init];
    helper.host = host;
    return helper;
}
```

```objc
// Return the current protection space for the host
- (NSURLProtectionSpace *) protectionSpace
{
    if (!_host) return nil;
    return [[NSURLProtectionSpace alloc] initWithHost:_host
        port:0 protocol:@"http" realm:nil authenticationMethod:nil];
}


// Return all the credentials for the supplied protection space
- (NSDictionary *) credentials
{
    if (!_host) return nil;

    NSDictionary *credentials =
        [[NSURLCredentialStorage sharedCredentialStorage]
            credentialsForProtectionSpace:self.protectionSpace];
    return credentials;
}


// Store a credential
- (void) storeCredential: (NSString *) value forKey: (NSString *) key
{
    if (!_host)
    {
        NSLog(@"Error: Cannot store credential for nil host");
        return;
    }

    NSURLCredential *credential =
        [NSURLCredential credentialWithUser:key password:value
            persistence: NSURLCredentialPersistencePermanent];
    [[NSURLCredentialStorage sharedCredentialStorage]
        setCredential:credential
        forProtectionSpace:self.protectionSpace];
}
// Store a credential but make it act as the default
- (void) storeDefaultCredential: (NSString *) value
    forKey: (NSString *) key
{
```

```objc
        if (!_host)
        {
            NSLog(@"Error: Cannot store credential for nil host");
            return;
        }
        
        NSURLCredential *credential =
            [NSURLCredential credentialWithUser:key password:value
                persistence: NSURLCredentialPersistencePermanent];
        [[NSURLCredentialStorage sharedCredentialStorage]
            setDefaultCredential:credential
            forProtectionSpace:self.protectionSpace];
}


// Return the default credential
- (NSURLCredential *) defaultCredential
{
    return [[NSURLCredentialStorage sharedCredentialStorage]
        defaultCredentialForProtectionSpace:self.protectionSpace];
}

// Remove a credential associated with a username (key)
- (void) removeCredential: (NSString *) key
{
    NSArray *keys = self.credentials.allKeys;
    if (![keys containsObject:key])
    {
        NSLog(@"Key %@ not found in credentials. \
            Skipping remove request.", key);
        return;
    }
    [[NSURLCredentialStorage sharedCredentialStorage]
        removeCredential:self.credentials[key]
        forProtectionSpace:self.protectionSpace];
}

// Remove all credentials associated with the protection space
- (void) removeAllCredentials
{
    NSArray *keys = self.credentials.allKeys;
```

```objc
        for (NSString *key in keys)
            [self removeCredential:key];
}
// Return a credential for a given user (key)
- (NSURLCredential *) credentialForKey: (NSString *) key
{
    if (!_host) return nil;
    return self.credentials[key];
}


// Return the password (value) for a given user (key)
- (NSString *) valueForKey:(NSString *)key
{
    NSURLCredential *credential = [self credentialForKey:key];
    if (!credential) return nil;
    return credential.password;
}


// Provide indexed access to the helper, e.g. helper[username]
// and helper[username] = value
- (id) objectForKeyedSubscript: (NSString *) key
{
    return [self valueForKey:key];
}
- (void) setObject: (NSString *) newValue
    forKeyedSubscript: (NSString *) aKey
{
    [self storeCredential: newValue forKey: aKey];
}
@end
```

获取这个秘诀的代码

要查找这个秘诀的完整示例项目，可以浏览 https://github.com/erica/iOS-6-Advanced-Cookbook，并进入第 5 章的文件夹。

5.2 秘诀：输入凭证

在涉及输入和获取凭证时，可以做许多工作，使用户的生活变得简单。主动填充密码是它的一部分。秘诀 5-2 演示了如何执行该任务。

秘诀 5-2 使用秘诀 5-1 中介绍的凭证存储机制，创建一个安全的密码输入表单。当它的视图出现时，秘诀将加载任何可用的默认凭证。然后，它将在用户名和密码框中设置它们。在利用 Done（完成）按钮关闭对话框之后，秘诀将存储任何更新。如果用户点按 Cancel（取消）而不是 Done（完成）按钮，将在不进行保存的情况下关闭对话框。

storeCredentials 方法总是设置默认的凭证。将把最近更新的用户名/密码对保存（或者重新保存）到密钥链，并将其标记为默认的。

这个秘诀还实现了几个文本框委托方法。如果用户点按键盘上的 Done（完成）按钮，将像他们点按了导航栏上的 Done（完成）按钮那样处理。文本框将受第一个响应者支配，存储数据，并且关闭密码输入视图控制器。

另一个文本框委托方法将寻找编辑的起点。无论何时用户编辑姓名框，都会使密码无效并清除它。与此同时，"应该改变字符"方法将检查用户名文本框中的任何文本。当它找到一个与新用户名匹配的凭证时，它将自动更新密码框。这允许用户为给定的服务在几个用户名之间切换，而不必在每次这样做时重新输入密码。

可以轻松地扩展这个秘诀，列出已知的用户名，提供对存储的凭证的一次点按式访问。

秘诀 5-2　密码输入视图控制器

```
@implementation PasswordBaseController
{
    UITextField *userField;
    UITextField *passField;
    CredentialHelper *helper;
}

- (id) initWithHost: (NSString *) hostName
{
    if (self = [super init])
    {
        _host = hostName;
        helper = [CredentialHelper helperWithHost:hostName];
    }
    return self;
}

- (void) listCredentials
{
```

```objc
        // Never log passwords in production code
        NSLog(@"Protection space for %@ has %d credentials:",
            _host, helper.credentialCount);
        for (NSString *userName in helper.credentials.allKeys)
            NSLog(@"%@: %@", userName, helper[userName]);
}

// Save the current credentials as default
- (void) storeCredentials
{
    if (!userField.text.length) return;
    [helper storeDefaultCredential:passField.text
        forKey:userField.text];
    [self listCredentials];
}

// Remove the currently-displayed credential
// from the credential storage
- (void) remove
{
    [helper removeCredential:userField.text];
    [self listCredentials];

    // Update GUI
    userField.text = @"";
    passField.text = @"";
    UIBarButtonItem *removeButton = self.navigationItem.leftBarButtonItems[1];
    removeButton.enabled = NO;
}

// Finish and store credentials
- (void) done
{
    [self storeCredentials];
    [self dismissViewControllerAnimated:YES completion:nil];
}

// Finish without updating credentials
- (void) cancel
{
```

```objc
    [self dismissViewControllerAnimated:YES completion:nil];
}


#pragma mark - Text Edits

// User tapping Done confirms changes. Store credentials
- (BOOL)textFieldShouldReturn:(UITextField *)textField
{
    [textField resignFirstResponder];
    [self done];
    return YES;
}


// Only enable Cancel on edits
- (void)textFieldDidBeginEditing:(UITextField *)textField
{
    UIBarButtonItem *cancelButton = self.navigationItem.leftBarButtonItems[0];
    cancelButton.enabled = YES;
}


- (BOOL)textFieldShouldClear:(UITextField *)textField
{
    // Empty the passfield upon a username clear
    if (textField == userField)
        passField.text = @"";
    return YES;
}


// Watch for known usernames during text edits
- (BOOL)textField:(UITextField *)textField
    shouldChangeCharactersInRange:(NSRange)range
    replacementString:(NSString *)string
{
    if (textField != userField) return YES;
    
    // Initially disable remove until there's a credential match
    UIBarButtonItem *removeButton =
        self.navigationItem.leftBarButtonItems[1];
    removeButton.enabled = NO;
```

```objc
    // Preemptively clear password field until there's a value for it
    passField.text = @"";

    // Calculate the target string that will occupy the username field
    NSString *username = [textField.text
        stringByReplacingCharactersInRange:range withString:string];
    if (!username) return YES;
    if (!username.length) return YES;

    // Always check if there's a matching password on file
    NSURLCredential *credential = [helper credentialForKey:username];
    if (!credential)
        return YES;

    // Match!
    passField.text = credential.password;
    removeButton.enabled = YES;

    // Never log passwords in production code!
    NSLog(@"Found match: %@: %@", username, passField.text);

    return YES;
}

#pragma mark - Load Defaults

- (void) viewWillAppear:(BOOL)animated
{
    // Disable the cancel button, there are no edits to cancel
    UIBarButtonItem *cancelButton =
        self.navigationItem.leftBarButtonItems[0];
    cancelButton.enabled = NO;

    // Disable the remove button, until a credential has been matched
    UIBarButtonItem *removeButton =
        self.navigationItem.leftBarButtonItems[1];
    removeButton.enabled = NO;

    NSURLCredential *credential = helper.defaultCredential;
    if (credential)
```

```objc
    {
        // Populate the fields
        userField.text = credential.user;
        passField.text = credential.password;
        
        // Enable credential removal
        removeButton.enabled = YES;
    }
}

- (void) loadView
{
    [super loadView];
    self.navigationItem.rightBarButtonItem =
        BARBUTTON(@"Done", @selector(done));
    self.navigationItem.leftBarButtonItems = @[
    BARBUTTON(@"Cancel", @selector(cancel)),
    BARBUTTON(@"Remove", @selector(remove)),
    BARBUTTON(@"List", @selector(listCredentials)),
    ];
    
    userField = [self textField];
    userField.placeholder = @"User Name";
    
    passField = [self textField];
    passField.secureTextEntry = YES;
    passField.placeholder = @"Password";
}
@end
```

获取这个秘诀的代码

要查找这个秘诀的完整示例项目，可以浏览 https://github.com/erica/iOS-6-Advanced-Cookbook，并进入第 5 章的文件夹。

5.3 秘诀：处理身份验证质询

一些 Web 站点是利用用户名和密码保护的。NSURLConnection 有助于通过响应站点的身份验证质询来访问页面。要测试身份验证，秘诀 5-3 连接到 http://ericasadun.com/Private，建立它就是为了用这种身份验证讨论。这个测试文件夹使用用户名 PrivateAccess 和密码 tuR7!mZ#eh。

要测试一条未经授权的连接(也就是说,你将被拒绝),可以把密码设置为 nil 或者一个无意义的字符串。当把密码设置为 nil 时,将给质询发送一个 nil 凭证,产生一个即时的失败。利用无意义的字符串,在服务器拒绝凭证之后质询将失败。

每次质询失败时,连接的 previousFailureCount 属性都会增加 1。可以轻松地检查这个计数,看看是否已经使一个质询失败了,如果是,就实现某种不能连接的处理程序。例如,你可能想请求试验一个不同的密码。

这个秘诀做了两件事来演示失败的和成功的身份验证,而这是你在自己的代码中不想做的:
- 它将在每次连接之后清除共享的 URL 缓存,以确保请求可以在成功之后失败。如果不这样做,缓存将会记住身份验证信息并继续连接,即使完全发送的是垃圾信息也会如此。不要把它放在生产代码中,因为用户期望身份验证持续存在。
- 这里使用的凭证是根据需要创建的,并且使用了持久性 None。此外,这种应用完全没有现实意义,除非你希望用户为每次 Web 页面访问都输入他们的凭证。这将不是成功的或者令人满意的浏览体验。

秘诀 5-3 利用 NSURLCredential 实例进行身份验证

```
- (void)connection:(NSURLConnection *)connection
    didReceiveData:(NSData *)data
{
    // Load the data into the view
    NSString *source = DATASTR(data);
    [webView loadHTMLString:source baseURL:_url];

    // Force clean the cache -- so you can "fail" after success
    [[NSURLCache sharedURLCache] removeAllCachedResponses];
}
- (void)connection:(NSURLConnection *) connection
    willSendRequestForAuthenticationChallenge:
        (NSURLAuthenticationChallenge *)challenge
{
    if (!challenge.previousFailureCount)
    {
        // Build a one-time use credential
        NSURLCredential *credential =
            [NSURLCredential credentialWithUser:@"PrivateAccess"
                password:_shouldFail ? @"foo" : @"tuR7!mZ#eh"
                persistence:NSURLCredentialPersistenceNone];
```

```objc
            [challenge.sender useCredential:credential
                   forAuthenticationChallenge:challenge];
        }
        else
        {
            // Stop challenge after first failure
            [challenge.sender cancelAuthenticationChallenge:challenge];
            [webView loadHTMLString:@"<h1>Failed</h1>" baseURL:nil];
        }
    }

- (BOOL)connectionShouldUseCredentialStorage:(NSURLConnection *)connection;
{
    NSLog(@"Being queried about credential storage. Saying no.");
    return NO;
}
```

获取这个秘诀的代码

要查找这个秘诀的完整示例项目，可以浏览 https://github.com/erica/iOS-6-Advanced-Cookbook，并进入第 5 章的文件夹。

5.4 秘诀：上传数据

秘诀 5-4 使用 POST 来创建多部分表单数据提交。它将把图像上传到 imgur.com 服务。imgur 提供了无需凭证的匿名图像上传，因此它提供了一个特别好的用于简单发布的示例。"匿名"指的是用户，而不是你这个开发人员。匿名上传允许把图像上传到 imgur，而无需把那些图像绑定到特定的用户账户。作为开发人员，你仍然需要注册 API，并用它发出你的请求。在 https://imgur.com/register/api_anon 上请求 API 键，确保把你的键添加到用于秘诀 5-4 的示例代码中。

秘诀 5-4 的质询将创建可以被 imgur 服务使用的正确格式化的主体。它实现了一个方法，可以通过键和值的字典生成表单数据。出于本示例的目的，那个字典中的对象被限制为字符串和图像。可以为其他数据类型扩展这个方法，只需利用不同的 MIME 类型更改内容类型字符串即可。显然，它不是创建这类数据的理想方式，但它多年来都可以完成工作。

这个秘诀使用一个同步请求来执行上传，处理它可能要花一分钟左右的时间。为了避免阻塞 GUI 更新，整个提交过程都嵌入在一个 NSOperation 子类中。操作封装了用于单个任务的代码和数据，使你能够异步地运行该任务。

使用 NSOperation 对象允许把它们提交给一个异步 NSOperationQueue。操作队列将会管理各个操作的执行。每个操作都会被设置优先级，并放入到队列，它将在那里以设置的优先级顺序执行。

无论何时子类化 NSOperation，都要确保实现一个 main 方法。在执行操作时将调用该方法。当 main 返回时，操作完成。

秘诀 5-4　上传图像到 imgur

```
#define NOTIFY_AND_LEAVE(MESSAGE) {[self bail:MESSAGE]; return;}
#define STRDATA(STRING) \
    ([STRING dataUsingEncoding:NSUTF8StringEncoding])
#define SAFE_PERFORM_WITH_ARG(THE_OBJECT, THE_SELECTOR, THE_ARG)\
    (([THE_OBJECT respondsToSelector:THE_SELECTOR]) ? \
        [THE_OBJECT performSelectorOnMainThread:THE_SELECTOR \
            withObject:THE_ARG waitUntilDone:NO] : nil)


// Form data constants
#define IMAGE_CONTENT(_FILENAME_) \
    @"Content-Disposition: form-data; name=\"%@\";\
    filename=\"_FILENAME_\"\r\nContent-Type: image/jpeg\r\n\r\n"
#define STRING_CONTENT \
    @"Content-Disposition: form-data; name=\"%@\"\r\n\r\n"
#define MULTIPART \
    @"multipart/form-data; boundary=------------0x0x0x0x0x0x0x0x"


@implementation ImgurUploadOperation

// Operation failure. Send an error off to the delegate.
- (void) bail: (NSString *) message
{
    SAFE_PERFORM_WITH_ARG(_delegate,
        @selector(handleImgurOperationError:), message);
}
// Create multipart data from a dictionary
- (NSData*)generateFormDataFromPOSTDictionary:(NSDictionary*)dict
{
    NSString *boundary = @"------------0x0x0x0x0x0x0x0x";
    NSArray *keys = [dict allKeys];
    NSMutableData *result = [NSMutableData data];

    for (int i = 0; i < keys.count; i++)
```

```objc
    {
        // Start part
        id value = dict[keys[i]];
        NSString *start =
            [NSString stringWithFormat:@"--%@\r\n", boundary];
        [result appendData:STRDATA(start)];
        
        if ([value isKindOfClass:[NSData class]])
        {
            // handle image data
            NSString *formstring =
                [NSString stringWithFormat:IMAGE_CONTENT(@"Cookbook.jpg"),
                    [keys objectAtIndex:i]];
            [result appendData:STRDATA(formstring)];
            [result appendData:value];
        }
        else
        {
            // all non-image fields assumed to be strings
            NSString *formstring =
                [NSString stringWithFormat:STRING_CONTENT,
                    [keys objectAtIndex:i]];
            [result appendData: STRDATA(formstring)];
            [result appendData:STRDATA(value)];
        }
        
        // End of part
        NSString *formstring = @"\r\n";
        [result appendData:STRDATA(formstring)];
    }
    
    // End of form
    NSString *formstring =
        [NSString stringWithFormat:@"--%@--\r\n", boundary];
    [result appendData:STRDATA(formstring)];
    return result;
}
- (void) main
{
    if (!_image)
```

```
        NOTIFY_AND_LEAVE(@"ERROR: Please set image before uploading.");

// Establish the post dictionary contents
NSMutableDictionary *postDictionary =
    [NSMutableDictionary dictionary];
postDictionary[@"key"] = IMGUR_API_KEY;
postDictionary[@"title"] = @"Random Image";
postDictionary[@"caption"] =
    @"Created by the iOS Developer's Cookbook";
postDictionary[@"type"] = @"base64";
postDictionary[@"image"] =
    [UIImageJPEGRepresentation(_image, 0.65)
        base64EncodedString];

// Create the post data from the post dictionary
NSData *postData =
    [self generateFormDataFromPOSTDictionary:postDictionary];

// Establish the API request.
NSString *baseurl = @"http://api.imgur.com/2/upload.json";
NSURL *url = [NSURL URLWithString:baseurl];
NSMutableURLRequest *urlRequest =
    [NSMutableURLRequest requestWithURL:url];
if (!urlRequest)
    NOTIFY_AND_LEAVE(@"ERROR: Error creating the URL Request");

[urlRequest setHTTPMethod: @"POST"];
[urlRequest setValue:MULTIPART
    forHTTPHeaderField: @"Content-Type"];
[urlRequest setHTTPBody:postData];

// Submit & retrieve results
NSError *error;
NSURLResponse *response;
NSData* result = [NSURLConnection
    sendSynchronousRequest:urlRequest
    returningResponse:&response error:&error];
if (!result)
{
    [self bail:[NSString stringWithFormat:
```

```
                    @"Submission error: %@", error.localizedFailureReason]];
            return;
        }
        // Success. Return results
        SAFE_PERFORM_WITH_ARG(_delegate,
            @selector(finishedImgurOperationWithData:), result);
}

// Helper method to return a pre-populated operation
+ (id) operationWithDelegate: (id <ImgurUploadOperationDelegate>) delegate andImage:
(UIImage *) image
{
    ImgurUploadOperation *op = [[ImgurUploadOperation alloc] init];
    op.delegate = delegate;
    op.image= image;
    return op;
}
@end
```

> **获取这个秘诀的代码**
> 要查找这个秘诀的完整示例项目，可以浏览 https://github.com/erica/iOS-6-Advanced-Cookbook，并进入第 5 章的文件夹。

5.5 秘诀：构建简单的 Web 服务器

Web 服务器提供了最干净的方式之一，用于把手机发出的数据提供给相同网络上的另一台计算机。不需要特殊的客户软件，任何浏览器都可以列出和访问基于 Web 的文件。最重要的是，Web 服务器只需要少数几个关键例程。必须建立服务，创建一个循环用于侦听请求（startServer），然后把那些请求传递给处理程序（handleWebRequest: ），响应请求的数据。秘诀 5-5 显示了一个 WebHelper 类，用于处理建立和控制基本的 Web 服务，提供在 iOS 设备屏幕上并发显示的相同图像。

循环例程使用低级套接字编程来建立侦听端口和捕获客户请求。当客户发出一个 GET 命令时，服务器将截获该请求，并把它传递给 Web 请求处理程序。处理程序可以分解它，通常会查找想要的数据文件的名称，但是在这个示例中，无论请求的特定细节是什么，它都将返回一幅图像。

秘诀 5-5 通过 Web 服务提供 iPhone 文件

```
#define SAFE_PERFORM_WITH_ARG(THE_OBJECT, THE_SELECTOR, THE_ARG)\
    (([THE_OBJECT respondsToSelector:THE_SELECTOR]) ? \
```

```objc
        [THE_OBJECT performSelectorOnMainThread:THE_SELECTOR \
            withObject:THE_ARG waitUntilDone:NO] : nil)
@implementation WebHelper
// Process the external request by sending an image
// (Customize this to do something more interesting.)
- (void) handleWebRequest: (int) fd
{
    // Request an image from the delegate
    if (!_delegate) return;
    if (![_delegate respondsToSelector:@selector(image)]) return;
    UIImage *image = (UIImage *)[_delegate performSelector:@selector(image)];
    if (!image) return;

    // Produce a jpeg header
    NSString *outcontent = [NSString stringWithFormat:
        @"HTTP/1.0 200 OK\r\nContent-Type: image/jpeg\r\n\r\n"];
    write (fd, [outcontent UTF8String], outcontent.length);

    // Send the data and close
    NSData *data = UIImageJPEGRepresentation(image, 0.75f);
    write (fd, data.bytes, data.length);
    close(fd);
}

// Listen for external requests
- (void) listenForRequests
{
    @autoreleasepool {
        static struct sockaddr_in cli_addr;
        socklen_t length = sizeof(cli_addr);

        while (1) {
            if (!isServing) return;

            if ((socketfd = accept(listenfd,
                (struct sockaddr *)&cli_addr, &length)) < 0)
            {
                isServing = NO;
                [[NSOperationQueue mainQueue]
                    addOperationWithBlock:^(){
```

```objc
                            SAFE_PERFORM_WITH_ARG(delegate,
                                @selector(serviceWasLost), nil);
                }];
                return;
            }
            [self handleWebRequest:socketfd];
        }
    }
}
// Begin serving data
- (void) startServer
{
    static struct sockaddr_in serv_addr;
    
    // Set up socket
    if((listenfd = socket(AF_INET, SOCK_STREAM,0)) < 0)
    {
        isServing = NO;
        SAFE_PERFORM_WITH_ARG(delegate,
            @selector(serviceCouldNotBeEstablished), nil);
        return;
    }
    
    // Serve to a random port
    serv_addr.sin_family = AF_INET;
    serv_addr.sin_addr.s_addr = htonl(INADDR_ANY);
    serv_addr.sin_port = 0;
    
    // Bind
    if (bind(listenfd, (struct sockaddr *)&serv_addr,
        sizeof(serv_addr)) <0)
    {
        isServing = NO;
        SAFE_PERFORM_WITH_ARG(delegate,
            @selector(serviceCouldNotBeEstablished), nil);
        return;
    }
    
    // Find out what port number was chosen.
    int namelen = sizeof(serv_addr);
```

```objc
    if (getsockname(listenfd, (struct sockaddr *)&serv_addr,
        (void *) &namelen) < 0) {
        close(listenfd);
        isServing = NO;
        SAFE_PERFORM_WITH_ARG(delegate,
            @selector(serviceCouldNotBeEstablished), nil);
        return;
    }

    chosenPort = ntohs(serv_addr.sin_port);

    // Listen
    if(listen(listenfd, 64) < 0)
    {
        isServing = NO;
        SAFE_PERFORM_WITH_ARG(delegate,
            @selector(serviceCouldNotBeEstablished), nil);
        return;
    }

    isServing = YES;
    [NSThread
        detachNewThreadSelector:@selector(listenForRequests)
        toTarget:self withObject:NULL];
    SAFE_PERFORM_WITH_ARG(delegate,
        @selector(serviceWasEstablished:), self);
}

- (void) stopService
{
    printf("Shutting down service\n");
    _isServing = NO;
    close(listenfd);
    SAFE_PERFORM_WITH_ARG(_delegate, @selector(serviceDidEnd), nil);
}

+ (id) serviceWithDelegate:(id)delegate
{
    if (![[UIDevice currentDevice] networkAvailable])
    {
```

```
            NSLog(@"Not connected to network");
            return nil;
        }
        
        WebHelper *helper = [[WebHelper alloc] init];
        helper.delegate = delegate ;
        [helper startServer];
        return helper;
    }
@end
```

> **获取这个秘诀的代码**
> 要查找这个秘诀的完整示例项目，可以浏览 https://github.com/erica/iOS-6-Advanced-Cookbook，并进入第 5 章的文件夹。

5.6 秘诀：OAuth 实用程序

OAuth 给许多 iOS 开发人员带来了无尽的烦恼。它是作为一项开放的标准开发的，用于保护用户凭证，它提供了一种安全的方式，在有限的基础上授权敏感的服务。许多流行的 API 都基于 OAuth 及其有争议的下一个版本 OAuth 2.0，其中后者仍然处在开发过程中。对于大多数开发人员来说，OAuth 过于繁琐、注重细节，并且难以使用。

Apple 利用其新的 Accounts 框架简化了 OAuth 的一些情况，但是在编写本书时，该框架仅限于 Facebook、Sina Weibo 和 Twitter。除此之外，私有的 OAuth 支持框架对于开发人员 API 调用仍然不受限制。

秘诀 5-6 介绍了许多实用程序，它们可能在你与 OAuth 打交道时提供帮助。它的代码可以回溯到多年前，并且是围绕 CommonCrypto 库构建的。它构建了一个 OAuthRequestSigner 类，它的职责是构建、签署和编码请求。这种实现难以编码成使用 HMAC-SHA1 签名。

秘诀 5-6 构成了秘诀 5-7 中讨论的所有签名请求的基础，在后一个秘诀中介绍了 OAuth 令牌交换过程。

> **注意：**
> 秘诀 5-6 的示例代码使用 imgur OAuth 凭证。在 https://imgur.com/register/api_oauth 上注册你自己的键。你将需要它们来编译和运行示例代码。

秘诀 5-6 基本的 OAuth 签名实用程序

```
#import <CommonCrypto/CommonHMAC.h>
```

```objc
@implementation OAuthRequestSigner
// Sign the clear text with the secret key
+ (NSString *) signClearText: (NSString *)text
      withKey: (NSString *) secret
{
    NSData *secretData = STRDATA(secret);
    NSData *clearTextData = STRDATA(text);

    //HMAC-SHA1
    CCHmacContext hmacContext;
    uint8_t digest[CC_SHA1_DIGEST_LENGTH] = {0};
    CCHmacInit(&hmacContext, kCCHmacAlgSHA1,
        secretData.bytes, secretData.length);
    CCHmacUpdate(&hmacContext,
        clearTextData.bytes, clearTextData.length);
    CCHmacFinal(&hmacContext, digest);

    // Convert to a base64-encoded result,
    // Thanks to Matt Gallagher's NSData category
    NSData *out = [NSData dataWithBytes:digest
        length:CC_SHA1_DIGEST_LENGTH];
    return [out base64EncodedString];
}

// RFC 3986
+ (NSString *) urlEncodedString: (NSString *) string
{
    NSString *result = (__bridge_transfer NSString *)
        CFURLCreateStringByAddingPercentEscapes(
            kCFAllocatorDefault, (__bridge CFStringRef)string,
            NULL, CFSTR(":/?#[]@!$&'()*+,;="),
            kCFStringEncodingUTF8);
    return result;
}

// Return url-encoded signed request
+ (NSString *) signRequest: (NSString *)
    baseRequest withKey: (NSString *) secret
{
```

```objc
    NSString *signedRequest = [OAuthRequestSigner
        signClearText:baseRequest withKey:secret];
    NSString *encodedRequest = [OAuthRequestSigner
        urlEncodedString:signedRequest];
    return encodedRequest;
}


// Return a nonce (a random value)
+ (NSString *) oauthNonce;
{
    CFUUIDRef theUUID = CFUUIDCreate(NULL);
    NSString *nonceString =
        (__bridge_transfer NSString *)CFUUIDCreateString(
            NULL, theUUID);
    CFRelease(theUUID);
    return nonceString;
}


// Build a token dictionary from a
// key=value&key=value&key=value string
+ (NSDictionary *) dictionaryFromParameterString:
    (NSString *) resultString
{
    if (!resultString) return nil;
    NSMutableDictionary *tokens = [NSMutableDictionary dictionary];
    NSArray *pairs = [resultString componentsSeparatedByString:@"&"];
    for (NSString *pairString in pairs)
    {
        NSArray *pair =
            [pairString componentsSeparatedByString:@"="];
        if (pair.count != 2) continue;
        tokens[pair[0]] = pair[1];
    }
    return tokens;
}

// Build a string from an oauth dictionary
+ (NSString *) parameterStringFromDictionary: (NSDictionary *) dict
{
    NSMutableString *outString = [NSMutableString string];
```

5.6 秘诀：OAuth 实用程序

```objc
    // Sort keys
    NSMutableArray *keys =
        [NSMutableArray arrayWithArray:[dict allKeys]];
    [keys sortUsingSelector:@selector(caseInsensitiveCompare:)];

    // Add sorted items to parameter string
    for (int i = 0; i < keys.count; i++)
    {
        NSString *key = keys[i];
        [outString appendFormat:@"%@=%@", key, dict[key]];
        if (i < (keys.count - 1))
            [outString appendString:@"&"];
    }
    return outString;
}

// Create a base oauth (header) dictionary
+ (NSMutableDictionary *) oauthBaseDictionary: (NSString *) consumerKey;
{
    NSMutableDictionary *dict = [NSMutableDictionary dictionary];
    dict[@"oauth_consumer_key"] = consumerKey;
    dict[@"oauth_nonce"] = [OAuthRequestSigner oauthNonce];
    dict[@"oauth_signature_method"] = @"HMAC-SHA1";
    dict[@"oauth_timestamp"] =
        [NSString stringWithFormat:@"%d", (int)time(0)];
    dict[@"oauth_version"] = @"1.0";
    return dict;
}
+ (NSMutableString *) baseRequestWithEndpoint: (NSString *) endPoint
    dictionary: (NSDictionary *)dict
    andRequestMethod: (NSString *) method
{
    NSMutableString *baseRequest = [NSMutableString string];
    NSString *encodedEndpoint =
        [OAuthRequestSigner urlEncodedString:endPoint];
    [baseRequest appendString:
        [NSString stringWithFormat:@"%@&%@&",
            method, encodedEndpoint]];
    NSString *baseParameterString =
```

```
            [OAuthRequestSigner parameterStringFromDictionary:dict];
    NSString *encodedParamString = 
            [OAuthRequestSigner urlEncodedString:baseParameterString];
    [baseRequest appendString:encodedParamString];
    return baseRequest;
}
@end
```

获取这个秘诀的代码

要查找这个秘诀的完整示例项目，可以浏览 https://github.com/erica/iOS-6-Advanced-Cookbook，并进入第 5 章的文件夹。

5.7 秘诀：OAuth 过程

使用 OAuth 访问服务前，一般需要 5 个步骤。这些步骤允许借助提供商对应用程序进行身份验证，并且确保用户正确地授权访问应用程序。下面几节详细说明了如何在标准的 OAuth 应用程序中执行这些步骤。

5.7.1 第 1 步：从 API 提供商请求令牌

首先，应用程序需要使用基本的 OAuth 令牌。应用程序给服务器的令牌端点（例如，https://api.imgur.com/oauth/request_token）发送一个请求。

这个请求包括 API 提供商提供的应用程序的独特消费者键（"oauth_consumer_key"），以及 OAuth 头部字典的标准元素。它们是一个特殊的时间（一个独特的无意义的字符串"oauth_nonce"）、一个时间戳（"oauth_timestamp"）、一个签名方法（"oauth_signature_method"，比如 HMAC-SHA1）和一个版本（"oauth_version"）。

可以从这些项目构造一个参数字符串，并利用应用程序的秘密密钥签署它。小贴士：确保给密钥追加"&"。如果密钥是"XYZZYPLUGH"，就签署"XYZZYPLUGH&"。这是由于签名密钥总是由应用程序（消费者）秘密密钥和最终用户（令牌）秘密密钥的组合组成的，并且在过程的这个阶段，令牌秘密密钥将保持为未定义。

给请求添加签名并把它发送给 API 提供商。秘诀 5-7 中的 requestTokens:方法演示了这个过程。

5.7.2 第 2 步：获取和存储令牌

一旦接收到经过身份验证的令牌请求，服务就可以在它的响应中发回许多项目。它们包括一个重要的数据对，确切地讲是一个用户令牌（"oauth_token"）和一个用户秘密令牌

("oauth_token_secret")。可以通过把服务返回的参数编码的字符串分解成它的成分键和值来提取这些值。秘诀 5-7 的 processTokens:方法显示了如何执行该任务。

由于用户尚未授权应用程序使用他的凭证，因此这些都是临时项目。它们允许应用程序移到这个过程的下一个阶段，但是不能走得更远。应用程序不能使用它们来发出一般的请求。在用户密钥链中安全地存储这些项目，而不要存储到用户默认的位置。

5.7.3　第 3 步：请求用户访问

用户必须现在访问 Web 页面，并给应用程序授权。你从一个 API 端点和第 2 步中返回的 OAuth 用户令牌构造 URL。该 URL 看起来如下，但是特定的 URL 将因提供商而异：

https://api.imgur.com/oauth/authorize?oauth_token= token

这个 URL 包括一个基本授权端点，它使用这个过程的第 2 步中返回的令牌。图 5-1 显示了这个授权屏幕的 imgur 版本。控制将传递给 Web 页面，并且应用程序负责从这个过程中监测和获取一个验证者。

在可能最简单的情况下，可以在 Safari 中打开 URL，或者指引用户访问一个 Web 站点，并输入 URL。不过，大多数 iOS 开发人员更喜欢把控制保留在应用程序内，并且应该提供一个应用程序中的 UIWebView，以显示身份验证屏幕，如图 5-1 所示。

5.7.4　第 4 步：获取一个 OAuth 验证者令牌

API 授权 Web 界面通过提供一个验证代码来结束它的交互，如图 5-2 所示。身份验证过程完全是在 HTML 中执行的，它位于标准的 Objective-C 交互之外。

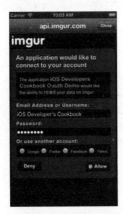

图 5-1　用户必须明确授权访问基于
OAuth 的应用程序，以允许 API 调用

图 5-2　用户必须明确授权访问基于
OAuth 的应用程序，以允许 API 调用

应用程序应该指示用户复制这段代码以便他们可以输入它，或者应该使用 JavaScript 直接从页面源代码中获取它，例如：

```objc
// Look for the "Your verification code is" pattern that imgur provides
- (void)webViewDidFinishLoad:(UIWebView *)aWebView
{
    // Retrieve HTML
    NSString *string = [webView
        stringByEvaluatingJavaScriptFromString:
            @"document.getElementById('content').innerHTML;"];

    // Check for code
    NSString *searchString = @"Your verification code is";
    if ([string rangeOfString:searchString].location == NSNotFound)
        return;

    // Split into lines
    NSArray *lines = [string componentsSeparatedByCharactersInSet:
        [NSCharacterSet newlineCharacterSet]];
    for (NSString *line in lines)
    {
        // Search each line for code
        if ([line rangeOfString:searchString].location == NSNotFound)
            continue;
        NSString *code = [self scanForCode:line];
        if (code)
        {
            // Store the credential
            CredentialHelper *helper =
                [CredentialHelper helperWithHost:HOST];
            [helper storeCredential:code forKey:@"oauth_verifier"];
        }
    }

    // Automatically close the window
    [self close];
}
```

可以看出，这段 HTML/Objective-C 的混杂代码很不美观、难以维护，如果 ATP 提供商改变任何用词，它就会遭受失败。这就是为什么一些提供商现在代之以提供 xAuth 访问的原因。xAuth 允许应用程序直接提供相同的凭证，用户通常可以把它们输入到 OAuth 身份验证 Web 页面中。这对于 iOS 应用程序很方便，因为不必显示 Web 页面，然后使用 JavaScript 提取结果。悲哀的是，许多提供商正在与 xAuth 渐行渐远。

你提供用户名（"x_auth_username"）、密码（"x_auth_password"）、OAuth 令牌，并为客户身份验证设置一种身份验证模式（"x_auth_mode"），比如 "client_auth"。这允许完全绕过 Web 页面身份验证阶段，并为应用程序的用户提供一种完美的体验。

并非所有的 API 提供商都会提供 xAuth 选项，有些提供商则倾向于限制访问范围以预防滥用。可以联系你的 API 提供商，看看 xAuth 是否是一个用于应用程序的选项。

5.7.5 第 5 步：利用访问令牌进行身份验证

使用第 4 步中返回的验证代码完成身份验证过程，如秘诀 5-7 的 authenticate:方法中所示。给 API 提供商提交最终的令牌请求，同时提供验证者作为请求的一部分。这个请求将使用一个与第 1 步中不同的端点，例如：

```
https://api.imgur.com/oauth/access_token
```

你的请求将包含代码（"oauth_verifier"）和第 2 步中返回的 OAuth 令牌。

签署请求的方式现在改变了。第 2 步中的第一个令牌请求只是利用应用程序的消费者秘密密钥签署的。从第 5 步开始，并从那里继续，所有的请求都必须同时利用消费者秘密密钥和 OAuth 秘密令牌（consumer_secret&user_token_secret）签署。更重要的是，这个步骤将返回新的 OAuth 秘密令牌。

这意味着第 2 步中返回的秘密令牌恰好会使用一次，以请求一个新的密钥对。这个步骤将返回新的 OAuth 令牌和新的 OAuth 秘密令牌。丢弃以前的值；利用这些新值替换它们。确保把它们存储到安全的密钥链中，而不要存储到用户默认的位置。应用程序现在将被授权，并准备好提交请求。

注意：

总是要使用密钥链保存 OAuth 令牌。永远不要把 OAuth 凭证存储到最终用户可读的位置，比如纯文本文件、数据库条目，或者用户的默认位置。几个众所周知的入侵方式允许访问这些令牌，在流行的社交网络服务上开放用户账户，从而进行有害的利用。可以通过保护用户的令牌来维护他们的利益，如秘诀 5-1 中所示。

秘诀 5-7　OAuth 过程

```
// Request initial tokens
- (BOOL) requestTokens: (NSString *) tokenEndpoint
```

```objc
{
    NSURL *endpointURL = [NSURL URLWithString:tokenEndpoint];
    
    // Create the preliminary (no token) dictionary
    NSMutableDictionary *dict = [self baseDictionary];
    if (!dict) return NO;
    // Create signature
    NSMutableString *baseRequest =
        [OAuthRequestSigner baseRequestWithEndpoint:tokenEndpoint
            dictionary:dict andRequestMethod:@"POST"];
    NSString *secretKey =
        [_consumerSecret stringByAppendingString:@"&"];
    dict[@"oauth_signature"] =
        [OAuthRequestSigner signRequest:baseRequest withKey:secretKey];
    
    // Produce the token request
    NSString *bodyString = [OAuthRequestSigner
        parameterStringFromDictionary:dict];
    NSMutableURLRequest *request =
        [NSMutableURLRequest requestWithURL:endpointURL];
    request.HTTPMethod = @"POST";
    request.HTTPBody = STRDATA(bodyString);
    [request setValue:@"application/x-www-form-urlencoded"
        forHTTPHeaderField:@"Content-Type"];
    
    // Request the tokens
    NSError *error;
    NSURLResponse *response;
    NSData *tokenData =
        [NSURLConnection sendSynchronousRequest:request
            returningResponse:&response error:&error];
    if (!tokenData)
    {
        NSLog(@"Failed to retrieve tokens: %@",
            error.localizedFailureReason);
        return NO;
    }
    return [self processTokenData:tokenData];
}
// Process and store tokens
```

```objc
- (BOOL) processTokens: (NSData *) tokenData
{
    NSString *tokenResultString = DATASTR(tokenData);

    // Check that we've received the right data
    NSRange range = [tokenResultString
        rangeOfString:@"oauth_token_secret"];
    if (range.location == NSNotFound)
    {
        NSLog(@"Failed to retrieve tokens: %@", tokenResultString);
        return NO;
    }
    // Convert the tokens
    NSDictionary *tokens = [OAuthRequestSigner
        dictionaryFromParameterString:tokenResultString];
    if (!tokens)
    {
        NSLog(@"Unable to process tokens: %@", tokenResultString);
        return NO;
    }

    // Store the tokens
    for (NSString *key in tokens.allKeys)
        credentialHelper[key] = tokens[key];

    return YES;
}

// Finish the end-user authentication step
- (BOOL) authenticate: (NSString *) accessEndpoint
{
    NSURL *endpointURL = [NSURL URLWithString:accessEndpoint];

    // This verifier invalidates after use
    NSString *access_verifier = credentialHelper[@"oauth_verifier"];
    if (!access_verifier)
    {
        NSLog(@"Error: Expected but did not find verifier");
        return NO;
    }
```

```objc
// Add the token and verifier
NSMutableDictionary *dict = [self baseDictionary];
if (!dict) return NO;
dict[@"oauth_token"] = credentialHelper[@"oauth_token"];
dict[@"oauth_verifier"] = credentialHelper[@"oauth_verifier"];

// Create signature
NSMutableString *baseRequest =
    [OAuthRequestSigner baseRequestWithEndpoint:accessEndpoint
        dictionary:dict andRequestMethod:@"POST"];
NSString *compositeKey = [NSString stringWithFormat:@"%@&%@",
    _consumerSecret, credentialHelper[@"oauth_token_secret"]];
dict[@"oauth_signature"] =
    [OAuthRequestSigner signRequest:baseRequest
        withKey:compositeKey];

// Build the request
NSString *bodyString =
    [OAuthRequestSigner parameterStringFromDictionary:dict];
NSMutableURLRequest *request =
    [NSMutableURLRequest requestWithURL:endpointURL];
request.HTTPMethod = @"POST";
request.HTTPBody = STRDATA(bodyString);
[request setValue:@"application/x-www-form-urlencoded"
    forHTTPHeaderField:@"Content-Type"];

// Place the request
NSError *error;
NSURLResponse *response;
NSData *resultData =
    [NSURLConnection sendSynchronousRequest:request
        returningResponse:&response error:&error];

// Check for common issues
if (!resultData)
{
    NSLog(@"Failed to retrieve tokens: %@",
        error.localizedFailureReason);
    return NO;
```

```
    }
    
    // Convert results to a string
    NSString *resultString = DATASTR(resultData);
    if (!resultString)
    {
        NSLog(@"Expected but did not get result string with tokens");
        return NO;
    }
    
    // Process the tokens
    NSDictionary *tokens = [OAuthRequestSigner
        dictionaryFromParameterString:resultString];
    if ([tokens.allKeys containsObject:@"oauth_token_secret"])
    {
        NSLog(@"Success. App is verified.");
        for (NSString *key in tokens.allKeys)
            credentialHelper[key] = tokens[key];
        
        // Clean up
        [credentialHelper removeCredential:@"oauth_verifier"];
        credentialHelper[@"authenticated"] = @"YES";
    }
    return YES;
}
```

获取这个秘诀的代码

要查找这个秘诀的完整示例项目,可以浏览 https://github.com/erica/iOS-6-Advanced-Cookbook,并进入第 5 章的文件夹。

5.8 小结

本章介绍了高级的网络支持技术。你学习了如何使用密钥链,处理安全身份验证质询,上传数据,使用基本的 OAuth 等。在继续学习下一章之前,要思考以下几点。

- 大多数 Apple 的联网支持都是通过低级的基于 C 语言的例程提供的。如果可以找到友好的 Objective-C 包装器来简化编程工作,可以考虑使用它。仅当在应用程序的最基本层面明确需要严格的联网支持时,才会出现缺陷,不过这种情况很少见。它们是极好的资源，在 Google 上就能搜索到它们。
- iOS 的系统密钥链使用起来极其简单，没有理由不利用它，尤其是在处理安全凭证时。要保护用户的利益，并且维护应用程序的声誉。
- 如果可以访问 xAuth（而不是 OAuth），就使用它。你一般将申请特定的应用程序，并且一些提供商需要截屏图、描述等。作为回报，你将获得更干净的界面，而不必费力地使用 Web 页面。

第 6 章 图像

图像是抽象的表示，存储构成图片的数据。本章介绍了 Cocoa Touch 图像，特别是 UIImage 类，并且讲述了在 iOS 上处理图像数据所需的全部专业知识。在本章中，你将学习如何在应用程序中加载、存储和修改图像数据。你将发现如何处理图像数据以创建特殊效果，如何逐个字节地访问图像，等等。

6.1 图像源

iOS 图像一般存储在设备上的几个明确定义的位置之一。这些源包括相册、应用程序包、沙盒、iCloud 和 Internet 等。下面回顾了你将在应用程序中使用的最常见的图像源。

- 相册：iOS 的相册包含相机胶卷（用于准备好相机的装置）、保存的图册、从用户的 Photo Stream 中同步的照片、计算机，或者使用相机连接工具包从另一种数码设备传输的图像。用户可以使用 UIImagePickerController 类提供的交互式对话框请求这个相册中的图像。该对话框允许用户逐个相册地浏览存储的照片，并选择他们想要处理的图像。
- 应用程序包：应用程序包中包括添加到 Xcode 项目中的静态图像。这些资源与应用程序的可执行文件、Info.plist 文件以及项目中包括的其他材料一起出现在应用程序包内（[NSBundle mainBundle]）。可以检查用于项目的 Target > Build Phases > Copy Bundle Resources 窗格，查看包括的文件。

  你的应用程序不能修改这些文件，但是它可以使用 UIImage 的方便的 imageNamed: 方法在应用程序内加载并显示它们。可以使用标准的文件命名约定（比如用于 Retina 显示屏的"@2x"），为特定于设备的分辨率选择以这种方式访问的图像。
- 沙盒：你的应用程序还可以把图像文件写入沙盒中，并根据需要读回它们。沙盒允许把文件存储到 Documents、Library 和 tmp 文件夹中。

  当应用程序在 Info.plist 文件中启用 UIFileSharingEnabled 时，可以从 iTunes 填充和访问顶级 Documents 目录。尽管沙盒外部的 iOS 的其他部分从技术上讲是可读的，iOS

还是明确指出这些区域不受 App Store 应用程序限制。

> **注意：**
> 当使用 Document 文件共享选项时，确保在 Library 文件夹中存储不应该与用户直接共享的任何特定于应用程序的文件。

- iCloud：Apple 的 iCloud 服务允许在共享的中心位置（存储在/private/var/mobile/Library/Mobile Documents/中的文件夹）存储文档，并从用户的所有计算机和 iOS 设备上访问它们。在 iOS 上，可能使用 UIDocument 类把 iCloud 图像加载到应用程序中。
- Internet：应用程序可以使用 URL 资源指向基于 Web 的文件，来下载图像。为了使其工作，iOS 需要一条活动的 Internet 连接，但是一旦建立连接，访问远程图像的数据就像访问本地存储的数据一样容易。
- 粘贴板：应用程序可以使用存储在系统粘贴板中的图像数据。难以把图像数据与其他任何类型的可粘贴数据区分开。可以查询 pasteboardTypes 取回该数据，这将返回一个统一类型标识符的数组，它们指定了粘贴板上目前可用的是哪些类型的数据。

  图像 UTI 通常具有 public.png、public.tiff、public.jpg 或类似的形式。每个应用程序都可以选择它是否能够基于那些 UTI 来处理粘贴板的内容。

  在第 2 章中可以阅读到关于系统粘贴板及其用法的更多知识。
- 共享数据：iOS 应用程序可以打开和显示由其他应用程序发送的图像文件。可以调用那些声明支持图像文件类型（通过在它们的 Info.plist 文件中定义一个 CFBundleDocumentTypes 数组）的应用程序，来打开那些文件。在第 2 章也详细讨论了共享数据。
- 生成的图像：在应用程序中不限于只使用预先构建的图像，可以根据需要自由地生成图像。UIKit 的图形调用允许通过代码创建图像，并根据需要构建 UIImage 对象。例如，可能像下面这样创建一个色块：

```
// Return a swatch with the given color
- (UIImage *) swatchWithColor:(UIColor *) color
    andSize: (CGFloat) side
{
    UIGraphicsBeginImageContext(CGSizeMake(side, side));
    CGContextRef context = UIGraphicsGetCurrentContext();
    [color setFill];
    CGContextFillRect(context, CGRectMake(0.0f, 0.0f, side, side));
    UIImage *image = UIGraphicsGetImageFromCurrentImageContext();
    UIGraphicsEndImageContext();
    return image;
}
```

把 Quartz 绘图调用放在开始和结束环境语句之间。通过调用 UIGraphicsGetImageFrom-CurrentImageContext()获取当前的图像。生成的图像就像任何其他的 UIImage 实例一样。可以在图像视图中显示它们，把它们传递给表格单元格以进行显示，以及把它们放入按钮中等。参考 Apple 的 *Quartz 2D Programming Guide*，学习关于 Quartz 绘图具体细节的更多知识。

6.2 读取图像数据

图像的文件位置控制着读取其数据的方式。可以想象一下，可以只使用一个诸如 UIImage 的 imageWithContentsOfFile:之类的方法加载图像，而无论它们的源是什么。在现实中，不能这样做。

例如，相册图片和它们的路径对于直接的应用程序访问是隐藏的，除非获取了一个 ALAsset URL。仅当用户明确授权应用程序可以访问时，Apple 的用户授予的特权才能确保被突破的隐私级别。即便如此，也必须使用 Asset Library 调用以读取那些文件，而不是 UIImage 调用。

存储在应用程序包中的图像的访问方式不同于那些存储在 Web 站点上或沙盒中的图像。每一类图像数据都需要一种有细微差别的方法以进行访问和显示。

下面总结了常见的图像读取方法，以及如何使用它们把图像数据读入到应用程序中。

6.2.1 UIImage 类的便捷方法

UIImage 类提供了一个简单的方法，用于加载存储在应用程序包中的任何图像。利用文件名（包括扩展名）调用 imageNamed:，例如：

    myImage = [UIImage imageNamed:@"icon.png"];

该方法将在应用程序的顶级文件夹中寻找具有所提供名称的图像。如果找到这样的图像，就会加载它，并由 iOS 缓存起来。这意味着图像是由该缓存管理的内存。

如果数据存储在沙盒中的另一个位置，则可代之以使用 imageWithContentsOfFile:。给它传递一个指向图像文件路径的字符串。该方法将不会像 imageNamed:那样缓存对象。

一般来讲，imageNamed:将预先做更多的工作。它不会懒惰地把文件解码为原始的像素并把它添加到缓存中，而是让你可以自由地进行内存管理。可以把它用于你将反复使用的图像，尤其是较小的图像。缓存较大的图像可能导致更多的重读开销，从而可能会耗尽内存。由于早期的 iOS 错误和早期的 iOS 设备上严格的内存限制，该方法具有不应有的坏名声，但是这些问题已经解决很长一段时间了，并且更新的设备带有大得多的 RAM。

named 和 contents 方法提供了超过其他图像加载方法的巨大优点。对于更高级别的设备（即 iPhone 4 及更高型号、第三代 iPad 及更高型号、第五代 iPod Touch 及更高型号，它们都具有更高像素密度的显示屏），可以使用这些方法自动在低分辨率图像与高分辨率图像之间做出选择。对于显示屏级别为 2.0 的情况，该方法首先将使用 "@2x" 命名提示搜索文件名。

1. 命名提示

调用 [UIImage imageNamed:@"basicArt.png"]，将在更高级别的设备上加载文件 basicArt@2x.png（如果该文件可用的话），并在 1.0 级别的设备上加载文件 basicArt.png。"@2x" 提示指示为 Retina 设备加载更高分辨率的艺术作品。这种命名模式绑定到了常规的图像加载和更微妙的应用程序启动。总体模式看起来如下。

```
<basename> <usage_specific_modifiers> <scale_modifier> <device_modifier>.png
```

下面说明了如何分解命名的成分。

- 基本名称指任何文件名，比如 basicArt、Default、icon 等。
- 特定于应用的修饰符用于启动和系统必需的项目，即 Default 图像、icon 图像、Spotlight 图像、Settings 图像等这些修饰符包含诸如方位提示（-Landscape、-Portrait、-Portrait-UpsideDown 等）、尺寸提示（-small、-72、-144、-50、-100、-568h 等）和 URL 模式之类的项目。比如说你的应用程序支持名为 xyz 的自定义模式，可以从 Mobile Safari 通过 xyz: whatever 启动它。可以把"-xyz"作为特定于应用的修饰符添加到 Default 图像中，例如，Default-xyz@2x.png。在通过该模式访问时，这允许应用程序利用此资源启动。
- 级别修饰符是可选的"@2x"。万一 Apple 引入了其他的级别，开发人员可能在这里添加新的修饰符。
- 设备修饰符指不同的设备家族，确切地讲是"~iphone"和"~ipad"。iPhone 5 和第五代 iPod Touch 属于"~iphone"家族，即使它们的 4″显示屏比 3.5″的 iPhone 4、iPhone 4S 和第四代 iPad 高。iPad mini 属于"~ipad"家族。

这些命名提示和多项目资源反映了 iOS 家族随着时间的推移其发展有多混乱。典型的应用程序现在带有一打以上的默认和图标图像。因此，这是 Apple 可能的演进领域，有希望转向向量图形（或者甚至是更好的解决方案），并远离多个文件，尤其是当新设备上带有更新的屏幕大小和分辨率时。

> **注意：**
> 在 Github 的 PDF-UIImage 集成上可以找到一些初始作品，它们是由第三方开发人员创建的。诸如 mindbrix/UIImage-PDF 之类的存储库尝试把图像移入可伸缩的向量资产。希望 Apple 最终重新设计图像表示，更长时间地消除使用这些成分修饰符的需要。

> **注意：**
> iOS 支持以下图像类型：PNG、JPG、JPEG、TIF、TIFF、GIF、BMP、BMPF、ICO、CUR、XBM 和 PDF。UIImage 类不会读取或显示 PDF 文件，可代之以使用 UIWebView。

6.2.2 查找沙盒中的图像

默认情况下，每个沙盒都包含 3 个文件夹：Documents、Library 和 tmp。用户生成的数据文件（包括图像）通常驻留在 Documents 文件夹中。这个文件夹的作用恰如其名所示，可以在这个目录中存储文档和访问它们。在这里保存程序创建或通过它浏览的文档文件数据。也可以使用 Documents 文件夹和 iTunes，允许基于桌面的用户直接访问数据。

Library 文件夹包含用户的默认设置以及程序的其他状态信息。可以把 Library 文件夹用于任何应用程序支持文件，这些文件是在连续的运行之间必须保留下来的，并且不打算用于一般的最终用户访问。

tmp 文件夹提供了一个位置，用于动态地创建短暂的（生存期较短的）文件。与 tmp 中不同，Documents 和 Library 中的文件不是短暂的。无论何时 iOS 进行同步，iTunes 都会备份 Documents 中的所有文件和 Library 中的大多数文件（不会备份 Library/Caches 中的项目，就像在 Mac 上一样）。与之相比，当 iOS 重新启动或者当它遇到设备上的存储空间不足的情况时，它将丢弃 tmp 中的任何文件。

下面显示了定位 Documents 文件夹的标准方式：

```
NSArray *paths = [NSSearchPathForDirectoriesInDomains(
    NSDocumentDirectory, NSUserDomainMask, YES);
return [paths lastObject];
```

还可以通过调用 **NSHomeDirectory()**，可靠地定位顶级沙盒文件夹。这允许向下一级导航到 Documents，这可以完全确保到达正确的目的地。下面的函数提供了一种方便的方式，用于返回 Documents 文件夹的路径：

```
NSString *documentsFolder()
{
    return [NSHomeDirectory()
        stringByAppendingPathComponent:@"Documents"];
}
```

要加载图像，可以把它的文件名追加到返回的路径上，并且告诉 UIImage 利用那些内容创建一幅新图像。这段代码将从顶级的文档文件夹中加载一个名为 image.png 的文件，并且返回一个利用该数据初始化的 UIImage 实例：

```
path = [documentsFolder() stringByAppendingPathComponent:@"image.png"];
return [UIImage imageWithContentsOfFile:path];
```

如果路径无效或者指向一种非图像资源，那么这个调用将返回 nil。

6.2.3 从 URL 加载图像

UIImage 类可以从 NSData 实例加载图像，但是不能直接从 URL 字符串或 NSURL 对象执行该操作。因此，要给 UIImage 提供已经从 URL 下载的数据。下面这个代码段将从 weather.com 异步下载最新的美国气象图，然后使用气象数据创建一幅新图像：

```
NSString *map =
    @"http://maps.weather.com/images/maps/current/curwx_720x486.jpg";
NSOperationQueue *queue = [[NSOperationQueue alloc] init];
[queue addOperationWithBlock: ^{
    // Load the weather data
    NSURL *weatherURL = [NSURL URLWithString:map];
    NSData *imageData = [NSData dataWithContentsOfURL:weatherURL];

    // Update the image on the main thread using the main queue
    [[NSOperationQueue mainQueue] addOperationWithBlock:^{
        UIImage *weatherImage = [UIImage imageWithData:imageData];
        imageView.image = weatherImage;
    }];
}];
```

首先，它将构造一个 NSURL 对象，然后它将创建一个利用该 URL 的内容初始化的 NSData 实例。返回的数据有助于构建 UIImage 实例，把它添加到图像视图中，负责在主线程上执行更新。

一种更详尽的实现将返回一幅占位符图像，在本地缓存获取的数据，并且当可以利用下载的资产替换占位符时将更新主线程。

6.2.4 从资产库中读取数据

资产 URL 允许直接访问存储在应用程序的沙盒外面的媒体。这种访问将无限期地持久存在，并且不会局限于由图像选择器返回的图像。典型的资产库 URL 看起来如下：

```
assets-library://asset/asset.JPG?id=553F6592-43C9-45A0-B851-28A726727436&ext=JPG
```

下面的代码段演示了如何访问图像资产。它创建了两个块，用于处理资产访问的两种情况：成功和失败。一旦获取某种资产，代码就会创建一种数据表示，把它存储到 CGImageRef，然后把它转换为标准的 UIImage：

```
// Retrieve an image from an asset URL
- (void) imageFromAssetURL: (NSURL *) assetURL into: (UIImage **) image
{
    ALAssetsLibrary *library = [[ALAssetsLibrary alloc] init];
```

```objc
ALAssetsLibraryAssetForURLResultBlock resultsBlock = ^(ALAsset *asset)
{
    ALAssetRepresentation *assetRepresentation =
        [asset defaultRepresentation];
    CGImageRef cgImage =
        [assetRepresentation CGImageWithOptions:nil];
    CFRetain(cgImage); // Thanks Oliver Drobnik. ARC weirdness
    if (image) *image = [UIImage imageWithCGImage:cgImage];
    CFRelease(cgImage);
};
ALAssetsLibraryAccessFailureBlock failure = ^(NSError *__strong error)
{
    NSLog(@"Error retrieving asset from url: %@",
        error.localizedFailureReason);
};

[library assetForURL:assetURL
    resultBlock:resultsBlock failureBlock:failure];
}
```

> **注意：**
> 在 The Core iOS Developer's Cookbook 一书的第 8 章 "常用控制器" 中讨论了用于从相册和从资产 URL 加载数据的秘诀。

6.3 秘诀：放入和填充图像

通常需要调整图像的大小以放入较小的空间中。可以调整图像视图的大小，但是它的内容可能不会像期望的那样更新。UIKit 提供了视图内容模式，它们准确指定了视图在更新其大小时如何调整内容。可以通过调整 contentMode 属性来设置它们。

最常用的内容模式如下。它们详细说明了视图（包括图像视图）调整它们的内容以匹配它们的当前大小的方式。

- 放入（**UIViewContentModeScaleAspectFit**）：在保留图像比例的同时调整它的大小，放入它，以便图像的每个部分保持可见。依赖于图像的宽高比，结果将是横向或纵向的，还需要某个额外的区域来装饰图像（参见图 6-1 的左上方）。
- 填充（**UIViewContentModeScaleAspectFill**）：冲压出图像的一部分以匹配可用的空间。这种方法将裁剪落在像素显示区域之外的任何元素，要么是顶部和底部，要么是左边和右边。图像通常居中显示，只有较小的尺寸会完全显示（参见图6-1，中上方）。

- 居中（UIViewContentModeCenter）：把图像以其自然比例直接放在图像的中心。大于像素显示区域的图像将被裁剪。那些较小的图像将根据需要进行装饰（参见图6-1，左上方）。类似的模式允许放置在视图的顶部、底部、左边、右边、左上方、右上方、左下方和右下方。
- 挤压（UIViewContentModeScaleToFill）：调整图像的宽高比，以将其放入可用的空间中。将填充所有的像素。

有时，在将图像重绘成较少的字节时，希望遵循内容规则。这允许在减少内存开销的同时模仿这种行为。

秘诀6-1显示了创建进行过模式调整的遵循目标大小的图像。这些方法是使用UIImage类别实现的，并且返回新的图像实例。你将传递一个目标大小和想要的内容模式。该秘诀将基于那个大小返回一幅新图像。

在这个秘诀中没有初始化环境背景。纵向或横向像素将保持透明。可以轻松地更新这段代码，在把图像绘制到它的目标矩形中之前提供一种后备的填充色。这里把它作为一个练习留给你完成。

图6-1 这些截屏图表现了操纵内容以填充可用空间的方式。放入（左上方）保留了原始的宽高比，可以根据需要利用额外的空间装饰图像。填充（中上方）确保每个可用的像素都会被填充，而只会裁剪那些落在框架以外的部分。居中（右上方）使用1:1的像素比例显示原始的图像，并从中心向外进行裁剪。挤压（左下方）把原始像素放入目标宽高比中，并在不保留宽高比的情况下根据需要进行缩放。原始图像出现在右下方

秘诀6-1 应用图像宽高比

```
// Calculate the destination scale for filling
CGFloat CGAspectScaleFill(CGSize sourceSize, CGRect destRect)
{
    CGSize destSize = destRect.size;
    CGFloat scaleW = destSize.width / sourceSize.width;
    CGFloat scaleH = destSize.height / sourceSize.height;
    return MAX(scaleW, scaleH);
}

// Calculate the destination scale for fitting
CGFloat CGAspectScaleFit(CGSize sourceSize, CGRect destRect)
```

```objc
{
    CGSize destSize = destRect.size;
    CGFloat scaleW = destSize.width / sourceSize.width;
    CGFloat scaleH = destSize.height / sourceSize.height;
    return MIN(scaleW, scaleH);
}

// Fit a size into another size
CGSize CGSizeFitInSize(CGSize sourceSize, CGSize destSize)
{
    CGFloat destScale;
    CGSize newSize = sourceSize;

    if (newSize.height && (newSize.height > destSize.height))
    {
        destScale = destSize.height / newSize.height;
        newSize.width *= destScale;
        newSize.height *= destScale;
    }

    if (newSize.width && (newSize.width >= destSize.width))
    {
        destScale = destSize.width / newSize.width;
        newSize.width *= destScale;
        newSize.height *= destScale;
    }

    return newSize;
}

// Destination rect by fitting
CGRect CGRectAspectFitRect(CGSize sourceSize, CGRect destRect)
{
    CGSize destSize = destRect.size;
    CGFloat destScale = CGAspectScaleFit(sourceSize, destRect);

    CGFloat newWidth = sourceSize.width * destScale;
    CGFloat newHeight = sourceSize.height * destScale;

    CGFloat dWidth = ((destSize.width - newWidth) / 2.0f);
```

```
        GFloat dHeight = ((destSize.height - newHeight) / 2.0f);

        CGRect rect = CGRectMake(dWidth, dHeight, newWidth, newHeight);
        return rect;
}

// Destination rect by filling
CGRect CGRectAspectFillRect(CGSize sourceSize, CGRect destRect)
{
        CGSize destSize = destRect.size;
        CGFloat destScale = CGAspectScaleFill(sourceSize, destRect);

        CGFloat newWidth = sourceSize.width * destScale;
        CGFloat newHeight = sourceSize.height * destScale;

        CGFloat dWidth = ((destSize.width - newWidth) / 2.0f);
        CGFloat dHeight = ((destSize.height - newHeight) / 2.0f);

        CGRect rect = CGRectMake(dWidth, dHeight, newWidth, newHeight);
        return rect;
}
// Return a new image created with a content mode
- (UIImage *) applyAspect: (UIViewContentMode) mode
        inRect: (CGRect) bounds
{
        CGRect destRect;

        UIGraphicsBeginImageContext(bounds.size);
        switch (mode)
        {
            case UIViewContentModeScaleToFill:
            {
                destRect = bounds;
                break;
            }
            case UIViewContentModeScaleAspectFill:
            {
                CGRect rect = CGRectAspectFillRect(self.size, bounds);
                destRect = CGRectCenteredInRect(rect, bounds);
                break;
```

```objc
        }
        case UIViewContentModeScaleAspectFit:
        {
            CGRect rect = CGRectAspectFitRect(self.size, bounds);
            destRect = CGRectCenteredInRect(rect, bounds);
            break;
        }
        case UIViewContentModeCenter:
        {
            CGRect rect = (CGRect){.size = self.size};
            destRect = CGRectCenteredInRect(rect, bounds);
            break;
        }
        default:
            break;
    }

    [self drawInRect:destRect];
    UIImage *newImage = UIGraphicsGetImageFromCurrentImageContext();
    UIGraphicsEndImageContext();
    return newImage;
}
```

获取这个秘诀的代码

要查找这个秘诀的完整示例项目，可以浏览 https://github.com/erica/iOS-6-Advanced-Cookbook，并进入第 6 章的文件夹。

6.4 秘诀：旋转图像

旋转图像的最容易的方式涉及对图像视图应用旋转变换。该方法不会影响图像数据，而只会影响它在屏幕上的表示。此外，变换还提供了额外的好处，即管理视图的自由触摸。当希望旋转数据本身时，可以把数据绘制到旋转的环境中，创建一幅变化的图像。

有许多原因要旋转图像。例如，由于用户握持设备的方式，可能需要调整一幅在某个角度拍摄的快照。或者，可能希望创建旋转和缩放过的子图像的大杂烩，以把它们合并到新图片中去。

秘诀 6-2 建立了一个 UIImage 类别，产生具有旋转内容的图像。可以利用一个角度偏移量调用它，并且它将返回一个旋转过的版本。在这个秘诀中，生成的图像的大小依赖于原始图像的几何形状（它的大小）以及对它应用的旋转角度。

例如，旋转 45° 的正方形图像大约每个尺寸要增大 40%（参见图 6-2），从而占据比原始图像大得多的空间。秘诀 6-1 通过应用一种仿射变换计算了这个新大小。仿射变换是把一个坐标系统中的点映射到另一个坐标系统中的几何操作。iOS 使用 CGAffineTransform 函数（在这里是 CGRectApplyAffineTransform），通过应用变换来转换矩形。

当然，可以裁剪旋转过的图像，切除边缘。此外，也可能想进行扩展以容纳整个图片。这个秘诀将保留更新的图像。它将操纵绘制图像的环境，调整它的大小，以便放入整个旋转过的版本。本章后面的秘诀将采用相反的方法，返回调整过大小以与原始版本匹配的图像。

最后，可以收缩结果，使它们适应可用的大小。这通常是最不想要的方法，因为改变的比例将使图像内容降级，并且在屏幕上展示突然收缩的图像时，可能会使用户产生混淆。

说到降级，如果需要保持旋转图像，就要确保总是从原始数据开始。每次旋转都会强制像素合并，并且不考虑裁剪的任何问题。对相同的数据应用多次旋转将产生损坏的结果。旋转原始图像到每个必需的程度，可以产生更干净的结果。

与秘诀 6-1 一样，未处理的像素将保持透明。此外，还可以利用你发现的适合需要的任何颜色和 Alpha 值填充环境的背景。

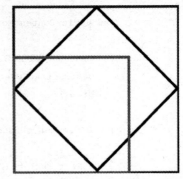

图 6-2　一旦进行了旋转，图像可能需要比它的原始大小大得多的空间。左下角较小的正方形代表初始的图像大小。在旋转后（旋转成中心的菱形），放入单独一幅图像所需的大小将增大到外部的大正方形。可以选择扩展最终的大小、收缩结果，或者裁剪图像的一部分以放入可用的空间内

秘诀 6-2　旋转图像

```
UIImage *rotatedImage(
    UIImage *image, CGFloat rotation)
{
    // Rotation is specified in radians
    // Calculate Destination Size
    CGAffineTransform t = CGAffineTransformMakeRotation(rotation);
    CGRect sizeRect = (CGRect) {.size = image.size};
    CGRect destRect = CGRectApplyAffineTransform(sizeRect, t);
    CGSize destinationSize = destRect.size;

    // Draw image
    UIGraphicsBeginImageContext(destinationSize);
```

```objc
CGContextRef context = UIGraphicsGetCurrentContext();

// Translate to set the origin before rotation
// to the center of the image
CGContextTranslateCTM(context,
    destinationSize.width / 2.0f,
    destinationSize.height / 2.0f);

// Rotate around the center of the image
CGContextRotateCTM(context, rotation);

// Draw the image into the rotated context
[image drawInRect:CGRectMake(
    -image.size.width / 2.0f,
    -image.size.height / 2.0f,
    image.size.width, image.size.height)];

// Save image
UIImage *newImage =
    UIGraphicsGetImageFromCurrentImageContext();
UIGraphicsEndImageContext();
return newImage;
}

@implementation UIImage (Rotation)
- (UIImage *) rotateBy: (CGFloat) theta
{
    return rotatedImage(self, theta);
}

+ (UIImage *) image: (UIImage *) image rotatedBy: (CGFloat) theta
{
    return rotatedImage(image, theta);
}
@end
```

获取这个秘诀的代码

要查找这个秘诀的完整示例项目，可以浏览 https://github.com/erica/iOS-6-Advanced-Cookbook，并进入第 6 章的文件夹。

6.5 秘诀：处理位图表示

尽管 Cocoa Touch 提供了与分辨率无关的优秀工具用于处理许多图像，有时还是需要处理构成图片基础的位，并逐位地访问该数据。例如，可能应用边缘检测或模糊例程，通过相对于实际的字节值卷积矩阵来计算它们的结果。

支持秘诀 6-3 的示例代码获取与某个点对应的字节，用户就是在这个点处触摸屏幕并利用此处的颜色更新导航栏。访问逐个像素的颜色允许项目从屏幕坐标中查找数据。

6.5.1 坐标系统之间的转换

要访问字节，必须把触摸点从图像视图的坐标转换为图像的坐标。通过把视图的内容模式设置为 UIViewContentModeScaleAspectFill，将使该任务变得稍微困难一点。如前一个秘诀所解释的，纵横填充意味着视图中的每个像素都会被填充，从中心向外扩展图像，并沿着边缘或其顶部和底部剪掉多余的材料。

下面的方法通过对视图的触摸点应用一种仿射变换来处理这种转换。这将把它从屏幕的坐标系缩放到图像的坐标系统，这需要执行几个步骤。首先必须对点重置中心，然后进行变换，再从图像的中心重置回原来的位置。仿射变换是针对视图的中心来工作的，因此触摸点的坐标必须适应这种改变。最终，中心—变换—重置中心的过程将产生一个属于图像数据数组的点，它遵循图像的左上角为原点(0,0)的规则：

```
- (void) handle: (CGPoint) aPoint
{
    // Set origin to the center of the view
    aPoint.x -= self.center.x;
    aPoint.y -= self.center.y;

    // Calculate the point in the image's view coordinate system
    CGFloat imageWidth = self.image.size.width;
    CGFloat imageHeight = self.image.size.height;
    CGFloat xScale = imageWidth / self.frame.size.width;
    CGFloat yScale = imageHeight / self.frame.size.height;
    CGFloat scale = MIN(xScale, yScale);
    CGAffineTransform t = CGAffineTransformMakeScale(scale, scale);
    CGPoint adjustedPoint = CGPointApplyAffineTransform(aPoint, t);

    // Reset the origin to the top-left corner of the image
    adjustedPoint.x += imageWidth / 2.0f;
```

```
    adjustedPoint.y += imageHeight / 2.0f;

    // Refresh the image data if needed (it shouldn't be needed)
    if (!_imageData)
        _imageData = self.image.bytes;

    // Retrieve the byte values at the given point -
    Byte *bytes = (Byte *)_imageData.bytes;
    CGFloat red = bytes[redOffset(adjustedPoint.x,
        adjustedPoint.y, imageWidth)] / 255.0f;
    CGFloat green = bytes[greenOffset(adjustedPoint.x,
        adjustedPoint.y, imageWidth)] / 255.0f;
    CGFloat blue = bytes[blueOffset(adjustedPoint.x,
        adjustedPoint.y, imageWidth)] / 255.0f;
    UIColor *color = [UIColor colorWithRed:red
        green:green blue:blue alpha:1.0f];

    // Update the nav bar to match the color at the user's touch
    _controller.navigationController.navigationBar.tintColor = color;
}
```

6.5.2 查找数据

位图没有自然的尺寸，它们只是数据块。因此，如果你想访问点(x,y)处的字节数据，就必须提供每一行的尺寸。每个像素的字节（ARGB 数据有 4 字节）开始于一个偏移量(y * rowWidth + x) * 4。因此，图像中的第一个像素开始于 0，并且会经过字节 3。第二个像素开始于 4，并延续到字节 7，等等。

每个通道都会占据 1 个字节。每个 4 字节的序列都包含 Alpha 值、红色、绿色和蓝色的级别。第一个字节保存 Alpha 值（不透明度）信息。第二个通道偏移 1 字节，绿色偏移 2 字节，蓝色则偏移 3 字节。4 个辅助函数允许该方法通过提供 x、y 和行宽度，把计算出的偏移量换算回字节数组中。下面的函数使用宽度 w 返回 ARGB 位图内的任何点(x,y)的偏移量。

```
NSUInteger alphaOffset(NSUInteger x, NSUInteger y, NSUInteger w)
    {return y * w * 4 + x * 4 + 0;}
NSUInteger redOffset(NSUInteger x, NSUInteger y, NSUInteger w)
    {return y * w * 4 + x * 4 + 1;}
NSUInteger greenOffset(NSUInteger x, NSUInteger y, NSUInteger w)
    {return y * w * 4 + x * 4 + 2;}
NSUInteger blueOffset(NSUInteger x, NSUInteger y, NSUInteger w)
```

```
{return y * w * 4 + x * 4 + 3;}
```

这些计算不需要高度；每一行的宽度允许确定一维缓冲区内的一个二维点。每个字节的范围在 0（0%或 0x00）~255（100%或 0xFF）之间。把它转换成一个浮点数并除以 255.0，以获取 ARGB 值。

6.5.3 在图像数据与位图数据之间转换

要构建位图，可以把图像绘制到位图环境中，然后获取字节作为一个 Byte *（aka uint8 *）缓冲区，秘诀 6-3 通过一个 NSData 对象存储它。bytesFromImage:类方法创建一个 RGB 颜色空间（备选的是 Gray 和 CMYK），分配位图数据，并创建用于绘制图像的位图环境。字节就是从该环境中获取的，放入 NSData 对象中，并会清理所有余下的 CG 对象。Data 对象负责正在处理的字节。

imageWithBytes:withSize:方法则执行相反的操作。你提供一个字节数组和想要的图像尺寸。该方法将利用此数据创建一个新的 CGImageRef，把它转换成一个 UIImage 实例，并返回那个图像。

在下一个秘诀中可以看到，来回摆弄位图表示允许给图片引入图像处理。

总是要确保在位图例程上运行 Xcode 的静态分析器。很容易忽视一次字节释放或自由调用，而分析器有助于查找错误。秘诀 6-3 的第一个例程（imageWithBytes:方法）就会释放传递给它的字节。

秘诀 6-3 在图像与位图之间转换

```
@implementation UIImage (Utils)
// Build image from bytes
+ (UIImage *) imageWithBytes: (Byte *) bytes withSize: (CGSize) size
{
    // Create a color space
    CGColorSpaceRef colorSpace = CGColorSpaceCreateDeviceRGB();
    if (colorSpace == NULL)
    {
        fprintf(stderr, "Error allocating color space\n");
        free(bytes);
        return nil;
    }

    // Create the bitmap context
    CGContextRef context = CGBitmapContextCreate(
        bytes, size.width, size.height, 8, size.width * 4,
        colorSpace, kCGImageAlphaPremultipliedFirst);
    if (context == NULL)
    {
        fprintf (stderr, "Error: Context not created!");
```

```objc
        free (bytes);
        CGColorSpaceRelease(colorSpace );
        return nil;
    }

    // Convert to image
    CGImageRef imageRef = CGBitmapContextCreateImage(context);
    UIImage *image = [UIImage imageWithCGImage:imageRef];

    // Clean up
    CGColorSpaceRelease(colorSpace );
    free(CGBitmapContextGetData(context)); // frees bytes
    CGContextRelease(context);
    CFRelease(imageRef);

    return image;
}

// Covert UIImage to byte array
+ (NSData *) bytesFromImage: (UIImage *) image
{
    CGSize size = image.size;

    CGColorSpaceRef colorSpace = CGColorSpaceCreateDeviceRGB();
    if (colorSpace == NULL)
    {
        fprintf(stderr, "Error allocating color space\n");
        return NULL;
    }

    void *bitmapData = malloc(size.width * size.height * 4);
    if (bitmapData == NULL)
    {
        fprintf (stderr, "Error: Memory not allocated!");
        CGColorSpaceRelease(colorSpace);
        return NULL;
    }

    CGContextRef context = CGBitmapContextCreate(
        bitmapData, size.width, size.height, 8, size.width * 4,
```

```
            colorSpace, kCGImageAlphaPremultipliedFirst);
    CGColorSpaceRelease(colorSpace );
    if (context == NULL)
    {
        fprintf (stderr, "Error: Context not created!");
        free (bitmapData);
        return NULL;
    }

    CGRect rect = (CGRect){.size = size};
    CGContextDrawImage(context, rect, image.CGImage);
    Byte *byteData = CGBitmapContextGetData (context);
    CGContextRelease(context);

    NSData *data = [NSData dataWithBytes:byteData
        length:(size.width * size.height * 4)];
    free(bitmapData);

    return data;
}

- (NSData *) bytes
{
    return [UIImage bytesFromImage:self];
}
@end
```

获取这个秘诀的代码

要查找这个秘诀的完整示例项目，可以浏览 https://github.com/erica/iOS-6-Advanced-Cookbook，并进入第 6 章的文件夹。

6.6 秘诀：基本的图像处理

Accelerate 框架是 iOS 的一个稍微有些新的增补功能，它提供了一些基于 C 语言的 API，允许应用程序使用基于向量和矩阵的算术运算执行数字信号处理（digital signal processing，DSP）。依据 Apple 的说法，它的数学函数支持"语音、声音、音频和视频处理，医学影像诊断，雷达信号处理，地震分析和科学数据处理"。这些例程是为高性能设计的，可以跨大多数 Apple 平台重用。Accelerate 框架提供了可移植、可靠和高度优化的代码。

考虑秘诀 6-4，它实现了类似于图 6-2 的图像旋转。它使用 Accelerate 框架代替 Core Graphics 环境调用。这种实现将把图像裁剪成原始大小，而不会保留旋转后的尺寸。

关于这个秘诀引人注目的是它的简单性（只需调用 ARGB 旋转例程）和簿记功能。数据被打包为一个 vImage_Buffer。图像缓冲区包含一个指针和几个字段，其中前者指向正在处理的数据，后者则用于存储图像的高度、宽度和每行上的字节数。在这种情况下，在一行上每个像素由 4 字节的 ARGB 数据组成。

可以分配输出缓冲区，并且负责在使用后解除分配。在这个秘诀中，调用秘诀 6-3 的 imageWithBytes:withSize:将自动负责解除分配，并返回一个 UIImage 实例。

利用 Accelerate 框架，可以执行大多数标准的图像处理例程，包括：缩放、剪切、直方图、伽玛值调整、卷积（参见秘诀 6-5）等。Accelerate 框架使得很容易对图像缓冲区应用标准例程。秘诀 6-3 和秘诀 6-4 演示了如何在 UIImage 实例的 UIKit 世界与 Accelerate 框架的基于 C 语言的函数的位图世界之间来去自如。

秘诀 6-4 利用 Accelerate 框架旋转图像

```
@implementation UIImage (vImage)

// Return a base buffer without bytes
- (vImage_Buffer) baseBuffer
{
    vImage_Buffer buf;
    buf.height = self.size.height;
    buf.width = self.size.width;
    buf.rowBytes = sizeof(Byte) * self.size.width * 4; // ARGB
    return buf;
}

// Return a buffer for the current image, with its bytes populated
- (vImage_Buffer) buffer
{
    vImage_Buffer buf = [self baseBuffer];
    buf.data = (void *)self.bytes.bytes;
    return buf;
}
// Perform rotation
- (UIImage *) vImageRotate: (CGFloat) theta
{
```

```
        vImage_Buffer inBuffer = [self buffer];
        vImage_Buffer outBuffer = [self baseBuffer];
        Byte *outData = (Byte *)malloc(
            outBuffer.rowBytes * outBuffer.height);
        outBuffer.data = (void *) outData;
        uint8_t backColor[4] = {0xFF, 0, 0, 0};

        vImage_Error error = vImageRotate_ARGB8888(
            &inBuffer, &outBuffer, NULL, theta, backColor, 0);
        if (error)
        {
            NSLog(@"Error rotating image: %ld", error);
            free(outData);
            return self;
        }

        // This frees the outData buffer
        return [UIImage imageWithBytes:outData withSize:self.size];
    }
@end
```

> **获取这个秘诀的代码**
> 要查找这个秘诀的完整示例项目，可以浏览 https://github.com/erica/iOS-6-Advanced-Cookbook，
> 并进入第 6 章的文件夹。

6.7 秘诀：图像卷积

许多常见的图像效果依赖于利用图像数据卷积矩阵。你可能希望锐化图像、平滑它，或者添加一种"浮雕"效果。创建一个二维数组，其中的值具有特定的特征，可以产生这些效果。下面的矩阵定义了一些像素操作，用于输出图 6-3 中所示的"浮雕"效果。

```
- (UIImage *) emboss
{
    static const int16_t kernel[] = {
        -2, -1, 0,
        -1, 1, 1,
        0, 1, 2};
    return [self convolve:kernel side:3];
}
```

图 6-3　凸出一幅图像将产生一种伪 3D 效果，其中的项目看上去好像向前挤压出屏幕。凸出的图像（底部）是通过相对于原始图像（顶部）卷积一个简单的矩阵产生的

卷积一幅图像意味着迭代地对所有像素应用一个矩阵。当给定位置右上方附近的像素较暗并且右下方的像素较亮时，这个特定的矩阵将产生最强烈的效果。由于这会在场景内的对象的右边缘和下边缘自然地发生，输入看起来就像"浮雕"，其中有一些项目似乎是从图像的余下部分中弹出的一样。图 6-3 中处理过的绵羊的右下部分比原始图像更醒目。

秘诀 6-5 演示了如何使用 Accelerate 框架执行卷积。在步骤中，它实际上与秘诀 6-4 完全相同。明显的区别是：你提供了一个要进行卷积的矩阵和一个用于规范化结果的除数。添加了两种常见的操作（锐化和 5×5 高斯模糊），以演示如何调用卷积方法，这对于在 iOS 6 以前安装的产品上执行模糊是一个有用的方法。

这些例程相当基础。可以使用 Core Image 应用复杂得多的图像增加，这将在下一个秘诀中讨论。

秘诀 6-5　利用 Accelerate 框架卷积图像

```
// Calculate the matrix sum for non-edge convolution
int32_t getDivisor(NSData *kernel)
{
    const int16_t *matrix = (const int16_t *)kernel.bytes;
    int count = kernel.length / sizeof(const int16_t);

    // Sum up the kernel elements
    int32_t sum = 0;
    for (CFIndex i = 0; i < count; i++)
        sum += matrix[i];
    if (sum != 0) return sum;
    return 1;
}
// Convolve an image against a kernel
- (UIImage *) vImageConvolve: (NSData *) kernel
```

```
    {
        vImage_Buffer inBuffer = [self buffer];
        vImage_Buffer outBuffer = [self baseBuffer];
        Byte *outData = (Byte *)malloc(
            outBuffer.rowBytes * outBuffer.height);
        outBuffer.data = (void *) outData;
        uint8_t backColor[4] = {0xFF, 0, 0, 0};

        const int16_t *matrix = (const int16_t *)kernel.bytes;
        uint32_t matrixSide = sqrt(kernel.length / sizeof(int16_t));
        int32_t divisor = getDivisor(kernel);

        vImage_Error error = vImageConvolve_ARGB8888(
            &inBuffer, &outBuffer, NULL, 0, 0, matrix,
            matrixSide, matrixSide, divisor,
            backColor, kvImageBackgroundColorFill);
        if (error)
        {
            NSLog(@"Error convolving image: %ld", error);
            free(outData);
            return self;
        }
        return [UIImage imageWithBytes:outData withSize:self.size];
    }

    - (UIImage *) sharpen
    {
        static const int16_t kernel[] = {
            0, -1, 0,
            -1, 8, -1,
            0, -1, 0
        };
        return [self convolve:kernel side:3];
    }

    - (UIImage *) gauss5
    {
        static const int16_t kernel[] = {
            1, 4, 6, 4, 1,
            4, 16, 24, 16, 4,
```

```
    6, 24, 36, 24, 6,
    4, 16, 24, 16, 4,
    1,  4,  6,  4, 1
};
return [self convolve:kernel side:5];
}
```

> **获取这个秘诀的代码**
>
> 要查找这个秘诀的完整示例项目，可以浏览 https://github.com/erica/iOS-6-Advanced-Cookbook，并进入第 6 章的文件夹。

6.8 秘诀：基本的 Core Image 处理

Core Image 滤镜允许快速处理图像。秘诀 6-6 通过对图像应用简单的褐色滤光镜，介绍了基本的 CI 过滤。iOS 6 提供了广泛的 CI 滤镜，它们随着时间的推移还在不断增多，可以通过查看内置的滤镜名称集来查询它们：

```
NSLog(@"%@", [CIFilter filterNamesInCategory: kCICategoryBuiltIn]);
```

滤镜包括颜色调整、几何变换、裁剪和合成。通常会把一个滤镜的结果馈送给下一个滤镜。例如，在创建一种图片中的图片的效果，可以缩放并平移一幅图像，然后把它合成到另一幅图像上。

每种滤镜都使用许多输入，包括源数据和参数，它们因滤镜而异。秘诀 6-6 使用了一种褐色滤光镜，其输入是 inputImage 和 inputIntensity。另一种透视滤镜会添加一些输入，它们指定了图像角落的位置。图 6-4 显示了这种透视滤镜实现的结果。调整 4 个角的输入将改变滤镜的"方向"。

在最新的"Core Image Filter Reference"（Core Image 滤镜参考）文档和 Apple 提供的 *Core Image Programming Guide* 中查找输入的具体信息。还可以调用 inputKeys 从滤镜中直接查询输入，这将返回滤镜使用的输入参数名称的数组。

图 6-4 透视滤镜将扭曲图像，就好像从另一个视角看它一样

几乎所有的滤镜都提供了默认设置，为你开始使用它们提供了一个基础。调用 setDefaults 加载这些值，允许开始进行测试。要特别注意 Apple 的指南，它详细说明了每种默认设置的含义

以及滤镜如何使用它。它还提供了一些图像示例，以便查看滤镜打算做什么。

顺便说一下，秘诀 6-6 实现了一个 coreImageRepresentation 属性，之所以提供它，是因为 CI/UIImage 集成仍然有点不可靠。当方法不能直接从图像获取一个 CIImage 实例时，它将从图像的 CGImage 表示构建这样一个示例。这种类型的解决办法有助于提醒你 Core Image 仍然是一种在 iOS 下不断演进的技术。

秘诀 6-6　Core Image 基础

```
- (CIImage *) coreImageRepresentation
{
    if (self.CIImage)
        return self.CIImage;
    return [CIImage imageWithCGImage:self.CGImage];
}

- (UIImage *) sepiaVersion: (CGFloat) intensity
{
    CIFilter *filter = [CIFilter
        filterWithName:@"CISepiaTone"
        keysAndValues: @"inputImage",
        self.coreImageRepresentation, nil];
    [filter setDefaults];
    [filter setValue:@(intensity) forKey:@"inputIntensity"];

    CIImage *output = [filter valueForKey:kCIOutputImageKey];
    if (!output)
    {
        NSLog(@"Core Image processing error");
        return nil;
    }

    UIImage *results = [UIImage imageWithCIImage:output];
    return results;
}

- (UIImage *) perspectiveExample
{
    CIFilter *filter = [CIFilter
        filterWithName:@"CIPerspectiveTransform"
        keysAndValues: @"inputImage",
```

```
        self.coreImageRepresentation,
        nil];
[filter setDefaults];
[filter setValue:[CIVector vectorWithX:180 Y:600]
    forKey:@"inputTopLeft"];
[filter setValue:[CIVector vectorWithX:102 Y:20]
    forKey:@"inputBottomLeft"];

CIImage *output = [filter valueForKey:kCIOutputImageKey];
if (!output)
{
    NSLog(@"Core Image processing error");
    return nil;
}
UIImage *results = [UIImage imageWithCIImage:output];
return results;
}
```

获取这个秘诀的代码

要查找这个秘诀的完整示例项目,可以浏览 https://github.com/erica/iOS-6-Advanced-Cookbook,并进入第 6 章的文件夹。

6.9 抓取基于视图的截屏图

有时,需要拍摄当前状态下的视图或窗口,或者把视图呈现到一幅图像中去。程序清单 6-1 详细说明了如何把视图绘制到图像环境中并获取 UIImage 实例。这段代码在工作时将把 Quartz Core 的 renderInContext 调用用于 CALayer 实例。它不仅可以产生视图的截屏图,而且可以产生视图所拥有的全部子视图的截屏图。

当然会有一些限制。不能制作整个窗口的截屏图(状态栏将会丢失),特别是在使用这段代码时,将不能制作视频或相机预览的截屏图。也不能抓取 OpenGL ES 视图。

在使用 Core Animation 创建从一种视图层次结构到另一种视图层次结构的过渡时,这种视图到图像的功能特别方便。

程序清单 6-1　制作视图的截屏图

```
UIImage *imageFromView(UIView *theView)
{
    UIGraphicsBeginImageContext(theView.frame.size);
```

```objc
        CGContextRef context = UIGraphicsGetCurrentContext();
        
        [theView.layer renderInContext:context];
        UIImage *theImage = 
            UIGraphicsGetImageFromCurrentImageContext();
        
        UIGraphicsEndImageContext();
        return theImage;
}

UIImage *screenShot()
{
        UIWindow *window = 
            [[UIApplication sharedApplication] keyWindow];
        return imageFromView(window);
}
```

6.10 绘制到 PDF 文件中

可以直接绘制到 PDF 文档中,就像绘制到图像环境中一样。在程序清单 6-2 中可以看到,可以使用 Quartz 绘图命令,比如 UIImage 的 drawInRect:和 CALayer 的 renderInContext:方法,以及其他 Quartz 函数的完整集合。

通常把每一幅图像都绘制到一个新的 PDF 页面中。在 PDF 文档中必须创建至少一个页面,如程序清单 6-2 所示。除了简单的呈现之外,还可以使用诸如 UIGraphicsSetPDFContextURL-ForRect()和 UIGraphicsSetPDFContextDestinationForRect()之类的函数向 PDF 中添加实时的链接,其中前者添加的链接将从定义的矩形中链接到外部 URL,后者添加的链接则将链接文档内部的内容。可以使用 UIGraphicsAddPDFContextDestinationAtPoint()创建目标点。

每个 PDF 文档都是利用自定义的字典构建的。如果想跳过字典,可以传递 nil,或者为 Author、Creator、Title、Password、AllowsPrinting 等分配值。这个字典的键列出在 Apple 的 CGPDFContext 参考文档中。

程序清单 6-2 把图像绘制到 PDF 文件中

```objc
+ (void) saveImage: (UIImage *) image
    toPDFFile: (NSString *) path
{
    CGRect theBounds = (CGRect){.size=image.size};
    UIGraphicsBeginPDFContextToFile(path, theBounds, nil);
```

```
{
    UIGraphicsBeginPDFPage();
    [image drawInRect:theBounds];
}
UIGraphicsEndPDFContext();
}
```

6.11 秘诀：倒影

本章中的最后两个秘诀将给图像添加一点超炫的效果，它们可以自然地适应由 Core Image 和 Accelerate 引入的特效。它们都是 QuartzCore 框架的一部分，用于创建、操纵和绘制图像。

倒影增强了屏幕上的对象的现实感。它们超越浮动在背景上的视图提供了一点额外的视觉情趣，这通常很流行。由于 CAReplicatorLayer 类，倒影变得特别容易实现。这个类的作用如其名称所示，它将复制视图的层，并且允许对复制的内容应用变换。

秘诀 6-7 显示了如何使用复制品的层构建可以自动映出内容的视图控制器。自定义的反射器视图类重写了 layerClass，以确保视图的层默认为复制品的层。把那个复制品的 instanceCount 设置为 2，使之把原始视图复制到第二个实例中。它的 instanceTransform 指定了如何操纵第二个实例并把它放在屏幕上：翻转、挤压，然后垂直移动到原始视图下方。这将在原始视图的底部创建一个较小的倒影。

由于是复制品，这个倒影将是完全"实时"的。对主视图的任何改变都将在倒影层中即时更新，可以使用本章配套的示例代码自己进行测试。可以添加滚动的文本视图、Web 视图、选项开关等。对原始视图的任何改变都会被复制，从而创建完全被动的倒影系统。

当"倒影"移动得距离原始视图更远时，良好的倒影将使用自然的损耗。这个秘诀添加了一个可选的渐变叠加（使用 usesGradientOverlay 属性），从视图的底部创建一种视觉急速下降的效果。这段代码添加渐变作为一个 CAGradientLayer。

这段代码在视图的底部与其倒影之间引入了任意的间隙。在这里，将其固定为 10 磅，如图 6-5 中所示，可以根据需要调整它。

为了使这个秘诀在视图控制器的环境内工作，它的子类必须做 3 件事。第一，把自定义的视图添加到控制器的背景

图 6-5 这种基于层的倒影将实时更新，以匹配它反射的视图。当像这样使用一个 ReflectionViewController 时，需要给控制器的反射 backdrop（而不是它的 view）添加一些子视图

中，而不是添加到它的视图中，例如，[self.backsplash add- Subview:cannedTable]。第二，确定调用 viewDidAppear 的父类的实现。最后，确保每次方位改变时通过为背景调用 setupReflectiong 来更新倒影。

秘诀 6-7　创建倒影

```
@implementation ReflectingView
// Always use a replicator as the base layer
+ (Class) layerClass
{
    return [CAReplicatorLayer class];
}

- (void) setupGradient
{
    // Add a new gradient layer to the parent
    UIView *parent = self.superview;
    if (!gradient)
    {
        gradient = [CAGradientLayer layer];
        CGColorRef c1 = [[UIColor blackColor]
            colorWithAlphaComponent:0.5f].CGColor;
        CGColorRef c2 = [[UIColor blackColor]
            colorWithAlphaComponent:0.9f].CGColor;
        [gradient setColors:[NSArray arrayWithObjects:
            (__bridge id)c1, (__bridge id)c2, nil]];
        [parent.layer addSublayer:gradient];
    }
    // Place the gradient just below the view using the
    // reflection's geometry
    float desiredGap = 10.0f; // gap between view and its reflection
    CGFloat shrinkFactor = 0.25f; // reflection size
    CGFloat height = self.bounds.size.height;
    CGFloat width = self.bounds.size.width;
    CGFloat y = self.frame.origin.y;

    [gradient setAnchorPoint:CGPointMake(0.0f,0.0f)];
    [gradient setFrame:CGRectMake(0.0f, y + height + desiredGap,
        width, height * shrinkFactor)];
    [gradient removeAllAnimations];
```

```objc
    [gradient setAnchorPoint:CGPointMake(0.0f,0.0f)];
    [gradient setFrame:CGRectMake(0.0f, y + height + desiredGap,
        maxDimension, height * shrinkFactor)];
    [gradient removeAllAnimations];
}

- (void) setupReflection
{
    CGFloat height = self.bounds.size.height;
    CGFloat shrinkFactor = 0.25f;

    CATransform3D t = CATransform3DMakeScale(1.0, -shrinkFactor, 1.0);

    // Scaling centers the shadow in the view.
    // Translate the results in shrunken terms
    float offsetFromBottom = height * ((1.0f - shrinkFactor) / 2.0f);
    float inverse = 1.0 / shrinkFactor;
    float desiredGap = 10.0f;
    t = CATransform3DTranslate(t, 0.0, -offsetFromBottom * inverse
        - height - inverse * desiredGap, 0.0f);

    CAReplicatorLayer *replicatorLayer =
        (CAReplicatorLayer*)self.layer;
    replicatorLayer.instanceTransform = t;
    replicatorLayer.instanceCount = 2;

    // Gradient use must be explicitly set
    if (usesGradientOverlay)
        [self setupGradient];
    else
    {
        // Darken the reflection when not using a gradient
        replicatorLayer.instanceRedOffset = -0.75;
        replicatorLayer.instanceGreenOffset = -0.75;
        replicatorLayer.instanceBlueOffset = -0.75;
    }
}
@end
```

> **获取这个秘诀的代码**
> 要查找这个秘诀的完整示例项目，可以浏览 https://github.com/erica/iOS-6-Advanced-Cookbook，并进入第 6 章的文件夹。

6.12 秘诀：发射器

引入发射器可以给应用程序增加一点活力。发射器将使用你调整的一套自定义的属性实时呈现微粒。秘诀 6-8 创建了在屏幕上跟随触摸移动的强烈的紫色微粒云。与秘诀 6-7 的倒影一样，发射器允许向应用程序中添加交互式的视觉兴趣。

创建发射器基于建立一个层，并把它添加到视图中。通过添加一个微粒发射器单元并调整它的特征，来建立发射器。在这里，每个微粒都基于一幅图像，在这个示例中，它是从示例代码中包括的 spark.png 文件中加载的。

除此之外，由你自己决定将多快地创建微粒，它们将存留多长时间，它们的速度、颜色等。这类秘诀是调教的喜悦。可以自由地探索每个参数如何影响当手指划过屏幕时所创建的微粒显示效果。

秘诀 6-8　添加发射器

```
@implementation SparkleTouchView
{
    CAEmitterLayer *emitter;
}

- (id) initWithFrame: (CGRect) aFrame
{
    if (!(self = [super initWithFrame:aFrame])) return self;

    return self;
}

- (void) touchesBegan:(NSSet *)touches withEvent:(UIEvent *)event
{
    float multiplier = 0.25f;

    CGPoint pt = [[touches anyObject] locationInView:self];

    //Create the emitter layer
```

```objc
    emitter = [CAEmitterLayer layer];
    emitter.emitterPosition = pt;
    emitter.emitterMode = kCAEmitterLayerOutline;
    emitter.emitterShape = kCAEmitterLayerCircle;
    emitter.renderMode = kCAEmitterLayerAdditive;
    emitter.emitterSize = CGSizeMake(100 * multiplier, 0);

    //Create the emitter cell
    CAEmitterCell* particle = [CAEmitterCell emitterCell];
    particle.emissionLongitude = M_PI;
    particle.birthRate = multiplier * 1000.0;
    particle.lifetime = multiplier;
    particle.lifetimeRange = multiplier * 0.35;
    particle.velocity = 180;
    particle.velocityRange = 130;
    particle.emissionRange = 1.1;
    particle.scaleSpeed = 1.0; // was 0.3
    particle.color = [[COOKBOOK_PURPLE_COLOR
        colorWithAlphaComponent:0.5f] CGColor];
    particle.contents =
        (__bridge id)([UIImage imageNamed:@"spark.png"].CGImage);
    particle.name = @"particle";

    emitter.emitterCells = [NSArray arrayWithObject:particle];
    [self.layer addSublayer:emitter];
}

- (void) touchesMoved:(NSSet *)touches withEvent:(UIEvent *)event
{
    CGPoint pt = [[touches anyObject] locationInView:self];

    // Disable implicit animations
    [CATransaction begin];
    [CATransaction setValue:(id)kCFBooleanTrue
        forKey:kCATransactionDisableActions];
    emitter.emitterPosition = pt;
    [CATransaction commit];
}

- (void) touchesEnded:(NSSet *)touches withEvent:(UIEvent *)event
```

```
{
    [emitter removeFromSuperlayer];
    emitter = nil;
}

- (void) touchesCancelled:(NSSet *)touches withEvent:(UIEvent *)event
{
    [self touchesEnded:touches withEvent:event];
}
@end
```

获取这个秘诀的代码

要查找这个秘诀的完整示例项目,可以浏览 https://github.com/erica/iOS-6-Advanced-Cookbook,并进入第 6 章的文件夹。

6.13 小结

本章介绍了处理图像的许多方式。你学习了生成图像、调整它们的大小、旋转它们,以及使用 Accelerate 和 Core Image 框架处理它们。在结束本章的学习之前,要思考以下几点,它们是关于你在本章中所看到的秘诀的。

- 在 UIImage 和位图表示之间移动允许逐字节地访问图像。这将使你能够执行图像处理,并针对内容测试触摸。在字节级别展示图像意味着总是能够完全控制图像内容,而不仅仅是一个不透明的图像对象。
- 加载和操纵对象的方式因源而异。应用程序包中的项目是使用一些方法调用导入的,它们不能用于在沙盒中、Internet 上或者系统资产库中发现的那些项目。在加载时,UIImage 类为显示此数据提供了统一的工具。
- Accelerate 框架可能很不美观,但是极其可靠。Core Image 框架仍然在 iOS 上不断演进,也许不能提供一种讨人喜欢的体验,尽管它随着时间的推移一直在变得越来越好。
- iOS 的 QuartzCore 库允许在较低层次上处理图形。它提供了一个轻量级的 2D 渲染系统,可以把这些调用用于你使用的任何类型的绘图操作,而无论是在 drawRect:中渲染视图还是自由地创建图像。

第 7 章
照相机

照相机把图像抬高到了下一个层次。它们使你能够把实时馈送和用户指示的快照集成到应用程序中,并且提供源于现实世界的原始数据。在本章中将会学习图像捕获。你将发现如何使用 Apple 提供的类来获取图片,以及你自己如何从零开始操纵图像。你将学习控制图像元数据,以及如何把实时馈送与高级过滤集成起来。本章从硬件的角度重点介绍了图像捕获。无论你是在打开照相机闪光灯还是在检测面孔,本章都会介绍 iOS 图像捕获技术的方方面面的知识。

7.1 秘诀:拍摄照片

iOS 提供了多种系统提供的视图控制器,允许开发人员利用最少量的编程来提供系统特性。iOS 图像选择控制器就是其中之一。它允许用户利用设备的内置照相机拍摄照片。由于在所有的 iOS 装置(确切地讲是老式的 iPod Touch 和 iPad 设备)上都不能使用照相机,因此首先要检查运行应用程序的系统是否支持使用照相机:

```
if ([UIImagePickerController isSourceTypeAvailable:
    UIImagePickerControllerSourceTypeCamera]) ...
```

规则是:永远不要为没有照相机的设备提供基于照相机的特性。尽管 iOS 6 是只为准备好照相机的设备开发的,Apple 还是没有承诺把它作为一项政策。它可能引入不带有照相机的新型号,这听起来似乎不太可能。在 Apple 给出另外的说法之前,假定可能会存在不带有照相机的系统,甚至会出现在现代的 iOS 版本之下。更进一步,假定这个方法可以准确地报告支持照相机的设备的状态,并且可以通过某种将来的系统设置禁用它的来源。

7.1.1 设置选择器

你以创建图片选择器的方式实例化了图像选择器的照相机版本,只需把源类型从库或相机胶卷改为照相机即可。与其他模式一样,可以通过设置 allowsEditing 属性,允许或禁止图像编辑作为照片拍摄过程的一部分。

图 7-1　图像选择控制器的照相机版本为拍摄照片提供了截然不同的用户体验

尽管设置相同，但是用户体验稍有差别（参见图 7-1）。照相机选择器提供了一个预览，它将在用户点按照相机图标以拍摄照片之后显示。这个预览允许用户重拍（Retake）照片或者原样使用（Use）照片。一旦他们点按了 Use，控制将传递到下一个阶段。如果启用了图像编辑，用户接下来就会这样做。如果没有启用它，控制将会转移给委托中标准的"完成选择"方法。

大多数现代设备都提供了不只一部照相机，iPhone 3GS（最后留存在 iOS 6 过时产品）则不然。指定 cameraDevice 属性，选择你想使用哪部照相机。后置的照相机总是默认的选择。

isCameraDeviceAvailable:类方法将查询照相机设备是否可用。下面这个代码段将检查前置的照相机是否可用，如果是，就选择它：

```
if ([UIImagePickerController isCameraDeviceAvailable:
    UIImagePickerControllerCameraDeviceFront])
    picker.cameraDevice = UIImagePickerControllerCameraDeviceFront;
```

对于可以通过 UIImagePickerController 类访问的照相机，下面给出了另外几点说明。

- 可以使用 isFlashAvailablForCameraDevice:类方法，查询设备使用闪光灯的能力。提供前置或后置设备常量。如果有闪光灯可用，该方法将返回 YES，否则就返回 NO。
- 当照相机支持闪光灯时，可以直接把 cameraFlashMode 属性设置为"自动"（UIImagePickerControllerCameraFlashModeAuto，这是默认设置）、"总是使用"（UIImagePickerControllerCameraFlashModeOn）或者"总是关闭"（UIImagePickerControllerCameraFlashModeOff）。如果选择"总是关闭"，那么无论周围的光线条件是什么，都会禁用闪光灯。
- 可以通过设置 cameraCaptureMode 属性，选择拍摄照片还是拍摄视频。选择器默认为拍摄照片模式。可以使用 availableCaptureModesForCameraDevice:测试设备能够使用什么模式，这将返回一个 NSNumber 对象的数组，其中每个对象都会编码一种有效的拍摄模式：即照片（UIImagePickerControllerCameraCaptureModePhoto）或视频（UIImagePickerControllerCameraCaptureModeVideo）。

7.1.2　显示图像

在处理照片时，始终要牢记图像大小。拍摄的图片（尤其是那些来自高分辨率照相机的图片）与屏幕大小相比可能相当大，甚至在 Retina 显示屏时代也是如此。这些通过前置视频照相机拍摄的照片将使用低质量传感器，并且比较小。

内容模式提供了一种视图解决方案，用于显示较大的图像。它们允许图像视图把它们的嵌入式图像缩放到可用的屏幕空间。考虑使用以下模式之一。

- UIViewContentModeScaleAspectFit 模式确保利用保留的宽高比显示完整的图像。可能在图像两侧或者顶部和底部填充空白的矩形，以保持它的宽高比。
- UIViewContentModeScaleAspectFill 模式将显示尽可能多的图像，同时将填满整个视图。可能会裁剪一些内容，以填满整个视图的边界。

7.1.3 把图像保存到相册

通过调用 UIImageWriteToSavedPhotosAlbum() 把拍摄的图像（或者实际上是任何 UIImage 实例）保存到相册。这个函数接受 4 个参数：第一个参数是要保存的图像；第二个和第三个参数指定回调目标和选择器，通常是主视图控制器和 image:didFinishSavingWithError:contextInfo:；第四个参数是可选的环境指针。无论使用哪种选择器，它都必须接受 3 个参数：图像、错误和指向传递的环境信息的指针。

秘诀 7-1 使用这个函数来演示如何拍摄一幅新图像，允许用户编辑它，然后把它保存到相册。这个调用将要求用户允许访问他们的图库。如果他们拒绝访问，尝试将会失败。

可以通过向项目的 Info.plist 文件中添加一个 **NSPhotoLibraryUsageDescription** 键，自定义 iOS 访问警报。这个条目将描述应用程序希望访问图库的原因，并将其显示在请求访问的对话框中。

秘诀 7-1　拍摄图片

```
// "Finished saving" callback method
- (void)image:(UIImage *)image
   didFinishSavingWithError: (NSError *)error
   contextInfo:(void *)contextInfo;
{
   // Handle the end of the image write process
   if (!error)
       NSLog(@"Image written to photo album");
   else
       NSLog(@"Error writing to photo album: %@", error.localizedFailureReason);
}


// Save the returned image
- (void)imagePickerController:(UIImagePickerController *)picker
   didFinishPickingMediaWithInfo:(NSDictionary *)info
{
    // Use the edited image if available
    UIImage __autoreleasing *image =
        info[UIImagePickerControllerEditedImage];
```

```objc
    // If not, grab the original image
    if (!image) image = info[UIImagePickerControllerOriginalImage];

    NSURL *assetURL = info[UIImagePickerControllerReferenceURL];
    if (!image && !assetURL)
    {
        NSLog(@"Cannot retrieve an image from selected item. Giving up.");
    }
    else if (!image)
    {
        NSLog(@"Retrieving from Assets Library");
        [self loadImageFromAssetURL:assetURL into:&image];
    }

    if (image)
    {
        // Save the image
        UIImageWriteToSavedPhotosAlbum(image, self,
            @selector(image:didFinishSavingWithError:contextInfo:), NULL);
        imageView.image = image;
    }
    [self performDismiss];
}
- (void) loadView
{

    self.view = [[UIView alloc] init];

    // Only present the "Snap" option for camera-ready devices
    if ([UIImagePickerController isSourceTypeAvailable:
        UIImagePickerControllerSourceTypeCamera])
        self.navigationItem.rightBarButtonItem =
            SYSBARBUTTON(UIBarButtonSystemItemCamera, @selector(snapImage));
}
```

获取这个秘诀的代码

要查找这个秘诀的完整示例项目，可以浏览 https://github.com/erica/iOS-6-Advanced-Cookbook，并进入第 7 章的文件夹。

7.2 秘诀：启用闪光灯

iPhone 4 是第一款提供内置 LED 照相机闪光灯的 iOS 设备。从那时起，就把闪光灯移植到了几种更新的设备上。在使用火炬模式（torch mode）的应用程序中可以很容易地控制该 LED。秘诀 7-2 演示了这种功能，允许用户打开和关闭灯光。如你所料，对于长时间使用来说，给"火炬"供电对电池不是特别友好，但是对于短期使用，则不是特别有害。

这个秘诀首先调查可用的机载照相机设备，它的数量在 0（无机载照相机）~2（对于当前的一代设备）之间。每部照相机可能会或者不会支持机载闪光灯。第五代 iPod Touch 会提供支持，第三代 iPad 则不然。要了解某个设备是否提供了准备好火炬模式的 LED 闪光灯，可以求助于 AVFoundation 框架。这个框架将给本章中的许多秘诀提供支持，它提供了一些编程界面，用于测试视听捕获并与之交互。

秘诀 7-2 查询当前单元，取回所有机载拍摄设备（即照相机）的列表。它将检查每个拍摄设备是否提供了"火炬"（即可以提供一般亮度的 LED 闪光灯），以及是否打开了"火炬"支持。

一旦找到了任何合乎需要的拍摄设备，它将把该信息存储到一个局部实例变量中，并且应用程序会展示一个简单的选项开关，它用于回调 toggleLightSwitch 方法。如果它没有找到一个准备好"火炬"的设备，秘诀的 supportsTorchMode 方法将返回 NO，并且应用程序实质上会关闭。

toggleLightSwitch 方法处理打开和关闭闪光灯的工作。该方法演示了这个过程的基本流程。在设置拍摄设备属性（包括聚焦、曝光度或者闪光灯的打开/关闭状态）之前，必须获得独占锁。

代码将锁定照相机设备，更新它的火炬模式，使之处于打开或关闭状态，然后解除锁定它。

通常，一旦完成了更新，总是要释放设备锁。持有锁而不释放它可能会影响其他的应用程序，它们可能正尝试访问相同的硬件。使用这个锁定/修改/解锁过程对拍摄设备的设置应用更新。

除了火炬模式之外，还可以检查某个单元是否支持基本的照相机闪光灯（hasFlash）、白平衡调整（isWhiteBalanceModeSupported:）、曝光度调整（isExposureModeSupported:）和自动聚焦模式（isFocusModeSupported:）。测试和更新这些其他的特性将遵循与秘诀 7-2 中相同的流程。

秘诀 7-2　控制火炬模式

```
@implementation TestBedViewController
{
    UISwitch *lightSwitch;
    AVCaptureDevice *device;
}

- (void) toggleLightSwitch
{
```

```objc
    // Lock the device
    if ([device lockForConfiguration:nil])
    {
        // Toggle the light on or off
        device.torchMode = (lightSwitch.isOn) ?
            AVCaptureTorchModeOn : AVCaptureTorchModeOff;

        // Unlock and proceed
        [device unlockForConfiguration];
    }
}

- (BOOL) supportsTorchMode
{
    // Survey all the onboard capture devices
    NSArray *devices = [AVCaptureDevice devices];
    for (AVCaptureDevice *aDevice in devices)
    {
        if ((aDevice.hasTorch) &&
            [aDevice isTorchModeSupported:AVCaptureTorchModeOn])
        {
            device = aDevice;
            return YES;
        }
    }
    // No torch-ready camera found. Return NO.
    return NO;
}

- (void) loadView
{
    self.view = [[UIView alloc] init];
    self.view.backgroundColor = [UIColor whiteColor];

    // Initially hide the light switch
    lightSwitch = [[UISwitch alloc] init];
    [self.view addSubview:lightSwitch];
    lightSwitch.alpha = 0.0f; // Alternatively, setHidden:

    // Only reveal the switch for torch-ready units
```

```
    if ([self supportsTorchMode])
        lightSwitch.alpha = 1.0f;
    else
        self.title = @"Flash/Torch not available";
}
@end
```

获取这个秘诀的代码

要查找这个秘诀的完整示例项目，可以浏览 https://github.com/erica/iOS-6-Advanced-Cookbook，并进入第 7 章的文件夹。

7.3 秘诀：访问 AVFoundation 照相机

AVFoundation 允许访问照相机缓冲机，而无需使用麻烦的图像选择器。它比选择器的速度更快并且响应性更好，但是没有提供那么友好的内置界面。例如，你可能想构建一个应用程序，评估照相机的内容是否增强了现实感，或者允许用户处理视频馈送。

AVFoundation 允许从照相机获取实时预览以及原始图像缓冲区数据，在你的开发工具箱中提供有价值的工具。要开始使用 AVFoundation，需要构建一个 Xcode 项目，它将使用相当多的框架。

下面列出了你应该考虑添加的框架以及它们在应用程序中扮演的角色，其中只有前 3 个框架是必需的。你可能想添加所有这些框架。额外的框架常用于 AVFoundation 视频馈送和图像捕获。

- AVFoundation——在 iOS 应用程序中管理和播放视听媒体：`<AVFoundation/AVFoundation.h>`。
- CoreVideo——通过逐帧控制播放和处理电影：`<CoreVideo/CoreVideo.h>`。
- CoreMedia——处理基于时间的 AV 资产：`<CoreMedia/CoreMedia.h>`。
- CoreImage——使用像素精度的、接近于实时的图像处理：`<CoreImage/CoreImage.h>`。
- ImageIO——添加和编辑图像元数据，包括 GPS 和 EXIF 支持：`<ImageIO/ImageIO.h>`。
- QuartzCore——添加 2D 图形渲染支持以及访问视图的层：`<QuartzCore/QuartzCore.h>`。

7.3.1 需要照相机

如果应用程序是围绕实时照相机馈送构建的，就不应该把它安装在任何没有提供内置照相机的设备上。可以编辑 Info.plist 文件，向 UIRequiredDeviceCapabilities 数组中添加一些项目，来强制执行这一点。可以添加 still-camera，声明相机（任何相机）是可用的。可以通过指定还必须有 auto-focus-camera、a front-facing-camera、camera-flash 和 video-camera，缩小设备型号和能力的范围。在第 1 章"特定于设备的开发"中进一步讨论了必需的项目。

7.3.2 查询和获取照相机

iOS 设备可能提供 0 部、1 部或 2 部照相机。对于具有多部照相机的设备,可以允许用户选择要使用哪部照相机,并在前置和后置之间切换。程序清单 7-1 显示了如何获取照相机的数量,检查给定的设备上是否有前置和后置照相机可用,以及如何为每个设备返回 AVCaptureDevice 实例。

程序清单 7-1　照相机

```
+ (int) numberOfCameras
{
    return [AVCaptureDevice devicesWithMediaType:AVMediaTypeVideo].count;
}

+ (BOOL) backCameraAvailable
{
    NSArray *videoDevices =
        [AVCaptureDevice devicesWithMediaType:AVMediaTypeVideo];
    for (AVCaptureDevice *device in videoDevices)
        if (device.position == AVCaptureDevicePositionBack) return YES;
    return NO;
}

+ (BOOL) frontCameraAvailable
{
    NSArray *videoDevices =
        [AVCaptureDevice devicesWithMediaType:AVMediaTypeVideo];
    for (AVCaptureDevice *device in videoDevices)
        if (device.position == AVCaptureDevicePositionFront) return YES;
    return NO;
}

+ (AVCaptureDevice *)backCamera
{
    NSArray *videoDevices =
        [AVCaptureDevice devicesWithMediaType:AVMediaTypeVideo];
    for (AVCaptureDevice *device in videoDevices)
        if (device.position == AVCaptureDevicePositionBack)
            return device;

    // Return whatever is available if there's no back camera
    return [AVCaptureDevice defaultDeviceWithMediaType:AVMediaTypeVideo];
```

```
+ (AVCaptureDevice *)frontCamera
{
    NSArray *videoDevices =
        [AVCaptureDevice devicesWithMediaType:AVMediaTypeVideo];
    for (AVCaptureDevice *device in videoDevices)
        if (device.position == AVCaptureDevicePositionFront)
            return device;

    // Return whatever device is available if there's no back camera
    return [AVCaptureDevice defaultDeviceWithMediaType:AVMediaTypeVideo];
}
```

7.3.3 建立照相机会话

在选择了要使用的设备并且获取了它的 **AVCaptureDevice** 实例之后，就可以建立一个新的照相机会话。程序清单 7-2 显示了涉及的步骤。会话包括创建捕获输入和捕获输出，前者用于从选择的照相机抓取数据，后者用于把缓冲区数据发送给它的委托。

程序清单 7-2 创建会话

```
// Create the capture input
AVCaptureDeviceInput *captureInput =
    [AVCaptureDeviceInput deviceInputWithDevice:device error:&error];
if (!captureInput)
{
    NSLog(@"Error establishing device input: %@", error);
    return;
}

// Create capture output.
char *queueName = "com.sadun.tasks.grabFrames";
dispatch_queue_t queue = dispatch_queue_create(queueName, NULL);
AVCaptureVideoDataOutput *captureOutput =
    [[AVCaptureVideoDataOutput alloc] init];
captureOutput.alwaysDiscardsLateVideoFrames = YES;
[captureOutput setSampleBufferDelegate:self queue:queue];

// Establish settings
```

```objc
NSDictionary *settings = [NSDictionary
    dictionaryWithObject:
        [NSNumber numberWithUnsignedInt:kCVPixelFormatType_32BGRA]
    forKey:(NSString *)kCVPixelBufferPixelFormatTypeKey];
[captureOutput setVideoSettings:settings];

// Create a session
_session = [[AVCaptureSession alloc] init];
[_session addInput:captureInput];
[_session addOutput:captureOutput];
```

在创建了照相机会话之后,就可以发送 startRunning 和 stopRunning 方法调用,分别用于开启和停止它,如下所示:

```objc
[session startRunning];
```

在运行时,会话将把常规的缓冲区更新发送给它的委托。在这里,可以捕获原始数据,并把它转换为更适合于图像处理工作的形式。这个方法将存储一个 **CIImage** 属性作为保留的属性。假定你在助手类中实现了捕获例程,就可以根据需要从应用程序中获取该数据。程序清单 7-3 显示了如何捕获输出。

程序清单 7-3　捕获输出

```objc
- (void)captureOutput:(AVCaptureOutput *)captureOutput
    didOutputSampleBuffer:(CMSampleBufferRef)sampleBuffer
    fromConnection:(AVCaptureConnection *)connection
{
    @autoreleasepool
    {
        // Transfer into a Core Video image buffer
        CVImageBufferRef imageBuffer =
        CMSampleBufferGetImageBuffer(sampleBuffer);
    // Create a Core Image result
    CFDictionaryRef attachments =
        CMCopyDictionaryOfAttachments(kCFAllocatorDefault,
            sampleBuffer, kCMAttachmentMode_ShouldPropagate);
    self.ciImage = [[CIImage alloc]
        initWithCVPixelBuffer:imageBuffer
        options:(__bridge_transfer NSDictionary *)attachments];
    }
}
```

Core Image 的 CIImage 类与 UIImage 相似，可以根据需要来回转换。可以利用一个 Core Image 实例初始化一个 UIImage 实例，如下所示：

```
UIImage *newImage = [UIImage imageWithCIImage:self.ciImage];
return newImage;
```

> **注意：**
> 在编写本书时，imageWithCIImage:方法还不能正确地工作。因此，在本章配套的示例代码中添加了解决方法。

捕获输出委托方法使用 Core Image 版本，这有以下两个原因。第一，它是 Core Video 像素缓冲区的自然的 API 接收者，只需单个调用即可转换到 CIImage。第二，它允许把捕获的图像与 Core Image 过滤和特性检测（比如人脸检测）集成起来。你很可能希望在捕获例程外面添加过滤器，因为它们可能延缓处理的速度；你将不希望在捕获的每一帧中添加过滤或检测。

7.3.4 切换照相机

在建立了会话之后，就可以切换照相机，而不必停止或取消那个会话。输入配置模式，执行改变，然后提交配置更新，如程序清单 7-4 所示。如果会话停止，你可能希望重新启动它，以使得用户清楚照相机实际上已经切换了。

程序清单 7-4　从可用的照相机中做出选择

```
- (void) switchCameras
{
    if (![CameraImageHelper numberOfCameras] > 1) return;

    _isUsingFrontCamera = !_isUsingFrontCamera;
    AVCaptureDevice *newDevice = _isUsingFrontCamera ?
        [CameraImageHelper frontCamera] : [CameraImageHelper backCamera];

    [session beginConfiguration];

    // Remove existing inputs
    for (AVCaptureInput *input in [session inputs])
        [_session removeInput:input];

    // Change the input
    AVCaptureDeviceInput *captureInput =
        [AVCaptureDeviceInput deviceInputWithDevice: newDevice error:nil];
```

```
        [_session addInput:captureInput];

        [_session commitConfiguration];
}
```

7.3.5 照相机预览

AVFoundation 的 **AVCaptureVideoPreviewLayer** 提供了实时图像预览层,可以把它嵌入到视图中。预览层既简单易用,又功能强大。程序清单 7-5 显示了如何创建新的预览层并把它插入到视图中,以匹配视图的框架。这里的视频拉伸方式(video gravity)默认为调整大小——外观,这将保留视频的宽高比,同时使之能够放入给定层的界限内。

可以搜索适当级别的层,从视图中获取预览层。

程序清单 7-5　嵌入和获取预览

```
- (void) embedPreviewInView: (UIView *) aView
{
    if (!_session) return;

    AVCaptureVideoPreviewLayer *preview =
        [AVCaptureVideoPreviewLayer layerWithSession: _session];
    preview.frame = aView.bounds;
    preview.videoGravity = AVLayerVideoGravityResizeAspect;
    [aView.layer addSublayer: preview];
}
- (AVCaptureVideoPreviewLayer *) previewInView: (UIView *) view
{
    for (CALayer *layer in view.layer.sublayers)
        if ([layer isKindOfClass:[AVCaptureVideoPreviewLayer class]])
            return (AVCaptureVideoPreviewLayer *)layer;

    return nil;
}
```

7.3.6 布置照相机预览

处理实时照相机预览的核心挑战之一是:无论设备方位是什么,都使视频指向上方。尽管照相机连接确实会处理一些方位问题,但是最容易的方式是直接处理预览层,并把它变换到正确的方位。程序清单 7-6 显示了如何执行该任务。

程序清单 7-6　添加照相机预览

```
- (void) layoutPreviewInView: (UIView *) aView
{
    AVCaptureVideoPreviewLayer *layer =
        [self previewInView:aView];
    if (!layer) return;

    UIDeviceOrientation orientation =
        [UIDevice currentDevice].orientation;
    CATransform3D transform = CATransform3DIdentity;
    if (orientation == UIDeviceOrientationPortrait) ;
    else if (orientation == UIDeviceOrientationLandscapeLeft)
        transform = CATransform3DMakeRotation(-M_PI_2, 0.0f, 0.0f, 1.0f);
    else if (orientation == UIDeviceOrientationLandscapeRight)
        transform = CATransform3DMakeRotation(M_PI_2, 0.0f, 0.0f, 1.0f);
    else if (orientation == UIDeviceOrientationPortraitUpsideDown)
        transform = CATransform3DMakeRotation(M_PI, 0.0f, 0.0f, 1.0f);

    layer.transform = transform;
    layer.frame = aView.frame;
}
```

7.3.7　照相机图像助手

秘诀 7-3 把本节中描述过的所有方法都结合进一个简单的助手类中。这个类提供了所需的全部基本特性，可用于查询照相机、建立会话、创建预览以及获取图像。要使用这个类，先要创建一个新实例，开始运行会话，或者还要把预览嵌入到主视图之一中。无论何时自动旋转了界面，都要确保再次布置预览，使之将保持正确的方位。

秘诀 7-3　用于照相机的助手类

```
@interface CameraImageHelper : NSObject
    <AVCaptureVideoDataOutputSampleBufferDelegate>

@property (nonatomic, strong) AVCaptureSession *session;
@property (nonatomic, strong) CIImage *ciImage;
@property (nonatomic, readonly) UIImage *currentImage;
@property (nonatomic, readonly) BOOL isUsingFrontCamera;
```

```
+ (int) numberOfCameras;
+ (BOOL) backCameraAvailable;
+ (BOOL) frontCameraAvailable;
+ (AVCaptureDevice *)backCamera;
+ (AVCaptureDevice *)frontCamera;

+ (id) helperWithCamera: (NSUInteger) whichCamera;

- (void) startRunningSession;
- (void) stopRunningSession;
- (void) switchCameras;

- (void) embedPreviewInView: (UIView *) aView;
- (AVCaptureVideoPreviewLayer *) previewInView: (UIView *) view;
- (void) layoutPreviewInView: (UIView *) aView;
@end
```

> **获取这个秘诀的代码**
>
> 要查找这个秘诀的完整示例项目，可以浏览 https://github.com/erica/iOS-6-Advanced-Cookbook，并进入第 7 章的文件夹。

7.4 秘诀：EXIF

EXIF（Exchangeable Image File Format，可交换图像文件格式）是由日本电子工业发展协会（Japan Electronic Industries Development Association，JEIDA）创建的。它提供了一组标准化的图像注释，包括照相机设置（比如快门速度和计量模式）、日期和时间信息，以及图像预览。其他元数据标准包括 IPTC（International Press Telecommunications Council，国际报业电信委员会）和 Adobe 的 XMP（Extensible Metadata Platform，可扩展的元数据平台）。

方位是目前唯一直接通过 UIImage 类展示的元信息。可以通过它的 imageOrientation 属性访问该信息，在下一节中可以看到，它具有与 EXIF 方位之间的对应关系，但是它的值使用不同的编号系统。

7.4.1 ImageIO

ImageIO 框架最初是在 iOS 4 中引入 iOS 中的。顾名思义，ImageIO 专门用于读写图像数据。使用"图像源"代替标准文件访问允许访问图像元数据属性。这些属性包括一些基本的值，比如图像的宽度和高度，以及一些更细致的信息，比如用于拍摄图像的照相机的快门速度和焦距。本书上一版的读者要求我添加一个秘诀，说明如何读取和更新 EXIF 和位置（GPS）元数据。秘诀

7-4 实现了一个自定义的 MetaImage 类，用于解决这个问题。

这个类封装了一个标准的 UIImage，并利用元数据属性字典增强它。这些属性源于图像，并展示那些值以供读写。为了创建那种透明度，必须在某个位置初始化属性。因此，在这个类中，当通过提供一幅图像构建实例时，类将生成用于该数据的基本类别。类似地，当通过一个文件创建实例时，它将读取与该图像存储在一起的任何元数据。

7.4.2 查询元数据

程序清单 7-7 演示了如何为 UIImage 和存储的文件源查询元数据。这两个函数中的第一个函数将从给定路径处的文件中读入图像属性字典。它将使用文件 URL 创建一个图像源，然后复制那个源中的属性，并把它们作为一个可变的字典返回。

第二个函数开始于一个图像实例，而不是一个文件。由于图像源必须处理数据，该方法将把 UIImage 实例转换为 JPEG 数据，它被选作为最常见的用例。这一次，将从图像数据生成图像源，但是余下的工作是完全相同的。该函数将复制源的属性，并把它们作为一个可变的字典返回。

从 UIImage 函数返回的数据预期是稀疏的。生成的图片不能提供关于白平衡、焦平面和计量的详细信息。用于 UIImage 实例的元数据包括它的深度、方位、像素尺寸和颜色模型。

程序清单 7-7　获取图像元数据

```
// Read image properties at a file path
NSMutableDictionary *imagePropertiesDictionaryForFilePath(
    NSString *path)
{
    CFDictionaryRef options = (__bridge CFDictionaryRef)@{
    };
    
    CFURLRef url = (__bridge CFURLRef) [NSURL fileURLWithPath:path];
    CGImageSourceRef imageSource =
        CGImageSourceCreateWithURL(url, options);
    if (imageSource == NULL)
    {
        NSLog(@"Error: Could not establish \
            image source for file at path: %@", path);
        return nil;
    }
    CFDictionaryRef imagePropertiesDictionary =
        CGImageSourceCopyPropertiesAtIndex(imageSource, 0, NULL);
    CFRelease(imageSource);
```

```objc
    return [NSMutableDictionary dictionaryWithDictionary:
        (__bridge_transfer NSDictionary *)
            imagePropertiesDictionary];
}

// Read the properties for an image instance
NSMutableDictionary *imagePropertiesFromImage(UIImage *image)
{
    CFDictionaryRef options = (__bridge CFDictionaryRef)@{
    };
    NSData *data = UIImageJPEGRepresentation(image, 1.0f);
    CGImageSourceRef imageSource =
        CGImageSourceCreateWithData(
            (__bridge CFDataRef) data, options);
    if (imageSource == NULL)
    {
        NSLog(@"Error: Could not establish image source");
        return nil;
    }
    CFDictionaryRef imagePropertiesDictionary =
        CGImageSourceCopyPropertiesAtIndex(imageSource, 0, NULL);
    CFRelease(imageSource);

    return [NSMutableDictionary dictionaryWithDictionary:
        (__bridge_transfer NSDictionary *)
            imagePropertiesDictionary];
}
```

7.4.3 包装 UIImage

尽管程序清单 7-7 显示了如何导入和获取元数据，秘诀 7-4 还是详细说明了 MetaImage 包装器实现如何把该数据导出回文件中。它的 writeToPath:方法使用图像目的地构建一个新文件，用于存储图像和元数据。

这种实现首先将把这种数据写到一个临时文件中，然后把它移入合适的位置。你可能想微调这些细节，以匹配正常的实践。与程序清单 7-7 一样，这个包装器方法将调用 ImageIO 的 Core Foundation 风格的 C 语言函数库，这就是为什么使用如此多的 ARC 风格的桥接的原因。

在这种实现中，两个类属性（gps 和 exif）展示嵌入在主要元数据中的子字典。它们被包装器类建立为可变的实例，允许直接更新字典值，准备好把元数据写到增强的图像文件中。下面这个示例说明了可能如何加载图像、调整它的元数据，以及把它存储到文件。

```
MetaImage *mi = [MetaImage newImage:image];
mi.exif[@"UserComment"] = @"This is a test comment";
[mi writeToPath:destPath];
```

两个字典属性都使用标准的键。这些键是在 ImageIO 框架中声明的，并且基于行业标准。参阅 Apple 的 CGImageProperties 参考资料，以了解相关的详细信息。这个文档定义了框架使用的图像特征，指定用于处理它们的关键常量，并且总结了通常如何使用它们。

秘诀 7-4　展示图像元数据

```
- (BOOL) writeToPath: (NSString *) path
{
    // Prepare to write to temporary path
    NSString *temporaryPath =
        [NSTemporaryDirectory() stringByAppendingPathComponent:
            [path lastPathComponent]];
    if ([[NSFileManager defaultManager]
        fileExistsAtPath:temporaryPath])
    {
        if (![[NSFileManager defaultManager]
            removeItemAtPath:temporaryPath error:nil])
        {
            NSLog(@"Could not establish temporary writing file");
            return NO;
        }
    }

    // Where to write to
    NSURL *temporaryURL = [NSURL fileURLWithPath:temporaryPath];
    CFURLRef url = (__bridge CFURLRef) temporaryURL;

    // What to write
    CGImageRef imageRef = self.image.CGImage;

    // Metadata
    NSDictionary *properties = [NSDictionary
        dictionaryWithDictionary:self.properties];
    CFDictionaryRef propertiesRef =
```

```objc
        (__bridge CFDictionaryRef) properties;

// UTI - See Chapter 2
NSString *uti = preferredUTIForExtension(path.pathExtension);
if (!uti) uti = @"public.image";
CFStringRef utiRef = (__bridge CFStringRef) uti;

// Create image destination
CGImageDestinationRef imageDestination =
    CGImageDestinationCreateWithURL(url, utiRef, 1, NULL);

// Save data
CGImageDestinationAddImage(imageDestination, imageRef, propertiesRef);
CGImageDestinationFinalize(imageDestination);

// Clean up
CFRelease(imageDestination);

// Move file into place
NSURL *destURL = [NSURL fileURLWithPath:path];

BOOL success;
NSError *error;

// Remove previous file
if ([[NSFileManager defaultManager] fileExistsAtPath:path])
{
    success = [[NSFileManager defaultManager]
        removeItemAtPath:path error:&error];
    if (!success)
    {
        NSLog(@"Error: Could not overwrite file properly. \
            Original not removed.");
        return NO;
    }
}
success = [[NSFileManager defaultManager]
    moveItemAtURL: temporaryURL toURL:destURL error:&error];
if (!success)
{
```

```
        NSLog(@"Error: could not move new file into place: %@",
            error.localizedFailureReason);
        return NO;
    }
    return YES;
}
```

获取这个秘诀的代码

要查找这个秘诀的完整示例项目，可以浏览 https://github.com/erica/iOS-6-Advanced-Cookbook，并进入第 7 章的文件夹。

7.5 图像方位

几何学可以表示处理 AVFoundation 照相机馈送的最困难的部分。与使用将为你自动处理这些事情的图像选择器不同，你必须与原始图像缓冲区打交道。你使用的照相机（前置或后置）与设备的方位将会影响图像数据的定向方式，如表 7-1 中所示。

表 7-1　把设备方位映射到图像方位

方位	Home 按钮	照相机	自然输出	镜像
Portrait	下	前置	左镜像	右
LandscapeLeft	右	前置	下镜像	下
PortraitUpsideDown	上	前置	右镜像	左
LandscapeRight	左	前置	上镜像	上
Portrait	下	后置	右	左镜像
LandscapeLeft	右	后置	上	上镜像
PortraitUpsideDown	上	后置	左	右镜像
LandscapeRight	左	后置	下	下镜像

表 7-1 中的"自然输出"列指定了由每个设备的方位与照相机的组合产生的方位。例如，设备是纵向握持的，使用前置照相机，产生左镜像的输出。iPhone 的最"自然的"设置是通过左边横向握持照相机创建的，也就是说，使 Home 按钮向右，并且使用后置照相机。这将创建一种面朝上的图像。

表 7-2 中对应的方位告诉 iOS 如何把 EXIF 值映射到 UIImage 设置。可以使用该数据把图像恢复到它们的正确表示。检查表中的第一个"F"，这个"F"是竖立的，并且它的叶子指向右边。具有这种方位的图像将具有与这种方位匹配的原始数据。像素数据从左上方开始并从左向右移

动,一行接着一行,最终到达右下方。

当用户使用其他方位拍摄照片或者使用前置照相机代替后置照相机时,存储的数据就像是从照相机的传感器捕获的一样。图片的左上角可能不会出现在图像数据的像素 0 处。例如,比如说用户使用前置照相机拍摄了一张照片,并且 Home 按钮位于右边。表 7-1 指示输出将是下镜像的。这是 UI 方位 5、EXIF 方位 4。原始的图片数据将不仅是颠倒的,而且还会被镜像。

知道底层数据表示(侧偏、镜像、颠倒等)如何映射到想要的表示(图像总是开始于左上方)对于图像处理是至关重要的,尤其对于自动检测例程则更是如此。表 7-2 中的最后一列显示了每个 UIImageOrientation 对应的 EXIF 方位。这种转换允许处理 Core Image。CI 就是使用 EXIF,而不是 UIImage 方位。

表 7-2 图像方位

方位	UIImageOrientation	对应的 EXIF
XXXXXX XX XXXX XX XX	UIImageOrientationUp 0	左上 1
XXXXXX XXXXXX XXXXXX XXXXXX XXXXXX	UIImageOrientationDown 1	右下 3
XX XXXXX XXXXXXXXXX	UIImageOrientationLeft 2	右上 6
XXXXXXXXXX XXXXXXXXXX XXXXXXXXXX	UIImageOrientationRight 3	左下 8
XXXXXX XXXXXX XXXXXX XXXXXX XXXXXX	UIImageOrientationUpMirrored 4	右上 2
XX XX XXXX XX XXXXXX	UIImageOrientationDownMirrored 5	左下 4

续表

方位	UIImageOrientation	对应的 EXIF
XXXXXXXXXX XXXXXX XX	UIImageOrientationLeftMirrored 6	左上 5
XXXXXXXXXX XXXXXXXXXX XXXXXXXXXX	UIImageOrientationRightMirrored 7	右下 7

下面的函数允许在 EXIF 与 UIImage 方位模式之间转换：

```
NSUInteger exifOrientationFromUIOrientation(
    UIImageOrientation uiorientation)
{
    if (uiorientation > 7) return 1;
    int orientations[8] = {1, 3, 6, 8, 2, 4, 5, 7};
    return orientations[uiorientation];
}

UIImageOrientation imageOrientationFromEXIFOrientation(
    NSUInteger exiforientation)
{
    if ((exiforientation < 1) || (exiforientation > 8))
        return UIImageOrientationUp;
    int orientations[8] = {0, 4, 1, 5, 6, 2, 7, 3};
    return orientations[exiforientation];
}
```

7.6 秘诀：Core Image 过滤

Core Image（CI）滤镜允许快速处理图像。秘诀 7-5 通过对利用机载照相机拍摄的图像应用简单的单色滤镜，引入了 CI 过滤。这个滤镜不适用于实时视频馈送，但是适用于单独的图像。单色滤镜可以把一幅 RGB 图像变换为单通道，然后应用单独一种颜色，它是你作为滤镜的输入指定的。

CADisplayLink 增强了这个秘诀。每次激活它时，snap 方法都会从其助手那里抓取当前的 ciImage（参考秘诀 7-3）。如果用户启用了过滤（使用一个简单的栏按钮开关），将创建新的滤镜，把图像设置为它的输入，把输出颜色调整为红色，然后获取并显示输出图像。

可以调整显示链接的帧间隔，在平滑表示与处理开销之间进行折衷。这是一个关键因素，尤其是在使用高分辨率的照相机时。增加帧间隔，以便在显示链接再次调用 snap 之前扩展必须传递的帧数。

秘诀 7-5 添加简单的 Core Image 滤镜

```objc
@implementation TestBedViewController

// Switch between cameras
- (void) switch: (id) sender
{
    [helper switchCameras];
}

// Enable/disable the filter
- (void) toggleFilter: (id) sender
{
    useFilter = !useFilter;
}

// Grab an image, optionally applying the filter,
// and display the results
- (void) snap
{
    // Orientation needed for CI image workaround
    UIImageOrientation orientation =
        currentImageOrientation(helper.isUsingFrontCamera, NO);

    if (useFilter) // monochrome - red
    {
        CIFilter *filter =
            [CIFilter filterWithName:@"CIColorMonochrome"];
        [filter setValue:helper.ciImage forKey:@"inputImage"];
        [filter setDefaults];
        [filter setValue:@1 forKey:@"inputIntensity"];
        [filter setValue:[CIColor colorWithRed:1.0f
            green:0.0f blue:0.0f] forKey:@"inputColor"];
        CIImage *outputImage = [filter valueForKey:kCIOutputImageKey];

        // Apply workaround to transform ciImage to UIImage
        if (outputImage)
            imageView.image = [UIImage imageWithCIImage:outputImage
                orientation:orientation];
        else NSLog(@"Missing image");
```

```
    }
    
    if (!useFilter) // no filter
    {
        // Convert using ciImage workaround
        imageView.image = [UIImage imageWithCIImage:helper.ciImage
            orientation:orientation];
    }
}
- (void) viewWillAppear:(BOOL)animated
{
    [super viewWillAppear:animated];
    
    // Establish the preview session
    helper = [CameraImageHelper helperWithCamera:kCameraBack];
    [helper startRunningSession];
    
    displayLink = [CADisplayLink displayLinkWithTarget:self
        selector:@selector(snap)];
    [displayLink addToRunLoop:[NSRunLoop currentRunLoop]
        forMode:NSRunLoopCommonModes];
}
@end
```

> **获取这个秘诀的代码**
> 要查找这个秘诀的完整示例项目，可以浏览 https://github.com/erica/iOS-6-Advanced-Cookbook，并进入第 7 章的文件夹。

7.7 秘诀：Core Image 人脸检测

人脸检测是在 iOS 5 中作为 iOS 的更绚丽的特性之一引入的。它比较容易实现，但是要使之正确却相当困难，这一切都可以归结为基本的几何学。获取特征集涉及创建新的人脸类型的检测器，以及查询它以获取检测到的特征。它在工作时，将返回 0 个或多个 CIFaceFeature 对象的数组，它们对应于在场景中发现的人脸的数量。

```
- (NSArray *) featuresInImage
{
    NSDictionary *detectorOptions = [NSDictionary
```

```
        dictionaryWithObject:CIDetectorAccuracyLow
            forKey:CIDetectorAccuracy];

    CIDetector *detector = [CIDetector
        detectorOfType:CIDetectorTypeFace context:nil
        options:detectorOptions];

    NSUInteger orientation = detectorEXIF(helper.isUsingFrontCamera, NO);
    NSDictionary *imageOptions =
        [NSDictionary dictionaryWithObject:
            [NSNumber numberWithInt:orientation]
            forKey:CIDetectorImageOrientation];
    return [detector featuresInImage:ciImage options:imageOptions];
}
```

有几个问题使之成为一项特别困难的任务。检测器依赖于知道正确的方位以解释传递给它的 CI 图像。它不能检测到偏斜或颠倒过来的人脸。这就是为什么该方法包括一个 detectorEXIF 函数调用的原因。问题是：至少在编写本书时，必需的检测器方位并不总是与现实的图像方位匹配，从而导致了如下的解决办法：

```
NSUInteger detectorEXIF(BOOL isUsingFrontCamera, BOOL shouldMirrorFlip)
{
    if (isUsingFrontCamera || deviceIsLandscape())
        return currentEXIFOrientation(isUsingFrontCamera,
            shouldMirrorFlip);

    // Only back camera portrait or upside down here.
    // Detection happens but the geometry is messed.
    int orientation = currentEXIFOrientation(!isUsingFrontCamera,
        shouldMirrorFlip);
    return orientation;
}
```

可以断定，这段代码有漏洞并且有错误，会有灾难等待发生。

第二个问题出在几何学上。Apple 没有提供 API，在把坐标映射到图像中的同时仍然保留图像方位信息。坐标(0,0)可能指左上角、左下角、右上角或右下角。更重要的是，镜像意味着坐标可能偏离左边或右边，或者偏离底部或顶部。数学运算留给你完成。进入程序清单 7-8 中，这些方法有助于把点和矩形结果从它们的原始数据坐标转换为规范化的图像坐标。

这里使用的 ExifOrientation 枚举没有构建到 iOS 中。定义它是为了匹配程序清单 7-8 中所示的自然的 EXIF 值。注意：没有针对 kRightTop 和 kLeftBottom 的情况，Core Image 检测器目前还不支持这两种 EXIF 方位。如果 Apple 更新了它的实现，就需要添加支持它们的情况。

程序清单 7-8　把几何结构从 EXIF 转换为图像坐标

```
CGPoint pointInEXIF(ExifOrientation exif, CGPoint aPoint, CGRect rect)
{
    switch(exif)
    {
        case kTopLeft:
            return CGPointMake(aPoint.x,
                rect.size.height - aPoint.y);
        case kTopRight:
            return CGPointMake(rect.size.width - aPoint.x,
                rect.size.height - aPoint.y);
        case kBottomRight:
            return CGPointMake(rect.size.width - aPoint.x,
                aPoint.y);
            case kBottomLeft:
                return CGPointMake(aPoint.x, aPoint.y);

        case kLeftTop:
            return CGPointMake(aPoint.y, aPoint.x);
        case kRightBottom:
            return CGPointMake(rect.size.width - aPoint.y,
                rect.size.height - aPoint.x);

        default:
            return aPoint;
    }
}

CGSize sizeInEXIF(ExifOrientation exif, CGSize aSize)
{
    switch(exif)
    {
        case kTopLeft:
        case kTopRight:
        case kBottomRight:
        case kBottomLeft:
            return aSize;
```

```objc
                case kLeftTop:
                case kRightBottom:
                    return CGSizeMake(aSize.height, aSize.width);
        }
}
CGRect rectInEXIF(ExifOrientation exif, CGRect inner, CGRect outer)
{
    CGRect rect;
    rect.origin = pointInEXIF(exif, inner.origin, outer);
    rect.size = sizeInEXIF(exif, inner.size);

    switch(exif)
    {
        case kTopLeft:
            rect = CGRectOffset(rect, 0.0f, -inner.size.height);
            break;
        case kTopRight:
            rect = CGRectOffset(rect, -inner.size.width,
                -inner.size.height);
            break;
        case kBottomRight:
            rect = CGRectOffset(rect, -inner.size.width, 0.0f);
            break;
        case kBottomLeft:
            break;
        case kLeftTop:
            break;
        case kRightBottom:
            rect = CGRectOffset(rect, -inner.size.width,
                -inner.size.height);
            break;
        default:
            break;
    }

    return rect;
}
```

秘诀 7-6 演示了如何在应用程序中执行人脸检测。这个方法将获取检测到的特征，利用阴

影矩形描绘检测到的人脸的界限，并利用圆圈突出显示 3 个
特定的特征（左眼、右眼和嘴巴）。图 7-2 显示了正在进行的
检测。

通过检测返回的 CIFaceFeature 数组的每个成员都可以报
告几个几何学上感兴趣的特征。这些包括人脸周围的矩形边
界，以及左眼、右眼和嘴巴的位置。这些子特征并不总会被创
建，因此在使用这些位置之前首先应该检测它们是否可用。秘
诀 7-6 演示了它们的测试和使用。

秘诀 7-6 还处理了由于 Core Image 的纵向 EXIF 的怪僻而
提出的几种解决办法。如果 Apple 更新了 Core Image 检测人
脸的方式，就要为余下的两种方位相应地调整代码。

图 7-2　自动检测人脸及其特征

秘诀 7-6　检测人脸

```
- (void) snap
{
    ciImage = helper.ciImage;
    UIImage *baseImage = [UIImage imageWithCIImage:ciImage];
    CGRect imageRect = (CGRect){.size = baseImage.size};
    
    NSDictionary *detectorOptions = [NSDictionary 
        dictionaryWithObject:CIDetectorAccuracyLow 
        forKey:CIDetectorAccuracy];
    
    CIDetector *detector = [CIDetector 
        detectorOfType:CIDetectorTypeFace 
        context:nil options:detectorOptions];
    
    ExifOrientation detectOrientation = 
        detectorEXIF(helper.isUsingFrontCamera, NO);
    
    NSDictionary *imageOptions = [NSDictionary 
        dictionaryWithObject:[NSNumber numberWithInt:detectOrientation] 
        forKey:CIDetectorImageOrientation];
    NSArray *features = 
        [detector featuresInImage:ciImage options:imageOptions];
    
    UIGraphicsBeginImageContext(baseImage.size);
```

```objc
    [baseImage drawInRect:imageRect];

    for (CIFaceFeature *feature in features)
    {
        CGRect rect = rectInEXIF(detectOrientation,
            feature.bounds, imageRect);
        if (deviceIsPortrait() && helper.isUsingFrontCamera) // workaround
        {
            rect.origin = CGPointFlipHorizontal(rect.origin, imageRect);
            rect.origin =
                CGPointOffset(rect.origin, -rect.size.width, 0.0f);
        }
        [[[UIColor blackColor] colorWithAlphaComponent:0.3f] set];
        UIBezierPath *path = [UIBezierPath bezierPathWithRect:rect];
        [path fill];

        if (feature.hasLeftEyePosition)
        {

            [[[UIColor redColor] colorWithAlphaComponent:0.5f] set];
            CGPoint position = feature.leftEyePosition;
            CGPoint pt = pointInEXIF(detectOrientation, position,
                imageRect);

            if (deviceIsPortrait() && helper.isUsingFrontCamera)
                pt = CGPointFlipHorizontal(pt, imageRect); // workaround

            UIBezierPath *path = [UIBezierPath bezierPathWithArcCenter:pt
                radius:30.0f startAngle:0.0f endAngle:2 * M_PI
                clockwise:YES];
            [path fill];
        }

        if (feature.hasRightEyePosition)
        {
            [[[UIColor blueColor] colorWithAlphaComponent:0.5f] set];
            CGPoint position = feature.rightEyePosition;
            CGPoint pt = pointInEXIF(detectOrientation, position,
                imageRect);
            if (deviceIsPortrait() && helper.isUsingFrontCamera)
```

```
            pt = CGPointFlipHorizontal(pt, imageRect); // workaround

        UIBezierPath *path = [UIBezierPath bezierPathWithArcCenter:pt
            radius:30.0f startAngle:0.0f endAngle:2 * M_PI
            clockwise:YES];

        [path fill];
    }
    if (feature.hasMouthPosition)
    {
        [[[UIColor greenColor] colorWithAlphaComponent:0.5f] set];
        CGPoint position = feature.mouthPosition;
        CGPoint pt = pointInEXIF(detectOrientation, position,
            imageRect);
        if (deviceIsPortrait() && helper.isUsingFrontCamera)
            pt = CGPointFlipHorizontal(pt, imageRect); // workaround

        UIBezierPath *path = [UIBezierPath bezierPathWithArcCenter:pt
            radius:30.0f startAngle:0.0f endAngle:2 * M_PI
                clockwise:YES];

        [path fill];
        }
    }

    imageView.image = UIGraphicsGetImageFromCurrentImageContext();
    UIGraphicsEndImageContext();
}
```

> **获取这个秘诀的代码**
>
> 要查找这个秘诀的完整示例项目,可以浏览 https://github.com/erica/iOS-6-Advanced-Cookbook,并进入第 7 章的文件夹。

7.8 秘诀：对实时馈送进行抽样

可以使用位图访问为用户创建实时响应。秘诀 7-7 对每幅图像的中心进行抽样，以获取一种主要的颜色。它在工作时，将提取 128 像素×128 像素的色块，并对它执行一些基本的统计图像处理。导航栏的颜色将持续更新，以显示抽样馈送的中心内的这种最流行的颜色。

该方法将把每个像素都从 RGB 转换为 HSB，使算法能够获取特有的色调。它将增大用于色调的直方图桶，并且会逐渐提高饱和度和亮度——最终将用累积量除以那个桶中的抽样次数，创建给定色调的平均饱和度和亮度。

抽样中最流行的色调将胜出。具有最大点击次数的色调（即模式）将构成新颜色的基础，并且利用平均饱和度和亮度修饰创建的颜色，然后指定该颜色对导航栏进行着色。

秘诀 7-7　分析位图样本

```
#define SAMPLE_LENGTH 128
- (void) pickColor
{
    // Retrieve the center 128x128 sample as bits
    UIImage *currentImage = helper.currentImage;
    CGRect sampleRect =
        CGRectMake(0.0f, 0.0f, SAMPLE_LENGTH, SAMPLE_LENGTH);
    sampleRect = CGRectCenteredInRect(sampleRect,
        (CGRect){.size = currentImage.size});
    UIImage *sampleImage =
        [currentImage subImageWithBounds:sampleRect];
    NSData *bitData = sampleImage.bytes;
    Byte *bits = (Byte *)bitData.bytes;

    // Create the histogram and average sampling buckets
    int bucket[360];
    CGFloat sat[360], bri[360];

    for (int i = 0; i < 360; i++)
    {
        bucket[i] = 0; // histogram sample
        sat[i] = 0.0f; // average saturation
        bri[i] = 0.0f; // average brightness
    }

    // Iterate over each sample pixel, accumulating hsb info
    for (int y = 0; y < SAMPLE_LENGTH; y++)
        for (int x = 0; x < SAMPLE_LENGTH; x++)
        {
            CGFloat r = ((CGFloat)bits[redOffset(x, y,
                SAMPLE_LENGTH)] / 255.0f);
            CGFloat g = ((CGFloat)bits[greenOffset(x, y,
```

```
                SAMPLE_LENGTH)] / 255.0f);
            CGFloat b = ((CGFloat)bits[blueOffset(x, y,
                SAMPLE_LENGTH)] / 255.0f);

            // Convert from RGB to HSV
            CGFloat h, s, v;
            rgbtohsb(r, g, b, &h, &s, &v);
            int hue = (hue > 359.0f) ? 0 : (int) h;

            // Collect metrics on a per-hue basis
            bucket[hue]++;
            sat[hue] += s;
            bri[hue] += v;
        }

// Retrieve the hue mode
int max = -1;
int maxVal = -1;
for (int i = 0; i < 360; i++)
{
    if (bucket[i] > maxVal)
    {
        max = i;
        maxVal = bucket[i];
    }
}
// Create a color based on the mode hue, average sat & bri
float h = max / 360.0f;
float s = sat[max]/maxVal;
float br = bri[max]/maxVal;

CGFloat red, green, blue;
hsbtorgb((CGFloat) max, s, br, &red, &green, &blue);

UIColor *hueColor = [UIColor colorWithHue:h saturation:s
    brightness:br alpha:1.0f];

// Display the selected hue
self.navigationController.navigationBar.tintColor = hueColor;
```

```
        free(bits);
}
```

> **获取这个秘诀的代码**
>
> 要查找这个秘诀的完整示例项目,可以浏览 https://github.com/erica/iOS-6-Advanced-Cookbook,并进入第 7 章的文件夹。

7.8.1 转换为 HSB

秘诀 7-8 依赖于使用程序清单 7-9 中的函数把它的颜色从 RGB 转换为 HSB。该函数是根据作者的博士生导师 Jim Foley 的一本具有重大影响的教科书改编而来的,该教科书是关于计算机图形学的,即 *Computer Graphics: Principles and Practice in C (2nd Edition)*(Addison-Wesley Professional,ISBN-13:9780201848403)。

程序清单 7-9 在 RGB 与 HSB 之间转换

```
void rgbtohsb(CGFloat r, CGFloat g, CGFloat b, CGFloat *pH,
    CGFloat *pS, CGFloat *pV)
{
    CGFloat h,s,v;

    // From Foley and Van Dam
    CGFloat max = MAX(r, MAX(g, b));
    CGFloat min = MIN(r, MIN(g, b));

    // Brightness
    v = max;

    // Saturation
    s = (max != 0.0f) ? ((max - min) / max) : 0.0f;
    if (s == 0.0f) {
        // No saturation, so undefined hue
        h = 0.0f;
    } else {
        // Determine hue distances from...
        CGFloat rc = (max - r) / (max - min); // from red
        CGFloat gc = (max - g) / (max - min); // from green
        CGFloat bc = (max - b) / (max - min); // from blue

        if (r == max) h = bc - gc; // between yellow & magenta
```

```
            else if (g == max) h = 2 + rc - bc; // between cyan & yellow
            else h = 4 + gc - rc; // between magenta & cyan

            h *= 60.0f; // Convert to degrees
            if (h < 0.0f) h += 360.0f; // Make non-negative
        }

        if (pH) *pH = h;
        if (pS) *pS = s;
        if (pV) *pV = v;
}
```

7.9 小结

本章介绍了使用 iOS 设备上的图像捕获硬件的代码级方式。你学习了如何检查机载设备以及如何微调设置，发现了 AVFoundation 框架如何在不使用标准的图像选择器的情况下提供对馈送的直接访问。你还探索了如何提样并处理那些馈送展示的数据。在结束本章的学习之前，要思考以下几点。

- OpenCV 计算机视觉库已被广泛连接到 iOS。可以考虑直接使用它的特性，而不要从头开始。你还可能想要探索 Accelerate 和 Core Image。
- 与这套 Cookbook 系列中任何其他的示例相比，我可能花了更多时间研究 AVFoundation 包装器类的过于繁琐的细节。仅仅只进行方位转换就花了几小时的时间。如果你发现了任何错误，可以与我联系，以便我可以在示例代码库上和将来的出版物中修正细节。
- 由于内置照相机的质量非常高，来自最新的 iOS 设备的馈送可能是巨大的。前置照相机提供的数据馈送一般比后置照相机的分辨率低。不要害怕进行子抽样，并且只使用图像中对应用程序有帮助的那些部分。

第 8 章 音频

iOS 设备是一个媒体大师,其内置的 iPod 特性可以专业地处理音频和视频。iOS SDK 向开发者展示了这种功能。一套丰富的类通过播放、搜索和录音简化了媒体处理。本章介绍了一些秘诀,使用那些类操纵音频、把媒体展示给用户,以及允许用户与媒体交互。你将看到如何构建音频播放器和录音机,还将发现如何浏览 iPod 库以及如何选择要播放的项目。你将在本章中学习的秘诀逐步地演示了如何把这些媒体丰富的特性添加到自己的应用程序中。

8.1 秘诀:利用 AVAudioPlayer 播放音频

AVAudioPlayer 类可以播放音频数据。作为 AVFoundation 框架的一部分,这个易于使用的类提供了众多特性,图 8-1 中突出显示了其中一些特性。利用这个类,可以加载音频、播放它、暂停它、停止它、监测平均和峰值声级、调整播放音量,以及设置和检测当前的播放时间。利用少许关联的开发成本即可使用所有这些特性。如你将看到的,AVAudioPlayer 类提供了一个稳定的 API。

8.1.1 初始化音频播放器

只需做少量的工作即可在代码中实现由 AVAudioPlayer 提供的音频播放特性。Apple 提供了一个比较简单的类,它为加载和播放文件进行了简化。

首先,分配一个播放器,并利用数据或者本地 URL 的内容初始化它。这个代码段使用一个文件 URL 指向一个音频文件,它将报告创建和设置播放器过程中涉及的任何错误。也可以使用 initWithData:error:利用已经存储在内存中的数据初始化播放器。当已经把数据读入到内存中时(比如在音频聊天期间)这将很方便,这样就无需读取存储在设备上的文件。

```
player = [[AVAudioPlayer alloc] initWithContentsOfURL:
    [NSURL fileURLWithPath:path] error:&error];

if (!player)
```

```
{
    NSLog(@"AVAudioPlayer could not be established: %@",
        error.localizedFailureReason);
    return;
}
```

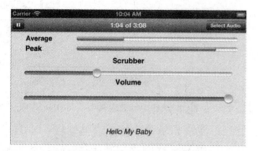

图 8-1　这幅截屏图中突出显示的特性是利用单个类 AVAudioPlayer 构建的。这个类提供了时间监测（在标题栏中间）、声级（平均和峰值）、擦除和音量滑块，以及播放/暂停控制（在标题栏右边）

在初始化播放器后，就准备好进行播放。可以调用 prepareToPlay，确保当准备好播放音频时，将尽可能快地开始播放。这个调用将预先加载播放器的缓冲区，并初始化音频播放硬件：

`[player prepareToPlay];`

可以随时调用 pause 暂停播放。暂停将冻结播放器的 currentTime 属性，然后可以再次调用 play 从那个位置恢复播放。

可以利用 stop 完全停止播放。停止播放将会撤销最初利用 prepareToPlay 建立的缓冲设置。不过，它不会把当前的时间设置回 0.0；可以再次调用 play，选择停止的位置，就像利用 pause 一样。在播放器重新加载其缓冲区时，可能会经历播放延迟。在切换新媒体之前，通常会停止播放。

8.1.2　监测音频级别

当打算监测音频级别时，首先要设置 meteringEnabled 属性。启用计量功能将允许在播放或录制音频时检查级别：

`player.meteringEnabled = YES;`

AVAudioPlayer 类提供了平均和峰值强度的反馈，可以在每个声道的基础上获取它。查询播放器，获取可用声道的数量（通过 numberOfChannels 属性），然后提供声道索引以请求每种强度级别。单声道信号使用声道 0，就像左声道用于立体声录制一样。

除了作为一个整体启用计量功能之外，每次希望测试声级时还需要调用 updateMeters；这个 AV 播放器方法将更新当前的测量声级。在这样做之后，使用 peakPowerForChannel:和 averagePowerForChannel:方法读取那些声级值。出现在本章后面的秘诀 8-6 显示了当它请求那些强度级别值时很可能将会在播放器中幕后发生的事情。可以看到代码请求测量声级值，然后提取峰值或平均强度。AVAudioPlayer 类隐藏了那些细节，简化了对这些值的访问。

AVAudioPlayer 以分贝（Decibel）度量强度，它是以浮点格式提供的。分贝使用对数标度来度量声强（sound intensity）。强度值的范围是从 0 dB（最高）到某个负值（表示低于最大强度）。

这个数字越低（并且它们都是负数），信号越弱。

```
int channels = player.numberOfChannels;
[player updateMeters];
for (int i = 0; i < channels; i++)
{
    // Log the peak and average power
    NSLog(@"%d %0.2f %0.2f",
        [player peakPowerForChannel:i],
        [player averagePowerForChannel:i]);
}

// Show average and peak values on the two meters for primary channel
meter1.progress = pow(10, 0.05f * [player averagePowerForChannel:0]);
meter2.progress = pow(10, 0.05f * [player peakPowerForChannel:0]);
```

要查询音频播放器增益（即它的音量），可以使用 volume 属性。这个属性返回一个 0.0~1.0 之间的浮点数，并且专门应用于播放器音量，而不是系统音量。可以设置这个属性并读取它。下面这个代码段可以用于目标—动作对，以便在用户操纵屏幕上的音量滑块时更新音量：

```
- (void) setVolume: (id) sender
{
    // Set the audio player gain to the current slider value
    if (player) player.volume = volumeSlider.value;
}
```

8.1.3　播放进度和擦除

属性 currentTime 和 duration 有助于监测音频的播放进度。要查找当前的播放百分比，可以用当前时间除以音频的总持续时间：

```
progress = player.currentTime / player.duration;
```

要擦除音频，也就是说，让用户可以在音轨内选择当前的播放位置，你可能希望暂停播放。尽管当前的 iOS 版本中的 AVAudioPlayer 类可以提供基于音频的擦除提示，但是这些提示起来可能比较愚蠢。你可能希望等待擦除完成，然后才开始在新位置播放音频。使用这个秘诀的示例代码测试实时擦除，就会了解到这个特性的相对粗糙的地方。

如果使擦除器基于标准的 UISlider，就要确保实现至少两个目标—动作对。对于第一个目标—动作项，可以利用 UIControlEventValueChanged 屏蔽 UIControlEventTouchDown。这些事件类型允许捕获用户擦除的开始以及值发生变化的任何时间。如果选择暂停擦除，就要通过暂停音频播放器来响应这些事件，并且为新选择的时间提供某种可视化反馈：

```
- (void) scrub: (id) sender
{
    // Pause the player -- optional
    [self.player pause];
    self.navigationItem.leftBarButtonItem =
        SYSBARBUTTON(UIBarButtonSystemItemPlay, @selector(play:));
    
    // Calculate the new current time
    player.currentTime = scrubber.value * player.duration;
    
    // Update the title with the current time
    self.title = [NSString stringWithFormat:@"%@ of %@",
        [self formatTime:self.player.currentTime],
        [self formatTime:self.player.duration]];
}
```

对于第二个目标——动作对,这个包含 3 个值的掩码(UIControlEventTouchUpInside | UIControl-EventTouchUpOutside | UIControlEventCancel)允许捕获释放事件和触摸中断。一旦释放,就将希望在通过用户擦除设置的新时间处开始播放:

```
- (void) scrubbingDone: (id) sender
{
    // resume playback here
    [self play:nil];
}
```

8.1.4 捕获播放的结束

通过设置播放器的委托并捕获 audioPlayerDidFinishPlaying:successfully:委托调用,来检测播放的结束。该方法能够非常好地清理任何细节,比如把 Pause(暂停)按钮恢复为 Play(播放)按钮。Apple 提供了几个专用于媒体播放的系统栏按钮项目:

- UIBarButtonSystemItemPlay
- UIBarButtonSystemItemPause
- UIBarButtonSystemItemRewind
- UIBarButtonSystemItemFastForward

Rewind(倒带)和 Fast Forward(快进)按钮提供了双箭头图标,它们通常用于把播放移到播放队列中的前一个或下一个项目上。也可以使用它们回到音轨的开头或者前进到它的末尾。不幸的是,Stop(停止)系统项目是一个 X,用于停止正在进行的加载操作,而不是许多消费者设备上用于停止播放或录音的标准实心方块。

秘诀 8-1 把所有这些部分组合在一起，创建在图 8-1 中看到的统一（虽然设计得有些恐怖）界面。在这里，用户可以选择音频、开始播放它、暂停它、调整它的音量以及擦除（scrub）等。

秘诀 8-1　利用 AVAudioPlayer 播放音频

```
// Set the meters to the current peak and average power
- (void) updateMeters
{
    // Retrieve current values
    [player updateMeters];

    // Show average and peak values on the two meters
    meter1.progress = pow(10, 0.05f * [player averagePowerForChannel:0]);
    meter2.progress = pow(10, 0.05f * [player peakPowerForChannel:0]);

    // And on the scrubber
    scrubber.value = (player.currentTime / player.duration);

    // Display the current playback progress in minutes and seconds
    self.title = [NSString stringWithFormat:@"%@ of %@",
        [self formatTime:player.currentTime],
        [self formatTime:player.duration]];
}

// Pause playback
- (void) pause: (id) sender
{
    if (player)
        [player pause];

    // Update the play/pause button
    self.navigationItem.leftBarButtonItem =
        SYSBARBUTTON(UIBarButtonSystemItemPlay, @selector(play:));

    // Disable interactive elements
    meter1.progress = 0.0f;
    meter2.progress = 0.0f;
    volumeSlider.enabled = NO;
    scrubber.enabled = NO;
```

```objc
    // Stop listening for meter updates
    [timer invalidate];
}

// Start or resume playback
- (void) play: (id) sender
{
    if (player)
        [player play];

    // Enable the interactive elements
    volumeSlider.value = player.volume;
    volumeSlider.enabled = YES;
    scrubber.enabled = YES;

    // Update the play/pause button
    self.navigationItem.leftBarButtonItem =
        SYSBARBUTTON(UIBarButtonSystemItemPause, @selector(pause:));

    // Start listening for meter updates
    timer = [NSTimer scheduledTimerWithTimeInterval:0.1f
        target:self selector:@selector(updateMeters)
        userInfo:nil repeats:YES];
}

// Update the volume
- (IBAction) setVolume: (id) sender
{
    if (player) player.volume = volumeSlider.value;
}

// Catch the end of the scrubbing
- (IBAction) scrubbingDone: (id) sender
{
    [self play:nil];
}

// Update playback point during scrubs
- (IBAction) scrub: (id) sender
{
```

```objc
    // Calculate the new current time
    player.currentTime = scrubber.value * player.duration;

    // Update the title, nav bar
    self.title = [NSString stringWithFormat:@"%@ of %@",
        [self formatTime:player.currentTime],
        [self formatTime:player.duration]];
}

// Prepare but do not play audio
- (BOOL) prepAudio
{
    NSError *error;
    if (![[NSFileManager defaultManager] fileExistsAtPath:path])
        return NO;

    // Establish player
    player = [[AVAudioPlayer alloc] initWithContentsOfURL:
        [NSURL fileURLWithPath:path] error:&error];
    if (!player)
    {
        NSLog(@"AVAudioPlayer could not be established: %@",
            error.localizedFailureReason);
        return NO;
    }

    [player prepareToPlay];
    player.meteringEnabled = YES;
    player.delegate = self;

    // Initialize GUI
    meter1.progress = 0.0f;
    meter2.progress = 0.0f;
    self.navigationItem.leftBarButtonItem =
        SYSBARBUTTON(UIBarButtonSystemItemPlay, @selector(play:));
    scrubber.enabled = NO;

    return YES;
}
// On finishing, return to quiescent state
```

```objc
- (void)audioPlayerDidFinishPlaying:(AVAudioPlayer *)player
    successfully:(BOOL)flag
{
    self.navigationItem.leftBarButtonItem = nil;
    scrubber.value = 0.0f;
    scrubber.enabled = NO;
    volumeSlider.enabled = NO;
    [self prepAudio];
}

// Select a media file
- (void) pick: (UIBarButtonItem *) sender
{
    self.navigationItem.rightBarButtonItem.enabled = NO;

    // Each of these media files is in the public domain via archive.org
    NSArray *choices = [@"Alexander's Ragtime Band*Hello My Baby*\
        Ragtime Echoes*Rhapsody In Blue*A Tisket A Tasket*In the Mood"
        componentsSeparatedByString:@"*"];
    NSArray *media = [@"ARB-AJ*HMB1936*ragtime*RhapsodyInBlue*Tisket*\
        InTheMood" componentsSeparatedByString:@"*"];

    UIActionSheet *actionSheet = [[UIActionSheet alloc]
        initWithTitle:@"Musical selections" delegate:nil
        cancelButtonTitle:IS_IPAD ? nil : @"Cancel"
        destructiveButtonTitle:nil otherButtonTitles:nil];
    for (NSString *choice in choices)
        [actionSheet addButtonWithTitle:choice];

    // See the Core Cookbook, Alerts Chapter for details on
    // the ModalSheetDelegate class
    ModalSheetDelegate *msd =
        [ModalSheetDelegate delegateWithSheet:actionSheet];
    actionSheet.delegate = msd;

    int answer = [msd showFromBarButtonItem:sender animated:YES];
    self.navigationItem.rightBarButtonItem.enabled = YES;

    if (IS_IPAD)
    {
```

```
            if (answer == -1) return; // cancel
            if (answer >= choices.count)
                return;
        }
        else
        {
            if (answer == 0) return; // cancel
            if (answer > choices.count)
                return;
            answer--;
        }

        // no action, if already playing
        if ([nowPlaying.text isEqualToString:[choices objectAtIndex:answer]])
            return;

        // stop any current item
        if (player)
            [player stop];

        // Load in the new audio and play it
        path = [[NSBundle mainBundle]
            pathForResource:[media objectAtIndex:answer] ofType:@"mp3"];
        nowPlaying.text = [choices objectAtIndex:answer];
        [self prepAudio];
        [self play:nil];
}
```

获取这个秘诀的代码

要查找这个秘诀的完整示例项目，可以浏览 https://github.com/erica/iOS-6-Advanced-Cookbook，并进入第 8 章的文件夹。

8.2 秘诀：循环音频

循环有助于展示环境背景音频，而不必在每次循环结束单独一遍播放时响应委托调用。可以使用循环多次播放一段音频，或者连续不断地播放它。秘诀 8-2 演示了一个只在展示特定的视频控制器期间播放的音频循环，它为该控制器提供一种背景音。

可以设置音频在重放结束前的播放次数。较高的次数（比如 999999）基本上相当于提供了无限次数的循环。例如，4 秒的循环将花费超过 1 000 小时的时间才能完全播放这么高的循环次数：

```
// Prepare the player and set the loops
[self.player prepareToPlay];
[self.player setNumberOfLoops:999999];
```

秘诀 8-2 为其主视图控制器使用循环的音频。无论何时其视图出现在屏幕上，都会在后台播放循环。希望你选择的循环不引人注目，为应用程序建立良好的情境，并从播放末尾平滑地过渡到开头。

这个秘诀使用一种淡出效果来引入和隐藏音频。当视图出现时，它将把循环渐渐出现在听觉中，并当视图消失时使循环渐渐淡出。它利用一个简单的方法实现这种效果。循环将遍历音量级别，在出现时是从 0.0 上升到 1.0，在消失时则是从 1.0 下降到 0.0。调用 NSThread 的内置睡眠功能将在不影响音频播放的情况下增加时间延迟（在每次音量改变之间增加十分之一秒）。

秘诀 8-2　通过循环创建环境音频

```
- (BOOL) prepAudio
{
    // Check for the file.
    NSError *error;
    NSString *path = [[NSBundle mainBundle]
        pathForResource:@"loop" ofType:@"mp3"];
    if (![[NSFileManager defaultManager]
        fileExistsAtPath:path]) return NO;

    // Initialize the player
    player = [[AVAudioPlayer alloc] initWithContentsOfURL:
        [NSURL fileURLWithPath:path] error:&error];
    if (!player)
    {
        NSLog(@"Could not establish AV Player: %@",
            error.localizedFailureReason);
        return NO;
    }

    // Prepare the player and set the loops large
    [player prepareToPlay];
    [player setNumberOfLoops:999999];
```

```
        return YES;
}

- (void) fadeIn
{
    player.volume = MIN(player.volume + 0.05f, 1.0f);
    if (player.volume < 1.0f)
        [self performSelector:@selector(fadeIn)
            withObject:nil afterDelay:0.1f];
}
- (void) fadeOut
{
    player.volume = MAX(player.volume - 0.1f, 0.0f);
    if (player.volume > 0.05f)
        [self performSelector:@selector(fadeOut)
            withObject:nil afterDelay:0.1f];
    else
        [player pause];
}

- (void) viewDidAppear: (BOOL) animated
{
    // Start playing at no-volume
    player.volume = 0.0f;
    [player play];

    // fade in the audio over a second
    [self fadeIn];

    // Add the push button
    self.navigationItem.rightBarButtonItem.enabled = YES;
}

- (void) viewWillDisappear: (BOOL) animated
{
    [self fadeOut];
}

- (void) push
```

```objc
{
    // Create a simple new view controller
    UIViewController *vc = [[UIViewController alloc] init];
    vc.view.backgroundColor = [UIColor whiteColor];
    vc.title = @"No Sounds";

    // Disable the now-pressed right-button
    self.navigationItem.rightBarButtonItem.enabled = NO;

    // Push the new view controller
    [self.navigationController
        pushViewController:vc animated:YES];
}

- (void) loadView
{
    [super loadView];
    self.navigationItem.rightBarButtonItem =
        BARBUTTON(@"Push", @selector(push));
    self.title = @"Looped Sounds";
    [self prepAudio];
}
```

获取这个秘诀的代码

要查找这个秘诀的完整示例项目，可以浏览 https://github.com/erica/iOS-6-Advanced-Cookbook，并进入第 8 章的文件夹。

8.3 秘诀：处理音频中断

当用户在音频播放期间接到电话时，该音频将渐渐消失，并会出现标准的接听/拒绝屏幕。当这发生时，AVAudioPlayer 委托将接收到 audioPlayerBeginInterruption:回调，如秘诀 8-3 中所示。音频会话将停用，并且播放器将暂停播放。直到中断结束之后，才能重新启动播放。

如果用户接听电话，应用程序委托将接收到一个 applicationWillResignActive:回调。打完电话之后，应用程序将重新启动（利用 applicationDidBecomeActive:回调）。如果用户拒听电话或者如果电话在无人接听的情况下结束，将代之以给委托发送 audioPlayerEndInterruption:。可以通过这个方法恢复播放。

如果在接听电话之后恢复播放是至关重要的，并且应用程序需要重新启动，那么就可以保存当前的时间,如秘诀 8-3 中所示。这个秘诀中的 viewDidAppear:方法(对大多数的恢复播放都会调用它)将在用户的默认设置中检查存储的中断值。当它找到一个这样的值时,将使用它设置恢复播放的当前时间。

这种方法考虑了应用程序重新启动,而没有考虑在打完电话后恢复播放。当用户接听电话时,你将不会接收到结束中断回调。

秘诀 8-3 为以后的选择存储中断时间

```
- (BOOL) prepAudio
{
    NSError *error;
    NSString *path = [[NSBundle mainBundle]
        pathForResource:@"MeetMeInSt.Louis1904" ofType:@"mp3"];
    if (![[NSFileManager defaultManager]
        fileExistsAtPath:path]) return NO;

    // Initialize the player
    player = [[AVAudioPlayer alloc]
        initWithContentsOfURL:[NSURL fileURLWithPath:path]
        error:&error];
    player.delegate = self;
    if (!player)
    {
        NSLog(@"Could not establish player: %@",
            error.localizedFailureReason);
        return NO;
    }

    [player prepareToPlay];

    return YES;
}
- (void)audioPlayerDidFinishPlaying:(AVAudioPlayer *)aPlayer
    successfully:(BOOL)flag
{
    // just keep playing
    [player play];
}

- (void)audioPlayerBeginInterruption:(AVAudioPlayer *)aPlayer
```

```objc
{
    // perform any interruption handling here
    fprintf(stderr, "Interruption Detected\n");
    [[NSUserDefaults standardUserDefaults]
        setFloat:[player currentTime] forKey:@"Interruption"];
}

- (void)audioPlayerEndInterruption:(AVAudioPlayer *)aPlayer
{
    // resume playback at the end of the interruption
    fprintf(stderr, "Interruption ended\n");
    [player play];

    // remove the interruption key. it won't be needed
    [[NSUserDefaults standardUserDefaults]
        removeObjectForKey:@"Interruption"];
}

- (void) viewDidAppear:(BOOL)animated
{
    [self prepAudio];
    // Check for previous interruption
    if ([[NSUserDefaults standardUserDefaults]
        objectForKey:@"Interruption"])
    {
        player.currentTime =
            [[NSUserDefaults standardUserDefaults]
                floatForKey:@"Interruption"];
        [[NSUserDefaults standardUserDefaults]
            removeObjectForKey:@"Interruption"];
    }
    // Start playback
    [player play];
}
```

> **获取这个秘诀的代码**
> 要查找这个秘诀的完整示例项目，可以浏览 https://github.com/erica/iOS-6-Advanced-Cookbook，并进入第 8 章的文件夹。

8.4 秘诀：录制音频

AVAudioRecorder 类简化了应用程序中的音频录制。它提供了与 AVAudioPlayer 相同的 API 友好性，以及类似的反馈属性。这两个类一起为许多标准的应用程序音频任务提供开发帮助。

通过建立一个 AVAudioSession 来开始录音。如果打算在同一个应用程序中在录音和播放之间切换，就要使用播放和录音会话，否则，可以使用一个简单的录音会话（通过 AVAudioSessionCategoryRecord）。

当具有会话时，可以检查它的 inputAvailable 属性。这个属性指示当前设备能够访问麦克风，它替换了 iOS 6 中不建议使用的 inputIsAvailable。如果计划部署 iOS 6 以前的框架，就要确保使用旧式的属性：

```
- (BOOL) startAudioSession
{
    // Prepare the audio session
    NSError *error;
    AVAudioSession *session = [AVAudioSession sharedInstance];

    if (![session
        setCategory:AVAudioSessionCategoryPlayAndRecord
        error:&error])
    {
        NSLog(@"Error setting session category: %@",
            error.localizedFailureReason);
        return NO;
    }

    if (![session setActive:YES error:&error])
    {
        NSLog(@"Error activating audio session: %@",
            error.localizedFailureReason);
        return NO;
    }

    // inputIsAvailable deprecated in iOS 6
    return session.inputAvailable;
}
```

秘诀 8-4 演示了创建会话之后的下一步操作。它将建立录音机，并提供用于暂停、恢复和停止录音的方法。

要开始录音，它将创建一个设置字典，并利用描述应该如何对录音进行抽样的键和值填充它。这个示例使用 8 000 次抽样/秒的单声道线性 PCM，这个抽样速率相当低。下面给出了几个自定义格式的示例。

- 对于单声道音频，把 AVNumberOfChannelsKey 设置为 1；对于立体声，则把它设置为 2。
- 在 iOS 上工作得很好的音频格式（AVFormatIDKey）包括 kAudioFormatLinearPCM（大文件）和 kAudioFormatAppleIMA4（紧凑的文件）。
- 标准的 AVSampleRateKey 抽样速率包括 8 000、11 025、22 050 和 44 100。
- 对于仅线性 PCM 的位深（AVLinearPCMBitDepthKey），可使用 16 位或 32 位。

代码分配一个新的 AVRecorder 实例，并利用文件 URL 和设置字典初始化它。在创建时，这段代码将设置录音机的委托并启用测量。AVAudioRecorder 实例的测量方式的工作原理像 AVAudioPlayer 实例的测量方式一样。在请求平均和峰值强度级别之前，必须更新测量仪：

```
- (void) updateMeters
{
    // Show the current power levels
    [self.recorder updateMeters];
    meter1.progress =
        pow(10, 0.05f * [recorder averagePowerForChannel:0]);
    meter2.progress =
        pow(10, 0.05f * [recorder peakPowerForChannel:0]);

    // Update the current recording time
    self.title = [NSString stringWithFormat:@"%
        [self formatTime:self.recorder.currentTime]];
}
```

这段代码还会跟踪录音的 currentTime。当暂停录音时，当前的时间将保持静止，直到恢复录音为止。基本上，当前的时间指示迄今为止的录音持续时间。

在准备好继续录音时，可以使用 prepareToRecord，并且利用 record 开始录音。发出 pause 可以暂停录音；再次调用 record 可以恢复录音。录音会选择它停止的地方。要完成录音，可使用 stop。这将产生一个对 audioRecorderDidFinishRecording:successfully:的回调，可以在这里清理界面，并完成任何录音细节。

要获取录音，可以在应用程序的 Info.plist 文件中把 UIFileSharingEnabled 键设置为 YES。这允许用户通过 iTunes 的特定的设备的 Apps 窗格访问所有录制的材料。

秘诀 8-4　利用 AVAudioRecorder 录制音频

```
- (void)audioPlayerDidFinishPlaying:(AVAudioPlayer *)player
    successfully:(BOOL)flag
{
    // Prepare UI for recording
    self.title = nil;
    meter1.hidden = NO;
    meter2.hidden = NO;
    {
        // Return to play and record session
        NSError *error;
        if (![[AVAudioSession sharedInstance]
            setCategory:AVAudioSessionCategoryPlayAndRecord
            error:&error])
        {
            NSLog(@"Error: %@", error.localizedFailureReason);
            return;
        }
        self.navigationItem.rightBarButtonItem =
            BARBUTTON(@"Record", @selector(record));
    }
}
- (void)audioRecorderDidFinishRecording:(AVAudioRecorder *)aRecorder
    successfully:(BOOL)flag
{
    // Stop monitoring levels, time
    [timer invalidate];
    meter1.progress = 0.0f;
    meter1.hidden = YES;
    meter2.progress = 0.0f;
    meter2.hidden = YES;
    self.navigationItem.leftBarButtonItem = nil;
    self.navigationItem.rightBarButtonItem = nil;

    NSURL *url = recorder.url;
    NSError *error;

    // Start playback
    player = [[AVAudioPlayer alloc] initWithContentsOfURL:url
```

```objc
        error:&error];
    if (!player)
    {
        NSLog(@"Error establishing player for %@: %@",
            url, error.localizedFailureReason);
        return;
    }
    player.delegate = self;

    // Change audio session for playback
    if (![[AVAudioSession sharedInstance]
        setCategory:AVAudioSessionCategoryPlayback
        error:&error])
    {
        NSLog(@"Error updating audio session: %@",
            error.localizedFailureReason);
        return;
    }

    self.title = @"Playing back recording...";
    [player prepareToPlay];
    [player play];
}

- (void) stopRecording
{
    // This causes the didFinishRecording delegate method to fire
    [recorder stop];
}

- (void) continueRecording
{
    // resume from a paused recording
    [recorder record];
    self.navigationItem.rightBarButtonItem =
        BARBUTTON(@"Done", @selector(stopRecording));
    self.navigationItem.leftBarButtonItem =
        SYSBARBUTTON(UIBarButtonSystemItemPause, self,
        @selector(pauseRecording));
}
```

```objc
- (void) pauseRecording
{
    // pause an ongoing recording
    [recorder pause];
    self.navigationItem.leftBarButtonItem =
        BARBUTTON(@"Continue", @selector(continueRecording));
    self.navigationItem.rightBarButtonItem = nil;
}

- (BOOL) record
{
    NSError *error;

    // Recording settings
    NSMutableDictionary *settings =
        [NSMutableDictionary dictionary];
    settings[AVFormatIDKey] = @(kAudioFormatLinearPCM);
    settings[AVSampleRateKey] = @(8000.0f);
    settings[AVNumberOfChannelsKey] = @(1); // mono
    settings[AVLinearPCMBitDepthKey] = @(16);
    settings[AVLinearPCMIsBigEndianKey] = @NO;
    settings[AVLinearPCMIsFloatKey] = @NO;

    // File URL
    NSURL *url = [NSURL fileURLWithPath:FILEPATH];

    // Create recorder
    recorder = [[AVAudioRecorder alloc] initWithURL:url
        settings:settings error:&error];
    if (!recorder)
    {
        NSLog(@"Error establishing recorder: %@",
            error.localizedFailureReason);
        return NO;
    }

    // Initialize delegate, metering, etc.
    recorder.delegate = self;
    recorder.meteringEnabled = YES;
```

```
            meter1.progress = 0.0f;
            meter2.progress = 0.0f;
            self.title = @"0:00";
            if (![recorder prepareToRecord])
            {
                NSLog(@"Error: Prepare to record failed");
                return NO;
            }

            if (![recorder record])
            {
                NSLog(@"Error: Record failed");
                return NO;
            }

            // Set a timer to monitor levels, current time
            timer = [NSTimer scheduledTimerWithTimeInterval:0.1f
                target:self selector:@selector(updateMeters)
                userInfo:nil repeats:YES];

            // Update the navigation bar
            self.navigationItem.rightBarButtonItem =
                BARBUTTON(@"Done", @selector(stopRecording));
            self.navigationItem.leftBarButtonItem =
                SYSBARBUTTON(UIBarButtonSystemItemPause, self,
                @selector(pauseRecording));

            return YES;
}

- (BOOL) startAudioSession
{
    // Prepare the audio session
    NSError *error;
    AVAudioSession *session = [AVAudioSession sharedInstance];

    if (![session setCategory:AVAudioSessionCategoryPlayAndRecord
            error:&error])
    {
        NSLog(@"Error setting session category: %@",
```

```
                error.localizedFailureReason);
            return NO;
        }

        if (![session setActive:YES error:&error])
        {
            NSLog(@"Error activating audio session: %@",
                error.localizedFailureReason);
            return NO;
        }
        return session.inputIsAvailable;
    }

    - (void) viewDidLoad
    {
        if ([self startAudioSession])
            self.navigationItem.rightBarButtonItem =
                BARBUTTON(@"Record", @selector(record));
        else
            self.title = @"No Audio Input Available";
    }
```

获取这个秘诀的代码

要查找这个秘诀的完整示例项目,可以浏览 https://github.com/erica/iOS-6-Advanced-Cookbook,并进入第 8 章的文件夹。

8.5 秘诀:利用 Audio Queues 录制音频

除了 AVAudioPlayer 类之外,Audio Queues 也可以在应用程序中处理录音和播放任务。在 AVAudioRecorder 类出现之前,需要 Audio Queues 来录音。直接使用队列有助于演示 AVAudioRecorder 类幕后所发生的事情。作为一种额外的好处,队列提供了对 HandleInputBuffer 中的原始底层音频的完全访问权限。你可以做各种有趣的事情,比如信号处理和分析。

秘诀 8-5 在 Audio Queue 级别录制音频,允许体验使用的 C 语言风格的函数和回调。这段代码在很大程度上基于 Apple 的示例代码,并且特别地展示了隐藏在 AVAudioRecorder 包装器背后的功能。

秘诀 8-5 的 setupAudioFormat:方法中使用的设置进行过测试,并且能够在 iOS 设备家族上可靠地工作。不过,在尝试自定义音频质量时,很容易混淆这些参数。如果没有正确地设置参数,队列可能会由于微不足道的反馈而失败。快速搜索一下 Google,即可提供丰富的设置示例。

秘诀 8-5　利用 Audio Queues 录音：Recorder.m 实现

```objc
// Write out current packets as the input buffer is filled
static void HandleInputBuffer (void *aqData,
    AudioQueueRef inAQ, AudioQueueBufferRef inBuffer,
    const AudioTimeStamp *inStartTime,
    UInt32 inNumPackets,
    const AudioStreamPacketDescription *inPacketDesc)
{
    RecordState *pAqData = (RecordState *) aqData;
    if (inNumPackets == 0 &&
        pAqData->dataFormat.mBytesPerPacket != 0)
        inNumPackets = inBuffer->mAudioDataByteSize /
            pAqData->dataFormat.mBytesPerPacket;

    if (AudioFileWritePackets(pAqData->audioFile, NO,
        inBuffer->mAudioDataByteSize, inPacketDesc,
        pAqData->currentPacket, &inNumPackets,
        inBuffer->mAudioData) == noErr)
    {
        pAqData->currentPacket += inNumPackets;
        if (pAqData->recording == 0) return;
        AudioQueueEnqueueBuffer (pAqData->queue, inBuffer,
            0, NULL);
    }
}
@implementation Recorder

// Set up the recording format as low quality mono AIFF
- (void)setupAudioFormat:(AudioStreamBasicDescription*)format
{
    format->mSampleRate = 8000.0;
    format->mFormatID = kAudioFormatLinearPCM;
    format->mFormatFlags = kLinearPCMFormatFlagIsBigEndian |
        kLinearPCMFormatFlagIsSignedInteger |
        kLinearPCMFormatFlagIsPacked;

    format->mChannelsPerFrame = 1; // mono
    format->mBitsPerChannel = 16;
    format->mFramesPerPacket = 1;
```

```objc
        format->mBytesPerPacket = 2;
        format->mBytesPerFrame = 2;
        format->mReserved = 0;
}


// Begin recording
- (BOOL) startRecording: (NSString *) filePath
{
    // Many of these calls mirror the process for AVAudioRecorder

    // Set up the audio format and the url to record to
    [self setupAudioFormat:&recordState.dataFormat];
    CFURLRef fileURL = CFURLCreateFromFileSystemRepresentation(
        NULL, (const UInt8 *) [filePath UTF8String],
        [filePath length], NO);
    recordState.currentPacket = 0;
// Initialize the queue with the format choices
OSStatus status;
status = AudioQueueNewInput(&recordState.dataFormat,
    HandleInputBuffer, &recordState,
    CFRunLoopGetCurrent(),kCFRunLoopCommonModes, 0,
    &recordState.queue);
if (status) {
    fprintf(stderr, "Could not establish new queue\n");
    return NO;
}


// Create the output file
status = AudioFileCreateWithURL(fileURL,
    kAudioFileAIFFType, &recordState.dataFormat,
    kAudioFileFlags_EraseFile, &recordState.audioFile);
if (status)
{
    fprintf(stderr, "Could not create file to record audio\n");
    return NO;
}


// Set up the buffers
DeriveBufferSize(recordState.queue, recordState.dataFormat,
    0.5, &recordState.bufferByteSize);
```

```
        for(int i = 0; i < NUM_BUFFERS; i++)
        {
            status = AudioQueueAllocateBuffer(recordState.queue,
                recordState.bufferByteSize, &recordState.buffers[i]);
            if (status) {
                fprintf(stderr, "Error allocating buffer %d\n", i);
                return NO;
            }
            status = AudioQueueEnqueueBuffer(recordState.queue,
                recordState.buffers[i], 0, NULL);
            if (status) {
                fprintf(stderr, "Error enqueuing buffer %d\n", i);
                return NO;
            }
        }


        // Enable metering
        UInt32 enableMetering = YES;
        status = AudioQueueSetProperty(recordState.queue,
            kAudioQueueProperty_EnableLevelMetering,
            &enableMetering,sizeof(enableMetering));
         if (status)
         {
             fprintf(stderr, "Could not enable metering\n");
             return NO;
         }


        // Start the recording
        status = AudioQueueStart(recordState.queue, NULL);
        if (status)
        {
            fprintf(stderr, "Could not start Audio Queue\n");
            return NO;
        }

        recordState.currentPacket = 0;
        recordState.recording = YES;
        return YES;
    }
```

```objc
// Return the average power level
- (CGFloat) averagePower
{
    AudioQueueLevelMeterState state[1];
    UInt32 statesize = sizeof(state);
    OSStatus status;
    status = AudioQueueGetProperty(recordState.queue,
        kAudioQueueProperty_CurrentLevelMeter, &state, &statesize);
    if (status)
    {
        fprintf(stderr, "Error retrieving meter data\n");
        return 0.0f;
    }
    return state[0].mAveragePower;
}

// Return the peak power level
- (CGFloat) peakPower
{
    AudioQueueLevelMeterState state[1];
    UInt32 statesize = sizeof(state);
    OSStatus status;
    status = AudioQueueGetProperty(recordState.queue,
        kAudioQueueProperty_CurrentLevelMeter, &state, &statesize);
    if (status)
    {
        fprintf(stderr, "Error retrieving meter data\n");
        return 0.0f;
    }
    return state[0].mPeakPower;
}

// There's generally about a one-second delay before the
// buffers fully empty
- (void) reallyStopRecording
{
    AudioQueueFlush(recordState.queue);
    AudioQueueStop(recordState.queue, NO);
    recordState.recording = NO;
```

```objc
            for(int i = 0; i < NUM_BUFFERS; i++)
            {
                AudioQueueFreeBuffer(recordState.queue,
                    recordState.buffers[i]);
            }
            
            AudioQueueDispose(recordState.queue, YES);
            AudioFileClose(recordState.audioFile);
}

// Stop the recording after waiting just a second
- (void) stopRecording
{
    [self performSelector:@selector(reallyStopRecording)
        withObject:NULL afterDelay:1.0f];
}

// Pause after allowing buffers to catch up
- (void) reallyPauseRecording
{
    if (!recordState.queue) {
        fprintf(stderr, "Nothing to pause\n"); return;}
    OSStatus status = AudioQueuePause(recordState.queue);
    if (status)
    {
        fprintf(stderr, "Error pausing audio queue\n");
        return;
    }
}

// Pause the recording after waiting a half second
- (void) pause
{
    [self performSelector:@selector(reallyPauseRecording)
        withObject:NULL afterDelay:0.5f];
}

// Resume recording from a paused queue
- (BOOL) resume
{
```

```objc
        if (!recordState.queue)
        {
            fprintf(stderr, "Nothing to resume\n");
            return NO;
        }

        OSStatus status = AudioQueueStart(recordState.queue, NULL);
        if (status)
        {
            fprintf(stderr, "Error restarting audio queue\n");
            return NO;
        }
        return YES;
}


// Return the current recording duration
- (NSTimeInterval) currentTime
{
    AudioTimeStamp outTimeStamp;
    OSStatus status = AudioQueueGetCurrentTime (
        recordState.queue, NULL, &outTimeStamp, NULL);
    if (status)
    {
        fprintf(stderr, "Error: Could not retrieve current time\n");
        return 0.0f;
    }

    // 8000 samples per second
    return outTimeStamp.mSampleTime / 8000.0f;
}

// Return whether the recording is active
- (BOOL) isRecording
{
    return recordState.recording;
}
@end
```

获取这个秘诀的代码

要查找这个秘诀的完整示例项目，可以浏览 https://github.com/erica/iOS-6-Advanced-Cookbook，并进入第 8 章的文件夹。

8.6 秘诀：利用 MPMediaPickerController 选择音频

MPMediaPickerController 类提供了与 UIImagePickerController 类的图像选择功能对应的音频功能。它允许用户从他们的音乐库中选择一个或多个项目，包括音乐、播客和音频书籍。iPod 风格的界面允许用户通过播放列表、艺术家、歌曲、专辑等浏览它们。

要使用这个类，可以分配一个新的选择器，并利用要使用的媒体种类初始化它。对于音频，可以选择 MPMediaTypeMusic、MPMediaTypePodcast、MPMediaTypeAudioBook 和 MPMediaType-AnyAudio。视频类型也可用于电影、电视剧、视频播客、音乐视频和 iTunesU。这些都是标志，可以把它们组合在一起构成一个掩码：

```
MPMediaPickerController *mpc = [[MPMediaPickerController alloc]
    initWithMediaTypes:MPMediaTypeAnyAudio];
mpc.delegate = self;
mpc.prompt = @"Please select an item";
mpc.allowsPickingMultipleItems = NO;
[self presentViewController:mpc animated:YES];
```

图 8-2 在这个多项选择媒体选择器中，将以灰色显示已经选择的项目。选择完成后，可以点按 Done 按钮。在正常的选择器元素上方将出现一个可选的提示区域（在这里是"Please Select an Item"）

设置一个符合 MPMediaPickerControllerDelegate 的委托，并且可以选择设置一个提示。该提示是出现在媒体选择器顶部的文本，如图 8-2 中所示。当选择允许进行多项选择时，标准选择器上的 Cancel 按钮将被标签 Done 所替代。通常，对话框会终结于用户点按乐曲时。利用多项选择，用户可以保持选择项目，直到他们按下 Done 按钮为止。选择的项目将更新为使用灰色文本。

mediaPicker:didPickMediaItems: 委托回调将处理用户选择的完成。作为参数传递的 MPMedia-ItemCollection 实例可以通过访问它的 items 来进行枚举。每个项目都是 MPMediaItem 类的成员，可以查询它的属性，如秘诀 8-6 中所示。秘诀 8-6 使用媒体选择器来选择多首乐曲。它将按艺术家和名称登记用户选择的项目。

参考 Apple 的 MPMediaItem 类文档，了解媒体项目的可用属性、它们返回的类型，以及它们是否可用于构造媒体属性谓词。在下一节中将讨论如何构建查询以及使用谓词。

记住：媒体项目属性不是 Objective-C 属性。必须使用 valueForProperty:方法获取它们，尽管很容易创建一个包装器类，该包装器类可以通过 Objective-C 方法调用把这些项目提供为真正的属性。

秘诀 8-6　从 iPod 库中选择音乐项目

```
- (void)mediaPicker: (MPMediaPickerController *)mediaPicker
    didPickMediaItems:(MPMediaItemCollection *)mediaItemCollection
{
    for (MPMediaItem *item in [mediaItemCollection items])
        NSLog(@"[%@] %@",
            [item valueForProperty:MPMediaItemPropertyArtist],
            [item valueForProperty:MPMediaItemPropertyTitle]);
    [self performDismiss];
}


- (void)mediaPickerDidCancel:(MPMediaPickerController *)mediaPicker
{
    if (IS_IPHONE)
        [self dismissViewControllerAnimated:YES completion:nil];
}
// Popover was dismissed
- (void)popoverControllerDidDismissPopover:
    (UIPopoverController *)aPopoverController
{
    popoverController = nil;
}


// Start the picker session
- (void) action: (UIBarButtonItem *) bbi
{
    MPMediaPickerController *mpc =
        [[MPMediaPickerController alloc]
            initWithMediaTypes:MPMediaTypeMusic];
    mpc.delegate = self;
    mpc.prompt = @"Please select an item";
    mpc.allowsPickingMultipleItems = YES;
    [self presentController:mpc];
}
```

> **获取这个秘诀的代码**
>
> 要查找这个秘诀的完整示例项目，可以浏览 https://github.com/erica/iOS-6-Advanced-Cookbook，并进入第 8 章的文件夹。

8.7 创建媒体查询

Media Queries 允许过滤 iPod 库内容，从而限制搜索的范围。表 8-1 列出了 MPMediaQuery 提供的 9 个类方法，它们可用于预定义的搜索。每种查询类型都会控制返回的数据组合，每个集合都按专辑、艺术家和音频书籍等组织为乐曲。

表 8-1 查询类型

类方法	是全局方法吗	过滤器	组类型
albumsQuery	否	MPMediaTypeMusic	MPMediaGroupingAlbum
artistsQuery	否	MPMediaTypeMusic	MPMediaGroupingArtist
audiobooksQuery	否	MPMediaTypeAudioBook	MPMediaGroupingTitle
compilationsQuery	否	MPMediaTypeAny | MPMediaItemPropertyIsCompilation	MPMediaGroupingAlbum
composersQuery	是	MPMediaTypeAny	MPMediaGroupingComposer
genresQuery	是	MPMediaTypeAny	MPMediaGroupingGenre
playlistsQuery	是	MPMediaTypeAny	MPMediaGroupingPlaylist
podcastsQuery	否	MPMediaTypePodcast	MPMediaGroupingPodcastTitle
songsQuery	否	MPMediaTypeMusic	MPMediaGroupingTitle

这种方法反映了 iTunes 在桌面上的工作方式。在 iTunes 中，选择一列用于组织结果，但是通过在应用程序的 Search（搜索）框中输入文本来执行搜索。

8.7.1 构建查询

使用专辑查询统计库中的专辑数量。下面这个代码段用于创建该查询，然后获取一个数组，其中每个项目都代表单独一个专辑。这些专辑项目是各个媒体项目的集合，一个集合可能包含一首或多首乐曲：

```
MPMediaQuery *query = [MPMediaQuery albumsQuery];
NSArray *collections = query.collections;
NSLog(@"You have %d albums in your library\n", collections.count);
```

许多 iOS 用户都具有大量的媒体集合，它们通常包含成百上千个专辑，更不用说单独的乐曲了。像这样的简单查询可能要花几秒钟的时间来运行，并且返回一个表示整个库的数据结构。

使用不同查询类型的搜索将返回按该类型组织的集合。可以使用类似的方法获取艺术家、歌曲、作曲家等的数量。

8.7.2　使用谓词

媒体属性谓词可以有效地过滤通过查询返回的项目。例如，你可能只想查找那些名称与短语"road"匹配的歌曲。下面的代码段创建了一个新的歌曲查询，并添加了一个过滤器谓词，用于搜索那个短语。该谓词是利用一个值（搜索短语）、一个属性（搜索歌曲名称）和一个比较类型（在这里是"包含"）。要实现精确匹配，可以使用 MPMedia-PredicateComparisonEqualTo；要实现子串匹配，可以使用 MPMediaPredicate-ComparisonContains。

```
MPMediaQuery *query = [MPMediaQuery songsQuery];

// Construct a title comparison predicate
MPMediaPropertyPredicate *mpp = [MPMediaPropertyPredicate
    predicateWithValue:@"road"
    forProperty:MPMediaItemPropertyTitle
    comparisonType:MPMediaPredicateComparisonContains];
[query addFilterPredicate:mpp];

// Recover the collections
NSArray *collections = query.collections;
NSLog(@"You have %d matching tracks in your library\n",
    collections.count);

// Iterate through each item, logging the song and artist
for (MPMediaItemCollection *collection in collections)
{
    for (MPMediaItem *item in [collection items])
    {
        NSString *song = [item valueForProperty:
            MPMediaItemPropertyTitle];
        NSString *artist = [item valueForProperty:
            MPMediaItemPropertyArtist];
        NSLog(@"%@, %@", song, artist);
    }
}
```

> **注意：**
> 如果你更愿意对媒体集合使用常规的谓词，而不是媒体属性谓词，我为此创建了一个 MPMediaItem 属性类别（http://github.com/erica/MPMediaItemProperties）。这个类别允许对集合应用标准的 NSPredicate 查询，比如那些由多项选择器返回的集合。

8.8 秘诀：使用 MPMusicPlayerController

Cocoa Touch 包括一个简单易用的音乐播放器类，它可以完美地处理媒体集合。MPMusicPlayerController 类不是一个视图控制器，尽管其名称暗示了这层意思。它没有提供用于播放音乐的屏幕元素。作为替代，它提供了一个抽象的控制器，用于处理音乐的播放和暂停。

当它的播放状态改变时，它将发布可选的通知。这个类提供了两个共享实例：iPodMusicPlayer 和 applicationMusicPlayer，总是要使用前者，它提供了可靠的状态改变反馈，你将希望以编程方式捕获它。

利用一个 MPMediaItemCollection 调用 setQueueWithItemCollection:，初始化播放器控制器：

```
[[MPMusicPlayerController iPodMusicPlayer]
    setQueueWithItemCollection: songs];
```

此外，还可以利用媒体查询来加载队列。例如，你可能设置一个与特定艺术家短语或艺术家查询匹配的 playlistsQuery，用于按给定的艺术家搜索歌曲。可以使用 setQueueWithQuery: 通过一个 MPMediaQuery 实例生成队列。

如果你想打乱播放的次序，可以给控制器的 shuffleMode 属性赋予一个值。可以选择 MPMusicShuffleModeDefault（它遵循用户的当前设置）、MPMusicShuffleModeOff（不打乱次序）、MPMusicShuffleModeSongs（打乱歌曲的次序）或 MPMusicShuffleModeAlbums（打乱专辑的次序）。对于音乐的 repeatMode，存在一组类似的选项。

在设置项目集合之后，可以播放、暂停或者跳过队列中的下一个项目，回到前一个项目，等等。要在不回到前一个项目的情况下倒带，可以发出 skipToBeginning。还可以在当前的播放项目内定位，前移或后移播放点。

秘诀 8-7 提供了一个简单的媒体播放器，显示了当前正在播放的歌曲（以及它的艺术作品，如果有的话）。在运行时，用户使用 MPMediaPickerController 选择一组项目。这个项目集合将返回并分配给播放器，它将开始播放这组项目。

一对观察者使用默认的通知中心来监控两个关键的改变：在当前项目改变时以及在播放状态改变时。要监控这些改变，必须手动请求通知。这允许在播放项目改变时利用新的"现在播放"信息更新界面：

```
[[MPMusicPlayerController iPodMusicPlayer]
    beginGeneratingPlaybackNotifications];
```

可以通过发出 endGeneratingPlaybackNotifications 来撤销这个请求，或者当应用程序自然地挂起或终止时可以简单地允许程序取消所有的观察者。由于这个秘诀使用了 iPod 音乐播放器，在离开应用程序后播放将会继续，除非明确地停止它。播放不会受到应用程序挂起或关闭的影响：

```
- (void) applicationWillResignActive: (UIApplication *) application
{
    // Stop player when the application quits
    [[MPMusicPlayerController iPodMusicPlayer] stop];
}
```

除了演示播放控制之外，秘诀 8-7 还展示了如何在播放期间显示专辑艺术。它使用了前一个秘诀中使用的相同类型的 MPItem 属性检索。在这里，它将查询 MPMediaItemPropertyArtwork，如果找到艺术作品，它将使用 MPMediaItemArtwork 把该艺术作品转换为给定大小的图像。

秘诀 8-7 利用 iPod 音乐播放器进行简单的媒体播放

```
#define PLAYER [MPMusicPlayerController iPodMusicPlayer]

#pragma mark PLAYBACK
- (void) pause
{
    // Pause playback
    [PLAYER pause];
    toolbar.items = [self playItems];
}

- (void) play
{
    // Restart play
    [PLAYER play];
    toolbar.items = [self pauseItems];
}

- (void) fastforward
{
    // Skip to the next item
    [PLAYER skipToNextItem];
}
```

```objc
- (void) rewind
{
    // Skip to the previous item
    [PLAYER skipToPreviousItem];
}


#pragma mark STATE CHANGES
- (void) playbackItemChanged: (NSNotification *) notification
{
    // Update title and artwork
    self.title = [PLAYER.nowPlayingItem
        valueForProperty:MPMediaItemPropertyTitle];
    MPMediaItemArtwork *artwork = [PLAYER.nowPlayingItem
        valueForProperty: MPMediaItemPropertyArtwork];
    imageView.image = [artwork imageWithSize:[imageView frame].size];
}


- (void) playbackStateChanged: (NSNotification *) notification
{
    // On stop, clear title, toolbar, artwork
    if (PLAYER.playbackState == MPMusicPlaybackStateStopped)
    {
        self.title = nil;
        toolbar.items = nil;
        imageView.image = nil;
    }
}


#pragma mark MEDIA PICKING
- (void)mediaPicker: (MPMediaPickerController *)mediaPicker
    didPickMediaItems:(MPMediaItemCollection *)mediaItemCollection
{
    // Set the songs to the collection selected by the user
    songs = mediaItemCollection;

    // Update the playback queue
    [PLAYER setQueueWithItemCollection:songs];

    // Display the play items in the toolbar
    [toolbar setItems:[self playItems]];
```

8.8 秘诀：使用 MPMusicPlayerController

```objc
    // Clean up the picker
    [self performDismiss];
;
}

- (void)mediaPickerDidCancel:(MPMediaPickerController *)mediaPicker
{
    // User has canceled
    [self performDismiss];
}

- (void) pick: (UIBarButtonItem *) bbi
{
    // Select the songs for the playback queue
    MPMediaPickerController *mpc = [[MPMediaPickerController alloc]
        initWithMediaTypes:MPMediaTypeMusic];
    mpc.delegate = self;
    mpc.prompt = @"Please select items to play";
    mpc.allowsPickingMultipleItems = YES;
    [self presentViewController:mpc];
}

#pragma mark INIT VIEW
- (void) viewDidLoad
{
    self.navigationItem.rightBarButtonItem = BARBUTTON(@"Pick",
        @selector(pick));
    // Stop any ongoing music
    [PLAYER stop];

    // Add observers for state and item changes
    [[NSNotificationCenter defaultCenter] addObserver:self
        selector:@selector(playbackStateChanged)
        name:MPMusicPlayerControllerPlaybackStateDidChangeNotification
        object:PLAYER];
    [[NSNotificationCenter defaultCenter] addObserver:self
        selector:@selector(playbackItemChanged)
        name:MPMusicPlayerControllerNowPlayingItemDidChangeNotification
        object:PLAYER];
```

```
    [PLAYER beginGeneratingPlaybackNotifications];
}
@end
```

> **获取这个秘诀的代码**
> 要查找这个秘诀的完整示例项目，可以浏览 https://github.com/erica/iOS-6-Advanced-Cookbook，并进入第 8 章的文件夹。

8.9 小结

本章介绍了用于处理音频媒体的许多方式，包括播放和录制。你看到一些秘诀使用了高级的 Objective-C 类，另外一些秘诀则使用了低级的 C 函数。你了解了媒体选择器、控制器等。在结束本章的学习之前，要思考以下几点。

- Apple 仍然在构建它的媒体播放类。随着时间的推移，它们正变得越来越强大。由于时间和篇幅限制，本章没有涉及许多给 iOS 上的音频提供动力的技术，包括：AVFoundation、OpenAL、Core Audio、Audio Units 或 Core MIDI。这些主题都很庞大，均能单独成书。有关它们的更多信息，参见 Chris Adamson 和 Kevin Avila 的图书 *Learning Core Audio: A Hands-On Guide to Audio Programming for Mac and iOS*（Addison-Wesley Professional）。

- 本章没有讨论 ALAssetsLibrary，尽管在本书中别的位置和 Core Cookbook 中介绍了使用这个类的示例。你可能想深入研究 MPMediaItemPropertyAssetURL 和 AVAssetExportSession 类，以便处理音频库资产。

- Audio Queue 提供了强大的低级音频例程，但是只有强者才能驾驭它们，它们也不适合于任何只想获得一个快速解决方案的人。如果需要 Audio Queue 引入的细粒度的音频控制，Apple 提供了大量的文档，可以帮助你实现自己的目标。

- 在使用录制特性创建本地资产时，确保在 Info.plist 文件中启用 UIFileSharingEnabled，以允许用户访问和管理他们通过 iTunes 创建的文件。

- MPMusicPlayerController 类介绍了一种简单的方式，用于同机载 iTunes 库中的音乐交互。确保对 AVAudioPlayer 和 MPMusicPlayerController 了如指掌，其中前者用于本地数据文件，后者则可以与用户的 iTunes 媒体集合交互。

第 9 章

连接到 Address Book

除了在任何计算机上看到的标准用户界面控件和媒体组件之外，iOS SDK 还提供了许多特定于 iOS 递送的受到紧密关注的开发人员解决方案。其中最有用的是 AddressBook 框架，它允许以编程方式访问和管理联系人数据库。本章将介绍 Address Book 并将演示如何在应用程序中使用它的框架。你将学到如何访问各个联系人的信息，如何修改和更新联系人信息，以及如何使用谓词只查找感兴趣的联系人。本章还将介绍 GUI 类，它们提供了交互式解决方案，用于挑选、查看和修改联系人。通过阅读本章，可以自下而上地了解 Address Book。

9.1 AddressBook 框架

iOS SDK 提供的不是一个而是两个 AddressBook 框架：AddressBook.framework 和 AddressBookUI.framework。顾名思义，它们在 iOS SDK 中占据着截然不同的位置。AddressBook 提供低级的 Core Foundation（基于 C 语言）对象和例程，用于从 iPhone 的机载数据库中访问联系人信息。AddressBookUI 则提供了高级的基于 Objective-C 的 UIViewController 浏览器对象，用于展示给用户。这两个框架都比较小，它们只提供了少数几个类和数据类型。

9.1.1 AddressBookUI

AddressBookUI 框架提供了几个预建的视图控制器，可以与机载联系人数据库交互。这些界面包括常规的人员选择器、联系人查看器和联系人编辑器。可以设置一个委托，然后把这些控制器压入到导航栈上或者以模态方式显示它们，如本章中的秘诀所示。

像这一整本书中讨论的其他专用的控制器一样，AddressBookUI 控制器不是通用的，也不灵活。Apple 打算让你像提供的那样使用它们，开发人员只需做很少的自定义工作或者根本不进行自定义。更重要的是，它们需要某种程度的低级编程技能。如你将在本章中所看到的，这些类以迂回的方式与底层的 Address Book 交互。

9.1.2 AddressBook 及其数据库

AddressBook 框架提供了对 iOS 的 Contacts 数据库的集中式访问。这个数据库存储关于各个联系人的信息，包括姓名、地址、电话号码、IM 详细信息等。如果使用 Core Foundation 风格的 API 构建，AddressBook 框架将允许从应用程序中查询、修改和添加联系人。

iOS 联系人数据驻留在用户的 Library 根文件夹中。在基于 Macintosh 的 iPhone 模拟器上，可以在~/Library/Application Support/iPhone Simulator/ iOS Version //Library/AddressBook 中自由地访问这些文件。你在这里找到的文件（确切地讲是 AddressBook.sqlitedb 和 AddressBookImages.sqlitedb）使用标准的 SQLite 来存储联系人信息，并且在后一个文件中，还可以选择存储联系人图像。

在 iOS 上，相同的文件存在于/var/mobile/Library/AddressBook 中，也就是说，位于应用程序沙盒之外，从而不能够直接检查它们。使用两个 AddressBook 框架，可以查询或修改用户的联系人信息，而不要尝试通过代码访问这些文件。

> **注意：**
> 默认情况下，从 OS X 10.7 开始，用户的 Library 文件夹就隐藏在 Finder 中。可以通过 Terminal 或者在使用 Finder 的 Go 菜单时按住 Option 键来访问它。此外，还可以在 Terminal 中输入 "chflags nohidden ~/Library/"，使该文件夹在 Finder 中可见。

9.1.3 记录

在 Core Foundation 基于 C 语言的 AddressBook 框架中，ABRecordRef 类型提供了核心联系人结构。这个记录将存储每个联系人的所有信息，包括姓名、电子邮件、电话号码和可选的图像等。每条记录都对应于一个完整的 Address Book 联系人，通过调用 ABPersonCreate()创建新记录：

```
ABRecordRef person = ABPersonCreate();
```

每条记录都存储在单个中心 Address Book 中。可以创建新的 Address Book 实例来访问这种数据，在 iOS 6 中操作方法已经发生了改变。在 iOS 6 之前，应该调用 ABAddressBookCreate()。现在，则要调用 ABAddressbookCreateWithOptions()，它接受两个参数。第一个参数是预留的，因此传递 NULL；第二个参数是一个 CFErrorRef 指针，例如，ABAddressBookCreateWithOptions(NULL, &errorRef)。

万一创建请求失败，可以轻松地把 CFErrorRef 强制转换为一个 NSError，以确定失败的原因。

尽管 ABAddressbookCreateWithOptions()函数具有这个名称，但它不会创建新的 Address Book；它将构建一个本地对象，从系统 Address Book 中加载它的数据。这个函数提供了主进入点，让你能够直接访问 Address Book 数据。更重要的是，直到创建 Address Book 或记录实例时，需要在应用程序中使用的许多常量仍然是未定义的。

　　Apple 在它的文档中与下面这段话一起提出了许多警告："直到调用了以下函数之一后才会定义这些常量的值：ABAddressBookCreate()、ABPersonCreate()、ABGroupCreate()。"

9.1.4　自定义的 ABStandin 类

　　无需每次访问数据时都创建一个 Address Book 实例，本章中的示例使用一个自定义的 ABStandin 类。这个类提供了几个基本的 Address Book 访问功能，并且存储一个静态共享的 ABAddressBookRef 实例。其实现如下：

```
// Shared reference
static ABAddressBookRef shared = NULL;

@implementation ABStandin

// C-function callback updates address book when changed
void addressBookUpdated(ABAddressBookRef reference,
    CFDictionaryRef dictionary, void *context)
{
    ABAddressBookRevert(reference);
}

// Return the current address book
+ (ABAddressBookRef) addressBook
{
    if (shared) return shared;

    // Create the new address book
    CFErrorRef errorRef;
    shared = ABAddressBookCreateWithOptions(NULL, &errorRef);
    if (!shared)
    {
        NSError *error = (__bridge_transfer NSError *)errorRef;
        NSLog(@"Error creating new address book object: %@",
            error.localizedFailureReason);
        return nil;
    }
    // Register for automatic updates when information changes
    ABAddressBookRegisterExternalChangeCallback(
        shared, addressBookUpdated, NULL);
```

```objc
    return shared;
}

// Load the current address book with updates
+ (ABAddressBookRef) currentAddressBook
{
    if (!shared)
        return [self addressBook];

    ABAddressBookRevert(shared);
    return shared;
}

// Thanks Frederic Bronner. Save the address book out
+ (BOOL) save: (NSError **) error
{
    CFErrorRef errorRef;
    if (shared)
    {
        BOOL success = ABAddressBookSave(shared, &errorRef);
        if (!success)
        {
            if (error)
                *error = (__bridge_transfer NSError *)errorRef;
            return NO;
        }
        return YES;
    }
    return NO;
}

// Test authorization status
+ (BOOL) authorized
{
    ABAuthorizationStatus status =
        ABAddressBookGetAuthorizationStatus();
    return (status == kABAuthorizationStatusAuthorized);
}

// Place access request
```

```objc
+ (void) requestAccess
{
    // Post a notification that the app is authorized
    if ([self authorized])
    {
        NSNotification *note = [NSNotification notificationWithName:
            kAuthorizationUpdateNotification object:@YES];
        [[NSNotificationCenter defaultCenter] postNotification:note];
        return;
    }

    // Build a completion handler for the next step
    ABAddressBookRequestAccessCompletionHandler handler =
    ^(bool granted, CFErrorRef errorRef){
        // Respond to basic error condition
        if (errorRef)
        {
            NSError *error = (__bridge NSError *) errorRef;
            NSLog(@"Error requesting Address Book access: %@",
                error.localizedFailureReason);
            return;
        }

        // Post notification on main thread for success/fail
        dispatch_async(dispatch_get_main_queue(), ^{
            NSNotification *note =
                [NSNotification notificationWithName:
                    kAuthorizationUpdateNotification
                    object:@(granted)];
            [[NSNotificationCenter defaultCenter]
                postNotification:note];
        });
    };
    ABAddressBookRequestAccessWithCompletion(shared, handler);
}

+ (void) showDeniedAccessAlert
{
    UIAlertView *alertView = [[UIAlertView alloc]
        initWithTitle:@"Cannot Access Contacts"
```

```
             message:@"Please enable in Settings > Privacy > Contacts."
             delegate:nil cancelButtonTitle:nil
             otherButtonTitles:@"Okay", nil];
        dispatch_async(dispatch_get_main_queue(), ^{
            [alertView show];
        });
    }
@end
```

在本书以前的版本中,Address Book 对象是动态创建的并且会自动释放。这一版中的代码假定单个静态实例总是可用,无论何时 Address Book 数据改变,它都会自动更新。一个名为 save: 的方法将利用对内存中的共享式 Address Book 实例所做的更新来更新持久的 Address Book 数据存储。

对这一版的重大改变表现在新的 iOS 6 限制上,它将限制访问用户的联系人。用户必须明确允许访问他们的 Address Book,并且一般不会再次询问他们何时会这样做(或者不这样做)。如果用户拒绝了访问(可以利用 ABAddressBookGetAuthorizationStatus()测试),必须指导他们如何授权访问。在这里,showDeniedAccessAlert 方法将把他们送往 Settings(设置)应用程序。

总是要提供某种指导,以使用户能够处理以前拒绝的权限,并且永远不要假定他们知道如何靠他们自己完成这个任务。用户会由于各种理由拒绝访问:错误点按的按钮、常规的不信任,或者由于他们没有注意到要求他们做什么。Apple 没有提供一个"再次询问"API。作为替代,可以让应用程序在启动时对当前的状态做出反应。

只有在这时才需要执行检查。如果应用程序在后台运行,更新 Settings 中的权限将自动使之停止。它将重新启动,而不是在下一次用户访问它时恢复运行。这个细节确保永远不会出现如下情形(至少在 iOS 6 中是这样):在应用程序运行以响应权限中的变化时必须更新 GUI。Apple 可能在将来改变这种行为。

这个类使用一个通知系统,用于更新应用程序的当前授权状态。这允许使用当前的状态来启用或禁用受影响的 GUI 元素。例如,当应用程序没有授权响应那些请求时,你将不希望提供一个 Add Contact(添加联系人)按钮:

```
[[NSNotificationCenter defaultCenter]
    addObserverForName:kAuthorizationUpdateNotification
    object:nil queue:[NSOperationQueue mainQueue]
    usingBlock:^(NSNotification *note)
{
    NSNumber *granted = note.object;
    [self enableGUI:granted.boolValue];
}];
```

9.1.5 查询 Address Book

可以使用单个函数调用，在 Address Book 的数据库中查询当前存储的对象数量。尽管名称指的是 Persons，但是这个方法可以统计所有的条目（包括业务），并返回一个简单的计数。

```
+ (int) contactsCount
{
    ABAddressBookRef addressBook = [ABStandin addressBook];
    return ABAddressBookGetPersonCount(addressBook);
}
```

9.1.6 包装 AddressBook 框架

由于 AddressBook 框架是基于 C 语言的，如果你更习惯使用 Objective-C，那么这个框架可能过于繁琐并且难以使用。为了应对这种情况，本章介绍了一个名为 ABContact 的简单的包装器。这个包装器类隐藏了类的 CF 性质，并利用一个 Objective-C 接口替换它。Objective-C 包装器简化了基于 C 语言的 Address Book 调用与正常的 Cocoa Touch 开发和内存管理之间的集成。在用于本章的示例代码库中可以找到它的完整来源，以及其他关联的包装器类。

ABContact 类隐藏了一个自动合成的内部 ABRecordRef，即与每条联系人记录对应的 CF 类型。包装器的其余部分只涉及生成一些属性和方法，使你能够进入 ABRecordRef，设置和访问它的子记录：

```
@interface ABContact : NSObject
@property (nonatomic, readonly) ABRecordRef record;
@end
```

第二个辅助类 ABContactsHelper 在全局级别包装 AddressBook 框架。它允许作为一个整体修改和查询数据库。例如，采用它的 contacts 方法。与记录统计一样，Address Book 也可以返回联系人的数组。可以通过调用 ABAddressBookCopyArrayOfAllPeople() 函数来获取单独的记录，它将返回一个 ABRecordRef 对象的 CF 数组。

下面的 ABContactsHelper 方法包装这个函数来获取一个记录数组。然后，它将把每条联系人记录添加到新的 ABContact 包装器中，并且返回一个利用 Objective-C 联系人实例填充的标准 NSArray：

```
+ (NSArray *) contacts
{
    ABAddressBookRef addressBook = [ABStandin addressBook];
    NSArray *thePeople =
        (__bridge_transfer NSArray *)
```

```
            ABAddressBookCopyArrayOfAllPeople(addressBook);
    NSMutableArray *array = [NSMutableArray array];
    for (id person in thePeople)
        [array addObject:[ABContact
            contactWithRecord:(__bridge ABRecordRef)person]];
    return array;
}
```

9.1.7 使用记录函数

绝大多数 ABRecordRef 函数都使用 ABPerson 前缀。这个前缀对应于 Macintosh 上提供的 ABPerson 类，但是在 iOS 上则没有提供它。因此，尽管函数调用是以 ABPerson 为中心的，但是受这些调用影响的所有数据实际上都是 ABRecordRef 实例。

当你注意到在 AddressBook 框架中使用相同的 ABRecordRef 结构来表示人员（各个联系人，无论是人员还是企业）和群组（联系人的集合，比如工作中的同事和私人朋友）时，这样做的原因就变得更清楚了。SDK 提供了 ABGroup 函数和 ABPerson 函数，使你能够添加成员和访问成员列表。在本节后面将学习关于群组的更多知识。

在本节前面你已经遇到过 ABPersonCreate()函数。如其前缀所指示的，它用于创建联系人记录。SDK 提供了差不多 24 个其他的以 ABPerson 命名的函数以及少量 ABGroup 和 ABRecord 函数。结合使用这些函数，允许设置和获取联系人内的值、比较记录、测试符合的数据等。

9.1.8 获取和设置字符串

每个 ABRecord 都会存储差不多 12 个简单的字符串值，用于表示人员的姓名、头衔、职业和组织。可以通过复制记录中的字段值来获取**单值字符串**（single-valued string）。单值意味着一个人只能拥有其中一个项目。一个人只具有一个名字、一个生日和一个昵称——至少就 AddressBook 框架来说是这样。它们与电话号码、电子邮件地址和社交网络身份形成了鲜明的对比，后面这些项目称为**多值**（multivalued）项目，可以自然地出现多次。

1. 获取字符串

如你所期望的，处理单值元素要比多值元素简单，只有一个项目要设置和获取。所有的单值元素都可以使用相同的 ABRecordCopyValue()函数访问。ABContact 包装器基于返回的数据类型特殊化了这种访问，比如用于字符串或日期。下面的方法将复制一个记录值，把它强制转换为字符串，然后返回此内容：

```
- (NSString *) getRecordString:(ABPropertyID) anID
{
```

```
    NSString *result =
        (__bridge_transfer NSString *)
            ABRecordCopyValue(_record, anID);
    return result;
}
```

要确定复制哪个值,将需要一个属性常量(ABPropertyID),用于标识记录中所请求的字段。下面的示例调用为名字和姓氏字段使用了标识符:

```
// Sample uses
- (NSString *) firstname
    {return [self getRecordString:kABPersonFirstNameProperty];}
- (NSString *) lastname
    {return [self getRecordString:kABPersonLastNameProperty];}
```

2. 字符串字段

下面列出了可以用这种方式获取的基于字符串的字段,它们的名称是自解释的。这些标识符在 ABPerson.h 头文件中被声明为常量整数,并且在 iOS 的更新换代过程中现在仍然保持稳定。每个项目都标识 ABRecordRef 记录中的单个基于字符串的字段:

- kABPersonFirstNameProperty
- kABPersonLastNameProperty
- kABPersonMiddleNameProperty
- kABPersonPrefixProperty
- kABPersonSuffixProperty
- kABPersonNicknameProperty
- kABPersonFirstNamePhoneticProperty
- kABPersonLastNamePhoneticProperty
- kABPersonMiddleNamePhoneticProperty
- kABPersonOrganizationProperty
- kABPersonJobTitleProperty
- kABPersonDepartmentProperty
- kABPersonNoteProperty

3. 设置字符串属性

事实证明,设置基于字符串的属性像获取它们一样简单。把你想设置的字符串强制转换为一个 CFStringRef。需要结合使用 Core Foundation 对象以及 Address Book 调用,然后使用

ABRecordSetValue()把数据存储到记录中。

这些调用不会更新 Address Book,它们只会改变记录内的数据。如果想存储用户的联系人信息,必须把该信息写回到 Address Book 中。如本节前面所示,可以使用 save:方法:

```
- (BOOL) setString: (NSString *) aString
    forProperty:(ABPropertyID) anID
{
    CFErrorRef errorRef = NULL;
    BOOL success = ABRecordSetValue(_record, anID,
        (__bridge CFStringRef) aString, &errorRef);
    if (!success)
    {
        NSError *error = (__bridge_transfer NSError *) errorRef;
        NSLog(@"Error: %@", error.localizedFailureReason);
    }
    return success;
}
// Examples of use
- (void) setFirstname: (NSString *) aString
    {[self setString: aString forProperty:
        kABPersonFirstNameProperty];}

- (void) setLastname: (NSString *) aString
    {[self setString: aString
        forProperty: kABPersonLastNameProperty];}
```

9.1.9 处理日期属性

除了你刚才看到的字符串属性之外,Address Book 还会存储 3 个关键的日期:可选的生日、创建记录的日期和上一次修改记录的日期。这些项目使用以下属性常量:

- kABPersonBirthdayProperty
- kABPersonCreationDateProperty
- kABPersonModificationDateProperty

可以用与字符串完全相同的方式访问这些项目,但是要强制转换的是 **NSDate** 实例,而不是 **NSString** 实例。不要尝试修改后两个属性,要允许 Address Book 为你处理它们。

```
// Return a date-time field from a record
- (NSDate *) getRecordDate:(ABPropertyID) anID
{
    return (__bridge_transfer NSDate *)
```

```objc
        ABRecordCopyValue(_record, anID);
}

// Get the contact's birthday
- (NSDate *) birthday
    {return [self getRecordDate:kABPersonBirthdayProperty];}

// Set a date-time field in a record
- (BOOL) setDate: (NSDate *) aDate forProperty:(ABPropertyID) anID
{
    CFErrorRef errorRef = NULL;
    BOOL success = ABRecordSetValue(_record, anID,
        (__bridge CFDateRef) aDate, &errorRef);
    if (!success)
    {
        NSError *error = (__bridge_transfer NSError *) errorRef;
        NSLog(@"Error: %@", error.localizedFailureReason);
    }
    return success;
}

// Set the contact's birthday
- (void) setBirthday: (NSDate *) aDate
    {[self setDate: aDate forProperty: kABPersonBirthdayProperty];}
```

9.1.10 多值记录属性

每个人都可能具有与他或她的联系人相关联的多个电子邮件地址、多个电话号码和多个重要的日期（除了生日单例之外）。ABPerson 使用一个多值结构来存储这些项目的列表。

每个多值项目实际上都是标签—值对的数组，在第一次开始使用这个特性时可能会觉得有点糊涂：

@[pair1, pair2, pair3,...]

进一步讲，ABContact 把这些标签/值对表示为两个项目的字典。其中值（value）指添加到 Address Book 中的实际项目，标签（label）指该项目的角色：

@[{label:label1, value:value1}, {label:label2, value:value2}, ...]

例如，一个人可能具有有一个工作电话、一个家庭电话和一个移动电话，其中每个角色（工作、家庭和移动）都是标签。实际的电话号码是与角色关联的值：

@[{label: *WorkPhone* , value: *303-555-1212* },

```
{label: HomePhone , value: 720-555-1212 }, ...]
```

尽管你可能认为 Apple 可以把该信息用作键/值对，这样使用单个字典就要容易得多，但这不是 AddressBook 的实现方式。

下面说明了如何创建 ABContact 字典来填充其中每个数组。可以传递值和标签，然后通过这些元素构建字典。在调用 ABMultiValueAddValueAndLabel()时将使用这些元素。

```
+ (NSDictionary *) dictionaryWithValue: (id) value
    andLabel: (CFStringRef) label
{
    NSMutableDictionary *dict = [NSMutableDictionary dictionary];
    if (value) dict[@"value"] = value;
    if (label) dict[@"label"] = (__bridge NSString *)label;
    return dict;
}
```

1. 多值标签

AddressBook 框架使用 3 个泛型标签，它们可以用于任何数据类。这 3 个标签是 kABWorkLabel、kABHomeLabel 和 kABOtherLabel。此外，还可以找到用于 URL、社交媒体（Social Media）、即时消息（Instant Messaging）、电话号码（Phone Number）、日期（Date）和关系（Relationship）的特定于类型的标签。其中每个项目都记录在 AddressBook 文档中，并且可能会随着时间的推移而改变。社交媒体元素（Twitter、Facebook、LinkedIn、Sina Weibo 等）都是最近添加到 iOS SDK 中的。

除了系统提供的标签之外，大多数多值项目现在都允许添加自定义的标签。例如，除了 "Friend"（系统标签）之外，还可以添加联系人的 "Receptionist"（自定义的标签）。下面的代码段中的方法调用是在 ABContact 包装器中实现的：

```
// System-supplied
[person addRelationItem:contact1.compositeName
    withLabel:kABPersonFriendLabel];
// Custom
[person addRelationItem:contact2.compositeName
    withLabel:(__bridge CFStringRef)@"Receptionist"];
```

> **注意：**
> 现在，在 iOS 6 中完全支持 "相关的名称"。kABPersonRelatedNamesProperty 常量有助于存储名字和它们的关系（例如，可能使用 kABPersonMotherLabel 把 Mary BallWashington 存储在 George Washington 的联系人中）。参见 ABPerson.h，了解预先定义的关系常量的完整列表。

2. 获取多值数组

从记录中通过它的属性标识符获取每个数组，返回一个 CFArrayRef，可以直接使用它或者桥接到一个 NSArray。使用的多值属性标识符如下：

- **kABPersonEmailProperty**
- **kABPersonPhoneProperty**
- **kABPersonURLProperty**
- **kABPersonDateProperty**
- **kABPersonAddressProperty**
- **kABPersonSocialProfileProperty**
- **kABPersonInstantMessageProperty**

每种多值类型在把数据存储回记录方面都起着重要作用。类型用于分配内存，以及确定记录内的每个字段的大小。

这些项目中的前 3 个项目（电子邮件、电话和 URL）存储 multistring，即字符串的数组。它们关联的类型是 kABMultiStringPropertyType。

下一个项目（即日期属性）使用 kABMultiDateTimePropertyType 存储日期的数组。

后 3 个项目（即地址、社会概况和即时消息属性）由字典的数组组成，并且使用 kABMultiDictionaryPropertyType。其中每个字典都具有它自己的复杂度和自定义的键（本节后面将讨论它们）。

3. 获取多值属性值

要获取其中任何属性的值的数组，可以从记录中复制属性（使用 **ABRecordCopyValue()**），然后将其分解为它的成分数组。Address Book 提供了一个函数，用于从属性中把数组复制到标准的 CFArrayRef 中：

```
- (NSArray *) arrayForProperty: (ABPropertyID) anID
{
    CFTypeRef theProperty =
        ABRecordCopyValue(_record, anID);
    if (!theProperty) return nil;
    NSArray *items = (__bridge_transfer NSArray *)
        ABMultiValueCopyArrayOfAllValues(theProperty);
    CFRelease(theProperty);
    return items;
}
```

尽管你可能认为利用这两个调用获取了所有的信息，实则不然。仅仅获取值不足以处理多值项目。如前所述，多值数组中存储的每个字典都使用一个标签和一个值。图 9-1 显示了 Address Book 的联系人页面的一部分。分组的项目使用标签来区分每个电子邮件地址、电话号码等的角色。这个联系人具有 3 个电话号码和 3 个电子邮件地址，它们都会显示一个标签，指示值的角色。

4. 获取多值标签

必须传送属性中的标签和值，以获取为每个多值属性存储的所有信息。下面的方法用于按索引复制每个标签，然后把它添加到一个标签数组中。标签和值一起构成了完整的多值集合：

图 9-1 多值项目由一个标签（例如，"main"、"Google" 和 "mobile" 用于这些电话号码，或者 "home"、"work" 和 "Google" 用于这些电子邮件地址）和一个值组成

```
- (NSArray *) labelsForProperty: (ABPropertyID) anID
{
    CFTypeRef theProperty = ABRecordCopyValue(_record, anID);
    if (!theProperty) return nil;
    NSMutableArray *labels = [NSMutableArray array];
    for (int i = 0;
        i < ABMultiValueGetCount(theProperty); i++)
    {
        NSString *label = (__bridge_transfer NSString *)
            ABMultiValueCopyLabelAtIndex(theProperty, i);
        [labels addObject:label];
    }
    CFRelease(theProperty);
    return labels;
}
```

像下面这样把获取的标签和值结合在一起，创建最终的字典的多值数组：

```
- (NSArray *) dictionaryArrayForProperty: (ABPropertyID) aProperty
{
    NSArray *valueArray = [self arrayForProperty:aProperty];
    NSArray *labelArray = [self labelsForProperty:aProperty];

    int num = MIN(valueArray.count, labelArray.count);
    NSMutableArray *items = [NSMutableArray array];
    for (int i = 0; i < num; i++)
```

```
    {
        NSMutableDictionary *md = [NSMutableDictionary dictionary];
        md[@"value"] = valueArray[i];
        md[@"label"] = labelArray[i];
        [items addObject:md];
    }
    return items;
}
```

9.1.11 存储多值数据

保存到多值对象中的工作方式与刚才学到的获取函数相同，但是过程是相反的。要把多值项目存储到记录中，可以把 Cocoa Touch 对象转换为记录可以处理的形式，即两个项目式的值-标签对。

1. 自定义的字典

尽管值和标签项目是作为单独的参数传递给 AddressBook 函数的，但是把它们一起存储在单个字典中也很方便。这使得当以后需要它们时，更容易一个接一个地获取它们。下面的方法测试字典的顺应性。

```
+ (BOOL) isMultivalueDictionary: (NSDictionary *) dictionary
{
    if (dictionary.allKeys.count != 2)
        return NO;
    if (!dictionary[@"value"])
        return NO;
    if (!dictionary[@"label"])
        return NO;
    return YES;
}
```

下面的方法期望这些自定义字典的数组，它们的对象对应于从原始的多值属性获取的值和标签。这段代码将遍历那个字典的数组，并把每个值和标签添加到可变的多值对象中，它是 ABMutableMultiValueRef 的一个实例，通过利用一种属性类型调用 ABMultiValueCreateMutable() 来创建它。

```
- (ABMutableMultiValueRef) copyMultiValueFromArray:
    (NSArray *) anArray withType: (ABPropertyType) aType
{
    ABMutableMultiValueRef multi = ABMultiValueCreateMutable(aType);
    for (NSDictionary *dict in anArray)
```

```objc
        {
            if (![ABContact isMultivalueDictionary:dict])
                continue;
            ABMultiValueAddValueAndLabel(
                multi, (__bridge CFTypeRef) dict[@"value"],
                (__bridge CFTypeRef) dict[@"label"], NULL);
        }
        return multi;
    }
```

提供作为方法参数的 **ABPropertyType** 指定存储什么类型的值。例如，可以使用 kABMultiStringPropertyType 利用字符串和标签填充电子邮件属性，或者使用 kABMultiDictionaryPropertyType 填充 Address Book。

9.1.12 处理多值项目

在使用像刚才所示的方法创建了多值项目之后，然后就可以通过设置一个记录值把它添加到联系人中。下面的方法使用标准的 **ABRecordSetValue()** 调用把一个多值项目赋予属性 ID。这个调用与设置单个日期或字符串属性的调用之间基本上没有什么区别。所有的工作都是在创建多值项目的过程中第一时间完成的，而不是在实际的赋值中完成的：

```objc
- (BOOL) setMultiValue: (ABMutableMultiValueRef) multi
    forProperty: (ABPropertyID) anID
{
    CFErrorRef errorRef = NULL;
    BOOL success = ABRecordSetValue(_record, anID, multi, &errorRef);
    if (!success)
    {
        NSError *error = (__bridge_transfer NSError *) errorRef;
        NSLog(@"Error: %@", error.localizedFailureReason);
    }
    return success;
}
```

可以从特殊化的设置器（比如这里所示的设置器）调用这个常规的赋值方法。该方法用于分配一个电子邮件字典的的数组，它首先将构建一个多值项目，然后设置值：

```objc
- (void) setEmailDictionaries: (NSArray *) dictionaries
{
    // kABWorkLabel, kABHomeLabel, kABOtherLabel
    ABMutableMultiValueRef multi =
        [self copyMultiValueFromArray:dictionaries
```

```
            withType:kABMultiStringPropertyType];
    [self setMultiValue:multi
            forProperty:kABPersonEmailProperty];
    CFRelease(multi);
}
```

9.1.13 地址、社会概况和即时消息属性

3 个属性（地址、社会概况和即时消息 ID）使用的是多值字典，而不是字符串或日期。这增加了一个额外的步骤，用于创建多值数组。在构建多值项目并将其添加到联系人记录中之前，必须填充一组字典，然后把它们连同其标签一起添加到数组中。图 9-2 展示了这个额外的层次。如图所示，电子邮件多值项目由标签-值对的数组组成，其中每个值都是一个字符串。与之相比，地址使用单独的字典，其中存放的是值项目。

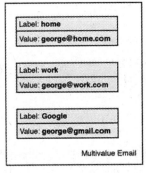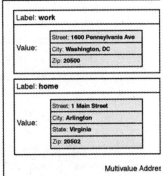

图 9-2 与为每个值存储一个字符串的多值电子邮件不同，多值地址包含整个地址字典

下面给出了一个示例，演示了在创建两个地址的多值项目时所涉及的步骤。这段代码构建字典，然后把它们连同它们的标签一起添加到基础数组中。这段代码创建的数组对应于图 9-2 右边显示的多值地址对象。

```
// Create the array that will store all the address
// value-label dictionaries
NSMutableArray *addresses = [NSMutableArray array];

// Create White House Address and add it to the array
NSDictionary *wh_addy = [ABContact
    addressWithStreet:@"1600 Pennsylvania Avenue"
    withCity:@"Washington, DC" withState:nil
    withZip:@"20500" withCountry:nil withCode:nil];
[addresses addObject:[ABContact dictionaryWithValue:wh_addy
    andLabel:kABWorkLabel]];

// Create a home address and add it to the array
NSDictionary *home_addy = [ABContact
    addressWithStreet:@"1 Main Street" withCity:@"Arlington"
```

```
            withState:@"Virginia" withZip:@"20502"
            withCountry:nil withCode:nil];
    [addresses addObject:[ABContact dictionaryWithValue:home_addy
            andLabel:kABHomeLabel]];
```

这段代码依靠便捷的方法来创建用于多值数组的地址字典和值/标签字典。下面的方法为这 3 种类型创建标签/值字典。注意怎样使用 Address Book 键常量预先定义每个字典的键。

```
+ (NSDictionary *) addressWithStreet: (NSString *) street
    withCity: (NSString *) city
    withState:(NSString *) state
    withZip: (NSString *) zip
    withCountry: (NSString *) country
    withCode: (NSString *) code
{
    NSMutableDictionary *md = [NSMutableDictionary dictionary];
    if (street) md[
        (__bridge NSString *) kABPersonAddressStreetKey] = street;
    if (city) md[
        (__bridge NSString *) kABPersonAddressCityKey] = city;
    if (state) md[
        (__bridge NSString *) kABPersonAddressStateKey] = state;
    if (zip) md[
        (__bridge NSString *) kABPersonAddressZIPKey] = zip;
    if (country) md[
        (__bridge NSString *) kABPersonAddressCountryKey] = country;
    if (code) md[
        (__bridge NSString *) kABPersonAddressCountryCodeKey] = code;
    return md;
}

+ (NSDictionary *) socialWithURL: (NSString *) url
    withService: (NSString *) serviceName
    withUsername: (NSString *) username
    withIdentifier: (NSString *) key
{
    NSMutableDictionary *md = [NSMutableDictionary dictionary];
    if (url) md[
        (__bridge NSString *) kABPersonSocialProfileURLKey] = url;
    if (serviceName) md[
        (__bridge NSString *) kABPersonSocialProfileServiceKey] = serviceName;
    if (username) md[
```

```objc
        (__bridge NSString *) kABPersonSocialProfileUsernameKey] = username;
    if (key) md[
        (__bridge NSString *) kABPersonSocialProfileUserIdentifierKey] = key;
    return md;
}

+ (NSDictionary *) imWithService: (CFStringRef) service
    andUser: (NSString *) userName
{
    NSMutableDictionary *im = [NSMutableDictionary dictionary];
    if (service) im[(__bridge NSString *)
        kABPersonInstantMessageServiceKey] = (__bridge NSString *) service;
    if (userName) im[(__bridge NSString *)
        kABPersonInstantMessageUsernameKey] = userName;
    return im;
}
```

9.1.14 Address Book 中的图像

Address Book 中的每条记录都与一幅可选的图像相关联。可以把图像数据复制到每条记录中，或者从中复制图像数据。ABPersonHasImageData()函数指示数据是否可供给定的记录使用。可以使用它来测试是否可以获取图像数据。

图像数据被存储为 Core Foundation 数据，它可以被桥接到 NSData。由于 UIImage 类完全支持把图像转换为数据以及通过数据创建图像，因此只需根据需要强制转换该数据即可。例如，可能使用 UIImagePNGRepresentation()函数，把 UIImage 实例转换为一种 NSData 表示。Address Book 支持 JPEG、JPG、BMP、PNG 和 GIF 数据。要执行相反的操作，可以使用 imageWithData:从 NSData 创建新图像。

Apple 在多个支持文档中详细说明了联系人的目标大小限制。特定于 iCloud 的文档（http://support.apple.com/kb/HT4489）把联系人限制为 25 000 个条目，其中每个条目可能包含 224 KB 的联系人照片（整个联系人被限制为总共 256KB）。对于整个联系人名册来说，其照片数据应该不超过 100MB，卡片信息则不超过 24MB。

```objc
// Retrieve image
- (UIImage *) image
{
    if (!ABPersonHasImageData(_record)) return nil;
    CFDataRef imageData = ABPersonCopyImageData(_record);
    if (!imageData) return nil;
```

```objc
    NSData *data = (__bridge_transfer NSData *)imageData;
    UIImage *image = [UIImage imageWithData:data];
    return image;
}

// Set image
- (void) setImage: (UIImage *) image
{
    CFErrorRef errorRef = NULL;
    BOOL success;
    
    if (image == nil) // remove
    {
        if (!ABPersonHasImageData(_record)) return;
        success = ABPersonRemoveImageData(_record, &errorRef);
        if (!success)
        {
            NSError *error = (__bridge_transfer NSError *) errorRef;
            NSLog(@"Error: %@", error.localizedFailureReason);
        }
        return;
    }
    
    NSData *data = UIImagePNGRepresentation(image);
    success = ABPersonSetImageData(
        _record, (__bridge CFDataRef) data, &errorRef);
    if (!success)
    {
        NSError *error = (__bridge_transfer NSError *) errorRef;
        NSLog(@"Error: %@", error.localizedFailureReason);
    }
    return;
}
```

9.1.15 创建、添加和删除记录

ABPersonCreate()函数返回一个新的 ABRecordRef 实例。这个记录存在于 Address Book 外面，表示一种独立的数据结构。迄今为止，本章中的所有示例都修改了单独的记录，但是没有任何示例实际地把记录存储到 Address Book 中。在查看下面这个便捷方法时要记住这一点，它返回一

个最近初始化的联系人:

```
+ (id) contact
{
    ABRecordRef person = ABPersonCreate();
    id contact = [ABContact contactWithRecord:person];
    CFRelease(person);
    return contact;
}
```

1. 写到 Address Book 中

要把新信息写到 Address Book 中, 需要执行两个步骤: 首先添加记录, 然后保存 Address Book。新的 iOS 开发人员通常会忘记第二个步骤, 这将导致 Address Book 好像会拒绝改变。这个方法将向 Address Book 中添加一个新的联系人, 但是没有保存改变。如前所示, 在添加了联系人以制作新的联系人标签之后, 要记得调用自定义的 **AddressBook** 类的 **save:** 方法:

```
+ (BOOL) addContact: (ABContact *) aContact
    withError: (NSError **) error
{
    ABAddressBookRef addressBook = [ABStandin addressBook];
    BOOL success;
    CFErrorRef errorRef = NULL;

    success = ABAddressBookAddRecord(
        addressBook, aContact.record, &errorRef);
    if (!success)
    {
        if (error)
            *error = (__bridge_transfer NSError *)errorRef;
        return NO;
    }
    return YES;
}
```

2. 删除记录

从 Address Book 中删除记录需要执行一个保存步骤, 就像添加记录一样。在删除记录后, 该记录仍将作为对象存在, 但是它将不再存储在 Address Book 数据库中。下面显示了如何从 Address Book 中删除记录。同样, 在此之后要调用[ABStandin save:&error]:

```
- (BOOL) removeSelfFromAddressBook: (NSError **) error
```

```
{
    CFErrorRef errorRef = NULL;
    BOOL success;
    
    ABAddressBookRef addressBook = [ABStandin addressBook];
    
    success = ABAddressBookRemoveRecord(
        addressBook, self.record, &errorRef);
    if (!success)
    {
        if (error)
            *error = (__bridge_transfer NSError *)errorRef;
        return NO;
    }
    
    return success;
}
```

9.1.16 搜索联系人

默认的 AddressBook 框架允许在全部记录中执行前缀搜索。这个函数返回一个记录的数组,这些记录的复合名称(通常在名字后面追加姓氏,但是对于颠倒了这种模式的国家,将进行本地化)与提供的字符串匹配:

```
NSArray *array = (__bridge_transfer NSArray *)
    ABAddressBookCopyPeopleWithName(addressBook, CFSTR("Eri"));
```

使用谓词和属性执行搜索更容易。结合使用自定义的 ABContact 类属性与 NSPredicate 实例,可以进行快速过滤,从而找到想要的匹配。下面的代码针对联系人的名字、中名、姓氏和昵称来匹配一个字符串。谓词使用属性名称来定义它将如何匹配或拒绝联系人。它将使用一个实例以及有差别的不区分大小写的匹配([cd]),比较每个字符串内的所有位置(contains),而不仅仅是开头(begins with):

```
+ (NSArray *) contactsMatchingName: (NSString *) fname
{
    NSPredicate *pred;
    NSArray *contacts = [ABContactsHelper contacts];
    pred = [NSPredicate predicateWithFormat:
        @"firstname contains[cd] %@ OR lastname contains[cd] %@ OR\
        nickname contains[cd] %@ OR middlename contains[cd] %@",
        fname, fname, fname, fname];
```

```
    return [contacts filteredArrayUsingPredicate:pred];
}
```

> **注意：**
> Apple 的 Predicate Programming Guide 提供了对谓词基础知识的全面介绍。它演示了如何创建谓词以及在应用程序中使用它们。

9.1.17 对联系人排序

有时，可能想以某种方式对联系人排序。下面的方法模仿了字符串排序，但它是使用 ABContact 类的复合名称属性来执行该任务的，该属性结合了名字、中名和姓氏。可以轻松地修改这个方法，使用其他键来排序：

```
- (BOOL) isEqualToString: (ABContact *) aContact
{
    return [self.compositeName isEqualToString:aContact.compositeName];
}

- (NSComparisonResult) caseInsensitiveCompare: (ABContact *) aContact
{
    return [self.compositeName
        caseInsensitiveCompare:aContact.compositeName];
}
```

9.1.18 处理群组

群组把联系人集中到相关的集合中，比如朋友、同事、酒友及其他自然的分组。每个群组仅仅只是另一个 ABRecord，但是带有几个特殊的属性。群组不存储名字、地址和电话号码。作为替代，它们存储一个指向其他联系人记录的引用。

1. 统计群组

可以通过获取群组记录，来统计当前 Address Book 中的群组数量，如这个方法中所示。不能像查询联系人数量那样直接查询群组的数量。下面的方法是 ABGroup 包装器类的一部分，为 Address Book 提供一个 Objective-C 包装器，比如 ABContact 包装 Address Book 联系人：

```
+ (int) numberOfGroups
{
    ABAddressBookRef addressBook = [ABStandin addressBook];
```

```
        NSArray *groups = (__bridge_transfer NSArray *)
            ABAddressBookCopyArrayOfAllGroups(addressBook);
    int ncount = groups.count;
    return ncount;
}
```

2. 创建群组

使用 ABGroupCreate()函数创建群组。这个函数以与 ABPersonCreate()相同的方式返回一个 ABRecordRef。它们的区别在于记录类型。对于群组，将把这个属性设置为 kABGroupType，而不是 kABPersonType：

```
+ (id) group
{
    ABRecordRef grouprec = ABGroupCreate();
    id group = [ABGroup groupWithRecord:grouprec];
    CFRelease(grouprec);
    return group;
}
```

3. 添加和删除成员

通过调用 ABGroupAddMember()和 ABGroupRemoveMember()，添加和删除群组的成员。这些调用只会影响记录，并且与以往一样，直到保存了 Address Book 之后，才会全局应用它们。

```
- (BOOL) addMember: (ABContact *) contact
    withError: (NSError **) error
{
    CFErrorRef errorRef = NULL;
    BOOL success;

    success = ABGroupAddMember(self.record,
        contact.record, &errorRef);
    if (!success)
    {
        if (error)
            *error = (__bridge_transfer NSError *)errorRef;
        return NO;
    }

    return YES;
}
```

```objc
- (BOOL) removeMember: (ABContact *) contact
    withError: (NSError **) error
{
    CFErrorRef errorRef = NULL;
    BOOL success;

    success = ABGroupRemoveMember(self.record,
        contact.record, &errorRef);
    if (!success)
    {
        if (error)
            *error = (__bridge_transfer NSError *)errorRef;
        return NO;
    }
    return YES;
}
```

4. 列出成员

群组的每个成员都是一个人，或者使用本章的 **ABContact** 术语来讲是一个联系人。下面这个方法将扫描群组的成员，并返回 **ABContact** 实例的数组，其中将利用群组成员的 **ABRecordRef** 初始化每个实例：

```objc
- (NSArray *) members
{
    NSArray *contacts = (__bridge_transfer NSArray *)
        ABGroupCopyArrayOfAllMembers(self.record);
    NSMutableArray *array =
        [NSMutableArray arrayWithCapacity:contacts.count];
    for (id contact in contacts)
        [array addObject:[ABContact contactWithRecord:
            (__bridge ABRecordRef)contact]];
    return array;
}
```

5. 群组名称

每个群组都有一个名称，它是可以设置和获取的主要的群组属性。它使用 kABGroupName-Property 标识符，其工作方式是像联系人属性一样：

```objc
- (NSString *) getRecordString:(ABPropertyID) anID
```

```objc
{
    return (__bridge_transfer NSString *)
        ABRecordCopyValue(_record, anID);
}

- (NSString *) name
{
    return [self getRecordString:kABGroupNameProperty];
}

- (void) setName: (NSString *) aString
{
    CFErrorRef errorRef = NULL;
    BOOL success;
    success = ABRecordSetValue(_record, kABGroupNameProperty,
        (__bridge CFStringRef) aString, &errorRef);
    if (!success)
    {
        NSError *error = (__bridge_transfer NSError *) errorRef;
        NSLog(@"Error: %@", error.localizedFailureReason);
    }
}
```

9.1.19 ABContact、ABGroup 和 ABContactsHelper

与本节内容配套的示例代码包括在这几节中演示过的 3 个包装器类的源代码。在全部讨论中显示的代码段突出说明了这些类中使用的技术。出于类的长度和篇幅考虑，在本节中省略了单个秘诀清单。在 github 代码库上用于本章的示例代码文件夹中可以找到完整的类。

自定义的 ABContact 类在某种程度上基于仅针对 Mac 的 ABPerson 类。它提供了一个 Cocoa Touch 接口，更多地针对内置的 Objective-C 2.0 属性，而不是 Apple 的 ABPerson 使用的 C 语言风格的属性查询。你在本节中迄今为止看到的所有特定于联系人的方法都源于这个类。

另一个名为 ABGroup 的类封装了用于 ABRecordRef 实例的所有群组功能。它提供了 Objective-C 访问，用于进行群组创建和管理。可以使用这个类构建新的类以及添加和删除成员。

最后一个类 ABContactsHelper 提供了特定于 Address Book 的方法。可以使用这个类搜索 Address Book、获取记录的数组等。尽管少量基本的搜索要涉及所有的名称和电话号码，但是可以为更复杂的查询轻松地扩展这个类，它宿主在 http://github.com/erica 上。

9.2 秘诀：搜索 Address Book

基于谓词的搜索既快速且有效。秘诀 9-1 构建在 ABContact 的谓词查询的基础之上，它将展示一个搜索表格，显示联系人的滚动表格。这个表格利用实时清单更新来响应用户的搜索栏查询。

在用户点按某一行时，这个秘诀将显示一个 **ABPersonViewController** 实例。这个类提供了联系人概览，显示给定记录的详细信息，类似于图 9-1 中所示的视图。要使用这个视图控制器，可以分配和初始化它，并设置它的 displayedPerson 属性。如你所期望的，这个属性将存储一个 ABRecordRef，可以轻松地从 ABContact 实例中获取它。

人员视图控制器提供了一个受限的委托。通过设置 personViewDelegate 属性，可以订阅 personViewController:shouldPerformDefault-ActionForPerson:方法。当用户选择视图中的某些项目（包括电话号码、电子邮件地址、URL 和地址）时，将触发该方法。如果该方法返回 YES，将会执行默认的动作（拨号、发送电子邮件等）；如果返回 NO，则会忽略它们。

秘诀 9-1 使用这个回调在调试控制台中显示所选项目的值，利用它可以方便地确定用户触摸的是什么以及他或她感兴趣的位置。

尽管可以与其他显示元素（比如联系人便条和铃声）交互，但是这些项目不会产生回调。

要扩展这个秘诀以允许进行编辑，可以把人员视图控制器的 allowsEditing 属性设置为 YES。这将提供 Edit 按钮，它出现在显示屏的右上方。点按这个按钮时，它将在人员视图中触发相同的编辑特性，就像在 Contacts 应用程序中通常所看到的那样。

秘诀 9-1 利用搜索选择和显示联系人

```
// Return the number of table sections
- (NSInteger)numberOfSectionsInTableView:(UITableView *)aTableView
{
    // One section for this example
    return 1;
}

// Return the number of rows per section
- (NSInteger)tableView:(UITableView *)aTableView
    numberOfRowsInSection:(NSInteger)section
{
    if (aTableView == self.tableView)
        return matches.count;
    // On cancel, matches are restored in the search bar delegate
    matches = [ABContactsHelper
```

```
            contactsMatchingName:searchBar.text];
    matches = [matches sortedArrayUsingSelector:
        @selector(caseInsensitiveCompare:)];
    return matches.count;

    matches = [matches sortedArrayUsingSelector:
        @selector(caseInsensitiveCompare:)];
    return matches.count;
}


// Produce a cell for the given index path
- (UITableViewCell *)tableView:(UITableView *)aTableView
    cellForRowAtIndexPath:(NSIndexPath *)indexPath
{
    // Dequeue or create a cell
    UITableViewCellStyle style = UITableViewCellStyleDefault;
    UITableViewCell *cell =
        [aTableView dequeueReusableCellWithIdentifier:@"BaseCell"];
    if (!cell)
        cell = [[UITableViewCell alloc]
            initWithStyle:style reuseIdentifier:@"BaseCell"];

    ABContact *contact = [matches objectAtIndex:indexPath.row];
    cell.textLabel.text = contact.compositeName;
    return cell;
}


- (BOOL)personViewController:
        (ABPersonViewController *) personViewController
    shouldPerformDefaultActionForPerson:(ABRecordRef)person
    property:(ABPropertyID)property
    identifier:(ABMultiValueIdentifier)identifierForValue
{
    // Reveal the item that was selected
    if ([ABContact propertyIsMultiValue:property])
    {
        NSArray *array =
            [ABContact arrayForProperty:property inRecord:person];
        NSLog(@"%@", [array objectAtIndex:identifierForValue]);
    }
```

```
        else
        {
            id object =
                [ABContact objectForProperty:property inRecord:person];
            NSLog(@"%@", [object description]);
        }

        return NO;
}

// Respond to user taps by displaying the person viewcontroller
- (void) tableView:(UITableView *)tableView
    didSelectRowAtIndexPath:(NSIndexPath *)indexPath
{
    ABContact *contact = [matches objectAtIndex:indexPath.row];
    ABPersonViewController *pvc =
        [[ABPersonViewController alloc] init];
    pvc.displayedPerson = contact.record;
    pvc.personViewDelegate = self;
    // pvc.allowsEditing = YES; // optional editing
    [self.navigationController
        pushViewController:pvc animated:YES];
}
```

获取这个秘诀的代码

要查找这个秘诀的完整示例项目，可以浏览 https://github.com/erica/iOS-6-Advanced-Cookbook，并进入第 9 章的文件夹。

9.3 秘诀：访问联系人图像数据

秘诀 9-2 扩展了秘诀 9-1，在可用时，它将把联系人图像缩略图添加到每个表格单元格中。它是通过从联系人的数据创建新缩略图来执行该任务的。当图像数据可用时，将把该图像呈现到缩略图上。当它不可用时，将把缩略图保持为空白。一旦绘制了图像，就会把它分配给单元格的 imageView。

图 9-3 显示了用于这个秘诀的界面。在这幅截屏图中，正在执行一种搜索（匹配字母 "e"）。与搜索匹配的每条记录都会显示它们的图像缩略图。

图 9-3 可以轻松地获取和显示与 Address Book 联系人关联的图像数据

注意在这个秘诀中创建和使用缩略图有多简单。只需几行代码,即可构建一种新的图像环境,在其中绘图(用于带有图像的联系人),并将其保存到 UIImage 实例。

秘诀 9-2 在表格单元格中显示 Address Book 图像

```
- (UITableViewCell *)tableView:(UITableView *)aTableView
    cellForRowAtIndexPath:(NSIndexPath *)indexPath
{
    // Dequeue or create a cell
    UITableViewCellStyle style = UITableViewCellStyleSubtitle;
    UITableViewCell *cell =
        [aTableView dequeueReusableCellWithIdentifier:@"BaseCell"];
    if (!cell)
        cell = [[UITableViewCell alloc]
            initWithStyle:style reuseIdentifier:@"BaseCell"];

    ABContact *contact = [matches objectAtIndex:indexPath.row];
    cell.textLabel.text = contact.compositeName;
    cell.detailTextLabel.text = contact.phonenumbers;

    CGSize small = CGSizeMake(48.0f, 48.0f);
    UIGraphicsBeginImageContext(small);
    UIImage *image = contact.image;
    if (image)
        [image drawInRect:(CGRect){.size = small}];
    cell.imageView.image = UIGraphicsGetImageFromCurrentImageContext();
    UIGraphicsEndImageContext();

    return cell;
}
```

> **获取这个秘诀的代码**
>
> 要查找这个秘诀的完整示例项目，可以浏览 https://github.com/erica/iOS-6-Advanced-Cookbook，并进入第 9 章的文件夹。

9.4 秘诀：选择人员

AddressBookUI 框架提供了人员选择控制器，它允许浏览整个联系人列表。其工作方式类似于单个联系人屏幕，但是不需要事先选择联系人。ABPeoplePickerNavigationController 类用于展示交互式浏览器，如图 9-4 所示。

图 9-4 iPhone 的人员选择器导航控制允许用户搜索联系人数据库，并选择人员或组织

在以模态方式展示控制器之前，要分配并显示它。确保设置 peoplePickerDelegate 属性，它允许利用视图捕获用户交互：

```
- (void) action: (UIBarButtonItem *) bbi
{
    ABPeoplePickerNavigationController *ppnc =
        [[ABPeoplePickerNavigationController alloc] init];
    ppnc.peoplePickerDelegate = self;
    [self presentViewController:ppnc animated:YES completion:nil];
}
```

在声明 ABPeoplePickerNavigationControllerDelegate 协议时，类必须实现下面 3 个方法。这些方法分别用于响应用户点按一个联系人、点按联系人的任何属性或者点按 Cancel 按钮。

- **peoplePickerNavigationController:shouldContinueAfterSelecting-Person:**——当用户点按一个联系人时，将具有两种选择。可以接受那个人作为最终的选择，并关闭模态视图；或者可以导航到单独的显示屏。如果只是选择人员，该方法将返回 NO。如果要继续显示单独的屏幕,该方法将返回 YES。第二个参数包含所选的人员,以防你想在选择任何 ABPerson 记录之后停止操作。

- **peoplePickerNavigationController:shouldContinueAfterSelecting-Person:property:identifier:**——直到用户前进到单独的联系人显示屏幕之后,才会调用该方法。然后,由你决定是把控制返回给程序(返回 NO),还是继续下面的操作(返回 YES)。你可以确定点按了哪个属性,并使用秘诀 9-1 中的代码获取它的值。尽管这个方法应该是可选的,但是在编写本书时不是这样。

- **peoplePickerNavigationControllerDidCancel:**——当用户点按 Cancel 按钮时,你仍然希望有机会关闭模态视图。这个方法将捕获取消事件,允许使用它执行关闭。

秘诀 9-3 展示了可能最简单的人员选择示例。它将展示选择器,并等待用户选择一个联系人。当用户这样做时,它将关闭选择器,并改变视图控制器的标题(在导航栏中),显示所选人员的复合名称。从主回调中返回 NO 意味着永远也不会调用属性回调。仍然必须在代码中包括它,因为全部 3 个方法都是必需的。

秘诀 9-3 选择人员

```
- (BOOL)peoplePickerNavigationController:
        (ABPeoplePickerNavigationController *)peoplePicker
    shouldContinueAfterSelectingPerson:(ABRecordRef)person
{
    self.title = [[ABContact contactWithRecord:person] compositeName];
    [self dismissViewControllerAnimated:YES completion:nil];
    return NO;
}

- (BOOL)peoplePickerNavigationController:
        (ABPeoplePickerNavigationController *)peoplePicker
    shouldContinueAfterSelectingPerson:(ABRecordRef)person
    property:(ABPropertyID)property
    identifier:(ABMultiValueIdentifier)identifier
{
    // required method that is never called in the people-only-picking
    [self dismissViewControllerAnimated:YES completion:nil];
    return NO;
}

- (void)peoplePickerNavigationControllerDidCancel:
    (ABPeoplePickerNavigationController *)peoplePicker
```

```
{
    [self dismissViewControllerAnimated:YES completion:nil];
}
- (void) action: (UIBarButtonItem *) bbi
{
    ABPeoplePickerNavigationController *ppnc =
        [[ABPeoplePickerNavigationController alloc] init];
    ppnc.peoplePickerDelegate = self;
    [self presentViewController:ppnc animated:YES completion:nil];
}
```

获取这个秘诀的代码

要查找这个秘诀的完整示例项目，可以浏览 https://github.com/erica/iOS-6-Advanced-Cookbook，并进入第 9 章的文件夹。

9.5 秘诀：限制联系人选择器属性

当需要用户选择某种属性（比如电子邮件地址）时，将不希望给用户展示人员的街道地址或传真号码。可以限制联系人的显示属性，以只显示你希望用户从中选择的那些项目。图 9-5 所示的选择器被限制为只能选择电子邮件。

图 9-5　人员选择器的显示属性允许选择要展示给用户哪些属性（在这里，只有电子邮件）

为了使之发生，可以通过把一个属性类型的数组提交给控制器来选择所显示的属性。设置选择器的 displayedProperties 属性。秘诀 9-4 提供了两个选择选项：一个用于电子邮件，另一个用于电话号码。尽管这些示例为属性数组使用单个属性，但是可以选择显示任意数量的属性。

秘诀 9-4 选择显示属性

```objc
- (BOOL)peoplePickerNavigationController:
        (ABPeoplePickerNavigationController *)peoplePicker
    shouldContinueAfterSelectingPerson:(ABRecordRef)person
{
    // Continue onto the detail screen
    return YES;
}

- (BOOL)peoplePickerNavigationController:
        (ABPeoplePickerNavigationController *)peoplePicker
    shouldContinueAfterSelectingPerson:(ABRecordRef)person
    property:(ABPropertyID)property
    identifier:(ABMultiValueIdentifier)identifier
{
    // Guaranteed to only be working with e-mail or phone here
    [self dismissViewControllerAnimated:YES completion:nil];
    NSArray *array =
        [ABContact arrayForProperty:property inRecord:person];
    self.title = (NSString *)[array objectAtIndex:identifier];
    return NO;
}

- (void)peoplePickerNavigationControllerDidCancel:
    (ABPeoplePickerNavigationController *)peoplePicker
{
    // Respond to cancel
    [self dismissViewControllerAnimated:YES completion:nil];
}

- (void) email: (UIBarButtonItem *) bbi
{
    ABPeoplePickerNavigationController *ppnc =
        [[ABPeoplePickerNavigationController alloc] init];
    ppnc.peoplePickerDelegate = self;
    [ppnc setDisplayedProperties:
        [NSArray arrayWithObject:@(kABPersonEmailProperty)]];
    [self presentViewController:ppnc animated:YES completion:nil];
}
```

```
- (void) phone: (UIBarButtonItem *) bbi
{
    ABPeoplePickerNavigationController *ppnc =
        [[ABPeoplePickerNavigationController alloc] init];
    ppnc.peoplePickerDelegate = self;
    [ppnc setDisplayedProperties:
        [NSArray arrayWithObject:@(kABPersonPhoneProperty)]];
    [self presentViewController:ppnc animated:YES completion:nil];
}
```

获取这个秘诀的代码

要查找这个秘诀的完整示例项目，可以浏览 https://github.com/erica/iOS-6-Advanced-Cookbook，并进入第 9 章的文件夹。

9.6 秘诀：添加和删除联系人

允许用户利用 ABNewPersonViewController 类创建新的联系人。这个视图控制器提供了一个编辑屏幕，简化了新 Address Book 条目的交互式创建。在分配并初始化视图控制器之后，首先要创建一个新联系人，并把它分配给 displayedPerson 属性。如果你愿意，可以先利用属性预先填充联系人的记录。

接下来，分配 newPersonViewDelegate，确保在委托类中声明 ABNewPersonViewController-Delegate 协议。委托将接收一个新的回调：newPersonViewController:didCompleteWithNewPerson:，这个回调是为选择和取消事件发送的。检查 person 参数，确定适用哪种情况，如秘诀 9-5 中所示。如果人员是 nil，用户就点按 Cancel，并且不应该把新联系人添加到 Address Book 中。如果用户在编辑了新联系人之后点按 Done，就会把联系人数据添加到 Address Book 中并保存它。

要从 Address Book 中删除联系人，首先要给用户展示标准的人员选择器导航控制器，并请求用户做出选择。当用户选择一个名字时，回调方法将要求他们在继续前确认删除。确认后，必须保存 Address Book。从 Address Book 中删除一条记录需要一个保存步骤，就像添加记录一样。

秘诀 9-5　使用新的人员视图控制器

```
#pragma mark NEW PERSON DELEGATE METHODS
- (void)newPersonViewController:
        (ABNewPersonViewController *)newPersonViewController
    didCompleteWithNewPerson:(ABRecordRef)person
{
```

```objc
        if (person)
        {
            ABContact *contact = [ABContact contactWithRecord:person];
            self.title = [NSString stringWithFormat:
                @"Added %@", contact.compositeName];
            
            NSError *error;
            BOOL success =
                [ABContactsHelper addContact:contact withError:&error];
            if (!success)
            {
                NSLog(@"Could not add contact. %@",
                    error.localizedFailureReason);
                self.title = @"Error.";
            }
            
            [ABStandin save:nil];
        }
        else
            self.title = @"Cancelled";
    
    [self.navigationController popViewControllerAnimated:YES];
}

#pragma mark PEOPLE PICKER DELEGATE METHODS
- (BOOL)peoplePickerNavigationController:
        (ABPeoplePickerNavigationController *)peoplePicker
    shouldContinueAfterSelectingPerson:(ABRecordRef)person
{
    [self dismissViewControllerAnimated:YES completion:nil];
    ABContact *contact = [ABContact contactWithRecord:person];
    
    NSString *query = [NSString stringWithFormat:
        @"Really delete %@?", contact.compositeName];
    if ([self ask:query])
    {
        self.title = [NSString stringWithFormat:@"Deleted %@",
            contact.compositeName];
        [contact removeSelfFromAddressBook:nil];
        [ABStandin save:nil];
```

    }

    return NO;
}

- (BOOL)peoplePickerNavigationController:
        (ABPeoplePickerNavigationController *)peoplePicker
    shouldContinueAfterSelectingPerson:(ABRecordRef)person
    property:(ABPropertyID)property
    identifier:(ABMultiValueIdentifier)identifier
{
    [self dismissViewControllerAnimated:YES completion:nil];
    return NO;
}

- (void)peoplePickerNavigationControllerDidCancel:
    (ABPeoplePickerNavigationController *)peoplePicker
{
    [self dismissViewControllerAnimated:YES completion:nil];
}

#pragma mark Respond to User Requests
- (void) add
{
    // create a new view controller
    ABNewPersonViewController *npvc =
        [[ABNewPersonViewController alloc] init];

    // Create a new contact
    ABContact *contact = [ABContact contact];
    npvc.displayedPerson = contact.record;

    // Set delegate
    npvc.newPersonViewDelegate = self;

    [self.navigationController
        pushViewController:npvc animated:YES];
}

- (void) remove
```
```

```
    {
        ABPeoplePickerNavigationController *ppnc =
            [[ABPeoplePickerNavigationController alloc] init];
        ppnc.peoplePickerDelegate = self;
        [self presentlViewController:ppnc animated:YES completion:nil];
    }
```

> **获取这个秘诀的代码**
>
> 要查找这个秘诀的完整示例项目，可以浏览 https://github.com/erica/iOS-6-Advanced-Cookbook，并进入第 9 章的文件夹。

9.7 修改和查看单独的联系人

ABPersonViewController 的 allowsEditing 属性允许用户编辑联系人。创建一个控制器，分配一个 Address Book 记录作为它的 displayedPerson，设置它的可编辑的属性，并显示它。下面显示了这个特性在回调方法中的典型示例。一旦解析到特定的联系人，应用程序就会创建新的人员展示，并把它压到导航栈上。

```
- (void)unknownPersonViewController:
       (ABUnknownPersonViewController *)unknownPersonView
    didResolveToPerson:(ABRecordRef)person
{
    // Handle cancel events
    if (!person) return;

    ABPersonViewController *abpvc = [[ABPersonViewController alloc] init];
    abpvc.displayedPerson = person;
    abpvc.allowsEditing = YES;
    abpvc.personViewDelegate = self;

    [self.navigationController pushViewController:abpvc animated:YES];
}
```

9.7.1 用户编辑

当把 allowsEditing 属性设置为 YES 时，在显示的人员展示的右上方将会出现一个 Edit 按钮。用户可能点按这个按钮，并修改联系人详细信息。编辑操作将自动传播到 Address Book。不需要明确添加调用以保存数据；Cocoa Touch 的 AddressBookUI 实现将自动为你处理它。与 Contacts

应用程序不同，它将不会在控制器底部以红色显示 Delete Contact（删除联系人）选项。

如果希望在不启用编辑的情况下允许用户查看联系人信息，可以把 allowsEditing 属性值切换为 NO，否则，可以保留原来的展示代码，而无需执行其他的修改。

9.7.2 委托方法

人员视图控制器只有一个委托回调方法。可以指定点按项目（比如电子邮件地址或 URL）是否应该基于那些点按执行相应的动作（也就是说打开 Mail 或者链接到 Safari）。根据需要返回 YES 或 NO。当设置为 YES 时，要知道控制可能传递到应用程序之外以响应这些动作：

```
#pragma mark PERSON DELEGATE
- (BOOL)personViewController:
        (ABPersonViewController *)personViewController
    shouldPerformDefaultActionForPerson:(ABRecordRef)person
    property:(ABPropertyID)property
    identifier:(ABMultiValueIdentifier)identifierForValue
{
    return NO;
}
```

9.8　秘诀："未知的"人员控制器

当具有某种信息（比如电子邮件地址或电话号码）但是还没有与之关联的联系人时，会发生什么事情？ABUnknownPersonViewController 允许把该信息添加到新的或现有的联系人中。这个类用于把已知的属性与未知的联系人相关联，下面介绍了它的工作方式。

分配并初始化视图控制器，然后创建记录并利用你喜欢的任何属性预先填充它。例如，秘诀 9-6 构建了一个随机的"怪物"图像。可以根据需要添加更多的项目，但是通常这个控制器用于添加一点单独的信息。未知的人员控制器专用于提供这种添加一个项目的功能。图 9-6 显示了这个秘诀的控制器的实际应用。

用户可以给现有的联系人添加图像，或者创建新的联系人。点按 Back 按钮将在不添加图像的情况下返回。

Create New Contact（创建新联系人）和 Add to Existing Contact（添加到现有联系人）按钮

图 9-6　可以利用未知的人员控制器给联系人添加图像，以及电子邮件地址、电话号码及其他基于文本的数据

是由控制器的 allowsAddingToAddressBook 属性控制的。当用户点按新联系人按钮时，将会出现一个表单，其中预先填充了所传递的记录中的属性。如果禁用这个添加属性，将不会显示这些按钮。

allowsAction 属性允许用户点按这个界面中的表单元素。这允许他们连接到电子邮件地址、连接到电话号码或者连接到 Web 页面等。通过点按任何元素，允许用户选择分支执行应用程序，比如执行调用或者转移到 Mobile Safari 等。这提供了一种方便的方式，用于把交互式联系人信息和 URL 添加到已经定义的视图控制器中。

控制器上的 alternateName 和 message 属性提供了用于填充名字和组织字段的文本。尽管可以利用记录中的数据填充这些字段，但是选项不会传送给联系人。因此，它们为提示用户提供了一种良好的方式，而不会产生任何副作用。

设置 unknownPersonViewDelegate 属性并且声明 ABUnknownPersonViewControllerDelegate 协议，以接收 unknownPersonViewController:didResolveToPerson:方法。当用户点按 Done（完成）按钮时将调用该方法，它允许取回保存数据的记录。它还提供了一个位置，其中可以用编程方式弹出视图控制器，从而知道用户利用对话框完成了交互。

> **注意：**
> 这个秘诀使用的 Monster ID 项目的艺术作品包含身体部位艺术作品的集合，可以把它们汇集到一起构成随机的图片。它是由 Andreas Gohr 开发的，并且受到了 Don Park 的 Web 贴画和组合式人物的启发。它是通过添加预先绘制的胳膊、腿和躯干等构建的，得到的合成图像将产生一个完整的生物。

秘诀 9-6　使用未知的控制器

```
#pragma mark UNKNOWN PERSON DELEGATE
- (void)unknownPersonViewController:
      (ABUnknownPersonViewController *)unknownPersonView
   didResolveToPerson:(ABRecordRef)person
{
   // Handle cancel events
   if (!person) return;

   ABPersonViewController *abpvc =
      [[ABPersonViewController alloc] init];
   abpvc.displayedPerson = person;
   abpvc.allowsEditing = YES;
   abpvc.personViewDelegate = self;
   [self.navigationController
```

```
            pushViewController:abpvc animated:YES];
}

#pragma mark PERSON DELEGATE
- (BOOL)personViewController:
        (ABPersonViewController *)personViewController
    shouldPerformDefaultActionForPerson:(ABRecordRef)person
    property:(ABPropertyID)property
    identifier:(ABMultiValueIdentifier)identifierForValue
{
    return NO;
}


- (BOOL)unknownPersonViewController:
        (ABUnknownPersonViewController *)personViewController
    shouldPerformDefaultActionForPerson:(ABRecordRef)person
    property:(ABPropertyID)property
    identifier:(ABMultiValueIdentifier)identifier
{
    return YES;
}

#pragma mark Action
- (void)assignAvatar
{
    ABUnknownPersonViewController *upvc =
        [[ABUnknownPersonViewController alloc] init];
    upvc.unknownPersonViewDelegate = self;

    ABContact *contact = [ABContact contact];
    contact.image = imageView.image;

    upvc.allowsActions = NO;
    upvc.allowsAddingToAddressBook = YES;
    upvc.message = @"Who looks like this?";
    upvc.displayedPerson = contact.record;

    [self.navigationController pushViewController:upvc animated:YES];
}
```

> **获取这个秘诀的代码**
> 要查找这个秘诀的完整示例项目,可以浏览 https://github.com/erica/iOS-6-Advanced-Cookbook,并进入第 9 章的文件夹。

9.9 小结

本章通过介绍 AddressBook 和 AddressBookUI 框架,描述了 Address Book 的核心功能。本章技术性很强,你在这一章中学习了它们是如何工作的。在结束本章的学习之前,要最后思考一下你刚才遇到的几个框架。

- 事实证明,低级的 Address Book 函数尽管是有用的,但是直接使用它们容易遭受挫折。与本章配套的几个辅助类可能有助于使事情变得更简单。所有的包装器都进行了更新,以适应 iOS 6 及其用户许可挑战。
- 访问和修改 Address Book 图像就像处理其他任何字段一样。提供图像数据,代替字符串、日期或多值数组。在联系人应用程序中,使用图像数据时不要犹豫。
- AddressBookUI 框架提供的视图控制器可以与底层的 AddressBook 例程无缝地协同工作。对于大多数常见的 Address Book 交互任务,无需开发你自己的 GUI。
- 你使用过 Vcard 吗?AddressBook 框架提供了两个特定于 Vcard 的函数,允许通过记录的数组创建 Vcard 数据(ABPersonCreateVCardRepresentationWithPeople())以及把 Vcard 数据转换为 ABRecordRef(ABPersonCreatePeopleInSourceWithVCardRepresentation())。
- 未知的人员控制器提供了一种极佳的方式,用于把特定的信息(比如公司的电子邮件地址、重要的 Web 站点等)存储到联系人中,同时允许用户决定在哪里(或者是否)存放该信息。

第 10 章
位置

位置具有深远的意义。快速计算正变得与计算的方式和计算的内容同样重要。iOS 终日忙个不停，整天都与它的用户一起旅行。Core Location（核心位置）给 iOS 注入了按需进行地理定位的功能。MapKit 添加了交互式的应用程序内的地图绘制功能，使用户能够查看和操纵加注释的地图。利用 Core Location 和 MapKit，可以开发应用程序，帮助用户与朋友会面、搜索本地资源，或者提供基于位置的个人信息流。本章介绍了这些可以感知位置的框架，并且说明了如何把它们集成进 iOS 应用程序中。

10.1 授权 Core Location

Apple 要求用户启用和授权 Core Location。它们是两个单独的设置，导致应用程序必须执行的两种单独的检查。用户可以在 Settings > Privacy > Location 中启用全局位置服务（参见图 10-1）。

10.1.1 测试位置服务

通过检查 locationServicesEnabled 类方法，测试是否启用了位置服务。这将返回一个布尔值，指示用户是否全局地退出了位置服务：

```
// Check for location services
if (![CLLocationManager locationServicesEnabled])
{
    NSLog(@"User has disabled location services");
    // Display instructions on enabling location services
    return;
}
```

用户可以逐个应用程序地授权位置服务，如图 10-2 所示。当应用程序第一次访问位置服务时，将会出现这个提示。用户可以通过点按 Don't Allow（不允许）或 OK（确定）按钮，来授权或拒绝访问。

图 10-1　用户可以在 Settings > Privacy > Location 中跨他们的设备启用和禁用位置服务。在启用 GPS 的设备上不需要 Wi-Fi

图 10-2　用户逐个应用程序地授权位置服务。这个提示只会出现一次，并且将把用户的选择存储到 Settings > General > Privacy > Application Name 中

万一用户拒绝访问，应用程序将孤立无援，只能指导他们如何手动恢复它。确切地讲，告诉用户在 Settings > Privacy > Location Services 中修改他们的授权选择。

可以用两种方式预防这种可能性。第一，可以在启动时使用 authorizationStatus 类方法测试授权状态。当"未确定"时，还不会提示用户做出决定。你可能希望显示某种预先检查的指导，解释他们为什么应该点按 OK 按钮：

```
if ([CLLocationManager authorizationStatus] ==
    kCLAuthorizationStatusNotDetermined)
{
    // Perform any instructions, e.g. tell user that the
    // app is about to request permission
}
```

第二，可以更新 Info.plist 文件，并添加一句较短的描述，说明将如何使用位置服务。把文本指定给 NSLocationUsageDescription 键，在图 10-2 中可以看到这个键的使用。自定义的描述（App demonstrates location capabilities to users（应用程序给用户演示位置能力））启用提示，更好地解释用户为什么应该授权。对于其他特权服务存在类似的键，包括访问图库、日历、联系人和提醒。

10.1.2　重置位置和隐私

在测试时，你可能想试验不同的情况，即用户授权访问、用户拒绝访问，以及用户授权访问然后撤回它等。你需要使用两个单独的 Settings（设置）窗格来产生这些场景。

要把当前的授权从打开切换到关闭状态或者执行相反的操作，可以访问 Settings > Privacy >

Location（参见图 10-3）。向下滚动屏幕，在那里可以找到一份请求位置访问的应用程序列表，并且在每个应用程序名称旁边会显示一个切换开关。

要把隐私设置恢复到它们的原始状态，即尚未显示授权提示之前，可以转到 Settings > General > Reset。点按 Reset Location & Privacy，然后点按 Reset Warnings（参见图 10-4）。这个选项将把隐私恢复到它的原始状态，即当应用程序下一次启动时将再次提示用户授权。测试应用程序的开发人员使用它特别方便。

图 10-3　在 Settings > Privacy > Location 中可以逐个应用程序地切换隐私设置。向下滚动屏幕，可以找到应用程序的列表

图 10-4　Reset Location & Privacy 按钮允许应用程序再次提示用户授权访问位置信息

10.1.3　检查用户权限

同一种授权状态检查既使你能够知道是否尚未提示用户，还使你能够测试应用程序是授予还是拒绝了权限。如果已授权，就可以开始 Core Location 更新：

```
// Check user permissions
if ([CLLocationManager authorizationStatus] ==
        kCLAuthorizationStatusDenied)
{
    NSLog(@"User has denied location services");
    return;
}


if ([CLLocationManager authorizationStatus] ==
```

```
                kCLAuthorizationStatusAuthorized)
{
    // The app need not query the user. It is already
    // authorized.
}

manager = [[CLLocationManager alloc] init];
[manager startUpdatingLocation];
```

> **注意：**
> 在 Settings > General > Date & Time 中，确保设备的日期、时间和时区设置正确，可以提高 GPS 的精度。选择 Set Automatically 即可。

10.1.4 测试 Core Location 特性

CLLocationManager 类提供了几个类方法，用于测试是否启用了 Core Location 特性。这些方法包括以下一些。

- **authorizationStatus** 指定用户是否为应用程序授权位置访问。如果用户同意允许应用程序使用位置，这个值将返回 YES。
- **locationServicesEnabled** 指示是否在设备上启用了位置服务。用户可以从 Settings > Location Services 中完全启用和禁用服务，或者可以逐个应用程序地启用/禁用服务。
- **significantLocationChangeMonitoringAvailable** 可以让应用程序知道设备是否可以提供低电量/低精度更新，它们是在设备上的原装电池容量改变时发生的。重要的位置改变监测提供了一种极佳的方式，只需做少量的工作即可粗略地跟踪车辆行进以及发现附近吸引人的事物。
- **headingAvailable** 确定位置管理器是否可以使用机载罗盘提供行进方向事件。
- **regionMonitoringAvailable** 指定当前设备上是否可以使用区域监测，regionMonitoringEnabled 指示用户是启用还是撤销了该特性。区域监测可以让应用程序定义要从固定区域移入和移出的地理区域。应用程序可以使用它将动作与真实的位置关联起来。例如，如果进入你的杂货店附近的区域，应用程序可能通知你挑选另外一夸脱牛奶。这称为**地理围栏**（geofencing）。

在尝试使用 Core Location 特性之前，总是要测试它们是否可用。其中一些限制是特定于设备的（例如，不能在第一代 iPhone 上使用 GPS），而另外一些限制则是由用户设置的。如果用户拒绝访问其中一些特性，应用程序将限制于仍然保持授权的任何特性（如果有的话）。在这些情况下，将在 App Store 批准过程中测试应用程序的行为。

> 注意：
> 使用 CLLocationCoordinate2DIsValid() 函数测试某个坐标是否提供了有效的纬度/经度对。纬度需要位于-90°~90°之间，经度则需要位于-180°~180°之间。

10.2 秘诀：Core Location 简介

Core Location 很容易使用，如下面的步骤所示。它们将引领你通过一个设置应用程序的过程，以请求代表正常使用的位置数据。这些步骤和秘诀 10-1 只提供了一个使用 Core Location 服务的示例，显示了如何查明用户的位置。

(1) 向项目中添加 Core Location 框架，并且可以选择编辑 Info.plist 文件，添加一个位置使用描述键。

(2) 测试 CLLocationManager 类的 locationServicesEnabled 类值，检查用户是否启用了 Core Location。用户可以选择从 Settings 应用程序中的 General > Location Services 中关闭 Core Location。

(3) 分配一个位置管理器。把管理器的委托设置为主视图控制器或应用程序委托，也可以选择设置其想要的距离过滤器和精度。距离过滤器指定最短的距离（以米为单位），设备在可以注册新的更新之前必须至少移动这段距离。例如，如果把距离设置为 5 米，那么直到设备移动那么远的距离之后，才能接收到新的事件。如果计划通过步行来执行测试，则可能希望减小那个数字。

精度属性指定你正在请求的精确程度。为了清楚起见，位置管理器不保证任何实际的精度。设置请求的精度将要求管理器（尝试）获取至少那个级别的精度。当不需要精确性时，管理器将使用任何可用的技术释放它的结果。

当需要精确性时，desiredAccuracy 属性将把那种需求通知给管理器。你将发现，对于步行和跑步应用程序，较高级别的精度特别重要。较低级别的精度可能适合于驾驶汽车，或者在较大的地理界限（比如城市、州和国家）内定位用户。

(4) 开始定位。告诉位置管理器开始更新位置。委托回调可让你知道何时找到了一个新位置，这可能要花许多秒或者多达一分钟的时间。

(5) 处理位置事件委托回调。你将处理两类回调：如果成功，就返回 CLLocation 数据（locationManager:didUpdateLocations:，它在 iOS 6 中替换了 locationManager:didUpdateToLocation:fromLocation:）；如果失败，就不返回数据（locationManager:didFailWithError:）。向代码中添加这些委托方法，以捕获位置更新。在秘诀 10-1 中，成功的位置将记录一种信息概述（description），其中包括当前的纬度和经度结果。

依赖于请求的精度，可能基于使用的不同位置方法和请求的精度接收到 3、4 个位置回调，因此要考虑到这种非线性情况。

(6) 开始移动并等待。当位置数据变得可用时，回调将异步到达。返回给应用程序的位置信息包括定位信息，以及可用于评估精确性的精度测量。

可能的话，要在设备上而不是在模拟器上测试 Core Location 应用程序。尽管模拟器现在可以支持各种位置场景，在设备上执行的测试将会提供最佳的结果。把秘诀 10-1 部署到设备上将允许在带着 iOS 设备步行或驾车时测试结果。如果在驾车时执行测试，那么你最好是作为乘客而不是驾驶员，这样可以获得最好的结果。

秘诀 10-1　使用 Core Location 获取纬度和经度

```
- (void)locationManager:(CLLocationManager *)manager
    didFailWithError:(NSError *)error
{
    if ([CLLocationManager authorizationStatus] ==
        kCLAuthorizationStatusDenied)
    {
        [self doLog:@"User has denied location services"];
        return;
    }

    [self doLog:@"Location manager error: %@",
        error.localizedFailureReason];
    return;
}
- (void)locationManager:(CLLocationManager *)manager
    didUpdateLocations:(NSArray *) locations
{
    [self doLog:@"%@\n",
        [[locations lastObject] description]];
}

- (void) startCL
{
    // Test for location services
    if (![CLLocationManager locationServicesEnabled])
    {
        [self doLog:@"User has disabled location services"];
        return;
    }
```

```
// Test for authorization
if ([CLLocationManager authorizationStatus] ==
    kCLAuthorizationStatusDenied)
{
    [self doLog:@"User has denied location services"];
    return;
}

manager = [[CLLocationManager alloc] init];
manager.delegate = self;
manager.desiredAccuracy = kCLLocationAccuracyBest;
manager.distanceFilter = 5.0f; // in meters
[manager startUpdatingLocation];
}
```

获取这个秘诀的代码

要查找这个秘诀的完整示例项目,可以浏览 https://github.com/erica/iOS-6-Advanced-Cookbook,并进入第 10 章的文件夹。

10.2.1 位置属性

由更新的位置回调返回的每个 CLLocation 实例都包含许多属性,它们描述了行进中的设备。位置对象可以通过 description 实例方法把它们的各种属性结合到单个文本结果中,如秘诀 10-1 中所使用的那样。此外,还可以逐个属性地取出每个值。位置属性包括:

- **altitude**:这个属性返回当前检测到的海拔高度。它返回一个海平面以上的浮点数(以米为单位)。笔者可以保证这个值的精度即使往好里说也是最低的,要小心谨慎地使用这些结果。

- **coordinate**:通过 coordinate 属性获取设备的检测到的地理位置。坐标是具有两个字段的结构:latitude 和 longitude,它们都存储一个浮点值。正的纬度值位于赤道的北面,负的纬度值则位于赤道的南面。正的经度值位于子午线的东面,负的经度值则位于子午线的西面。

- **course**:使用 course 值确定设备移动的普通方向。这个值粗略接近于行进的方向,其中 0 表示 North(北)、90 表示 East(东)、180 表示 South(南)、270 表示 West(西)。为了获得更好的精度,可以使用行进方向(CLHeading 实例)而不是路线。行进方向通过磁力计提供了对磁性的和真实的 North 读数的访问。

- **horizontalAccuracy**:这个属性指示当前坐标的精度(即不确定性或测量误差,以米为单位）。可以把返回的坐标视作圆的中心，并把水平精度视作它的半径，真正的设备位置将出现在那个圆内的某个位置。圆越小，位置越精确。圆越大，它将越不精确。负的精度值指示测量失败。

- **verticalAccuracy**：这个属性提供了与水平精度对应的海拔高度。它返回与海拔高度的真实值相关的精度，理论上讲，它可能在海拔高度减去那个数量到海拔高度加上那个数量之间变化。在实际中，海拔高度的读数极不准确，垂直精度通常与现实情况也没有多少关系。
- **speed**：这个值返回设备的速度，以米/秒为单位。
- **timestamp**：这个属性确定位置测量发生的时间。它返回一个 NSDate 实例，该实例被设置为当 Core Location 确定位置时的时间。

10.2.2 跟踪速度

由每个 CLLocation 实例返回的内置 speed 属性允许随着时间的推移跟踪设备的速度。它将以米/秒为单位报告这个值。可以通过把那个值乘以 3.6，轻松地计算以公里/小时（KPH）为单位的速度。如果以英里/小时为单位，则要乘以 2.23693629。

```
- (void)locationManager:(CLLocationManager *)manager
    didUpdateLocations: (NSArray *) locations
{
    CLLocation *newLocation = [locations lastObject];

    // If a speed is detected, log that data in miles per hour
    if (newLocation.speed > 0.0f)
    {
        NSString *speedFeedback = [NSString stringWithFormat:
            @"Speed is %0.1f MPH, %0.1f KPH",
            2.23693629 * newLocation.speed,
            3.6 * newLocation.speed];
        NSLog(@"%@", speedFeedback);
    }
}
```

10.3 秘诀：地理围栏

CLRegion 对象利用某个半径定义了以给定的点为中心的地理区域。无论何时用户越过其边界，区域都允许注册事件。可以调用 initCircularRegionWithCenter:radius:identifier: 创建新的区域，然后可以通过 startMonitoringForRegion:把该区域传递给位置管理器。为你想监测的每个区域调用这个方法。每个标识符都必须是独特的；否则，新区域将覆盖任何旧区域。

一旦越过了区域边界，位置管理器委托就会接收到 locationManager:didEnterRegion:或 locationManager:didExitRegion:回调。如果应用程序没有运行，在注册越过边界的事件时系统将重新启动它，并且 UIApplicationLaunchOptionsLocationKey 将出现在启动选项字典中。落在边界内的第一个成功的位置锁将触发回调。

这意味着越过边界信号并不是实时的，并且不是 100%可靠。你可能会遇到如下情形：用户进入某个区域，但是由于在那个时间内没有找到位置，从而导致在用户再次走出那个区域时没有注册回调。换句话说，要倍加小心地使用这个特性。

秘诀 10-2 演示了你可能需要完成的几个基本任务。listMonitoredRegions 方法用于列出应用程序当前监测的所有区域。每个应用程序一次可以注册最多 20 个区域，这些区域将不依赖于每次运行而持续存在，因此结果可能会反映以前添加的区域。

第二个方法 clearMonitoredRegions 用于中止监测，它可以为与应用程序关联的每个区域停止正在进行的监测。它将遍历位置管理器的 monitoredRegions 属性中的项目，请求停止监测它所发现的每个区域。通过清除这些区域，可以确保当设备进入以前监测的任何区域时应用程序将不会重新启动。

第三个方法是 markAndMonitor。这个方法使用 50 米的半径在当前位置设置一道围栏。这个范围最容易使用汽车进行测试，尽管作为乘客比作为驾驶员测试这个特性将更好一些。该方法将基于现有区域的数量生成它的区域名称，你必须确保区域名称将不会与现有的注册名称相冲突。

最后两个方法实现了用于进入和离开区域的回调。在这里，最低限度地实现了简单地记录更新。关注区域监测的应用程序自然会提供更健壮、更有趣的实现。

秘诀 10-2　使用 Core Location 创建地理围栏

```
// List all regions registered to the current location manager
- (void) listMonitoredRegions
{
    for (CLRegion *eachRegion in [manager monitoredRegions])
        [self doLog:@"Region: %@", eachRegion];
}

// Remove all region monitors
- (void) clearMonitoredRegions
{
    for (CLRegion *eachRegion in [manager monitoredRegions])
```

```objc
    {
        [self doLog:@"Stopping monitor for %@", eachRegion];
        [manager stopMonitoringForRegion:eachRegion];
    }
}

// Create a new region monitor at the current location
- (void) markAndMonitor
{
    if (!mostRecentLocation)
    {
        [self doLog:@"No location. Sorry"];
        return;
    }

    [self doLog:@"Setting Geofence"];
    NSString *geofenceName = [NSString stringWithFormat:
        @"Region #%d", manager.monitoredRegions.count + 1];
    CLRegion *region = [[CLRegion alloc]
        initCircularRegionWithCenter:
            mostRecentLocation.coordinate
        radius:50.0f identifier:geofenceName];
    [manager startMonitoringForRegion:region];
}

// Callback for entering region
- (void) locationManager:(CLLocationManager *)manager
    didEnterRegion:(CLRegion *)aRegion
{
    [self doLog:@"Entered region %@", aRegion.identifier];
}

// Callback for departing region
- (void) locationManager:(CLLocationManager *)manager
    didExitRegion:(CLRegion *)aRegion
{
    [self doLog:@"Leaving region %@", aRegion.identifier];
}
```

> **获取这个秘诀的代码**
> 要查找这个秘诀的完整示例项目，可以浏览 https://github.com/erica/iOS-6-Advanced-Cookbook，并进入第 10 章的文件夹。

10.4 秘诀：使用行进方向值跟踪"North"

机载位置管理器可以返回一个计算的 course 值，指示当前的行进方向，即 North（北）、South（南）、Southeast（东南）等。这些值采用 0~360 之间的浮点数的形式，其中 0° 指示 North（北）、90° 指示 East（东）等。这个计处的值源于随着时间的推移跟踪用户的位置。更新的设备具有更好的方式确定用户的行进路线。最新的设备提供了一个机载磁力计，它可以返回磁性的 North 值和真实的 North 值。

并非所有的 iOS 设备都支持行进方向；尽管此时，大多数现代设备都会支持。磁力计最初是在 iPhone 3GS 中推出的。在预订行进方向回调之前，每个设备都要测试是否具有这种能力。如果位置管理器可以生成行进方向事件，headingAvailable 属性将返回 YES。使用这个结果来控制 startUpdatingHeading 请求：

```
if (CLLocationManager.headingAvailable)
    [manager startUpdatingHeading];
```

Cocoa Touch 允许过滤行进方向回调，就像对距离回调所做的那样。把位置管理器的 headingFilter 属性设置为最小的角度改变，将其指定为一个浮点数。例如，如果直到设备旋转了至少 5° 之后，才希望接收到反馈，则可以把该属性设置为 5.0。所有的行进方向值都使用度，其值在 0.0~360.0 之间。要把行进方向值转换为弧度，可以除以 180.0，然后乘以 π。

行进方向回调返回一个 CLHeading 对象。可以查询行进方向的两个属性：magneticHeading 和 trueHeading。前者返回磁性 North 的相对位置，真正的 North 总是指向地理上的北极。磁性 North 对应于地球磁场的极点，它将随着时间的推移而改变。iPhone 使用一个计算的偏移量（称为**磁偏角**（declination））来确定两者之间的差值。

在启用磁力计的设备上，将会提供磁性行进方向更新，即使用户在 Settings 应用程序中关闭了位置更新也会如此。更重要的是，不会提示用户授权使用行进方向数据。磁性行进方向信息不能危及用户隐私，因此它应该可供应用程序自由地使用。

只能把 trueHeading 属性用于位置检测。iPhone 需要设备的位置，以计算确定真实的 North 所需的磁偏角。磁偏角因地理位置而异。洛杉矶的磁偏角不同于珀斯的，珀斯的又不同于莫斯科和伦敦的，等等。一些位置不能使用磁力计读数。某些异常的区域（比如苏必利尔湖中的米奇皮科滕岛和美国新墨西哥州的格兰茨）提供了铁矿床和熔岩流，它们会干扰正常的磁性罗盘使用。

金属和磁源（比如计算机、汽车或冰箱）也可能会影响磁力计。App Store 中的几个"金属检测器"应用程序利用了这种特性。

headingAccuracy 属性提供了一个误差值。这个数字指示实际的行进方向所属于的一个加减范围。较小的误差条指示更准确的读数，负值表示读取行进方向中的误差。

可以使用 CLHeading 属性 x、y 和 z 沿着 x 轴、y 轴和 z 轴获取原始磁性值。这些值是以微特斯拉为单位测量的，并且规范化到 Apple 指定的-128~128 的范围（基于标准的位数学运算，实际的范围更可能是-128~127）。每个轴值都表示距离设备的内置磁力计跟踪的磁力线的偏移量。

秘诀 10-3 使用 CLHeading 数据利用一个箭头指针旋转较小的图像视图，旋转可以确保箭头总是指向 North。图 10-5 显示了实际的界面。

秘诀 10-3　检测 North 的方向

```
// Allow calibration
- (BOOL)locationManagerShouldDisplayHeadingCalibration:
    (CLLocationManager *)manager
{
    return YES;
}


// Respond to heading updates
- (void)locationManager:(CLLocationManager *)manager
    didUpdateHeading:(CLHeading *)newHeading
{
    CGFloat heading = -M_PI * newHeading.magneticHeading / 180.0f;
    imageView.transform = CGAffineTransformMakeRotation(heading);
}
- (void) startCL
{
    if (![CLLocationManager locationServicesEnabled])
    {
        [self doLog:@"User has disabled location services"];
        return;
    }

    if ([CLLocationManager authorizationStatus] == kCLAuthorizationStatusDenied)
    {
        [self doLog:@"User has denied location services"];
        return;
    }
```

```
manager = [[CLLocationManager alloc] init];
manager.delegate = self;
if ([CLLocationManager headingAvailable])
    [manager startUpdatingHeading];
else
    imageView.alpha = 0.0f;
}
```

图 10-5　iPhone 的内置磁力计和秘诀 10-4 中的代码确保这个箭头总是指向 North

获取这个秘诀的代码

要查找这个秘诀的完整示例项目，可以浏览 https://github.com/erica/iOS-6-Advanced-Cookbook，并进入第 10 章的文件夹。

10.5　秘诀：前向和反向地理编码

反向地理编码（reverse geocoding）这个短语意指把纬度和经度信息转变成人类可识别的地址信息；正向地理编码（forward geocoding）则接受用户可读的地址，并把它转换成纬度和经度值。MapKit 提供了正向和反向地理编码器例程，作为它的 CLGeocoder 类的一部分。

CLPlacemark 类存储所有这类信息。它的 location 属性包含纬度和经度的 coordinate 值；它的现实世界的属性（country、postalCode、locality 等）存储人类可读的元素。可以在任何一个方向上（从现实世界的地址或者坐标）完全填充这一个类。

秘诀 10-4 演示了如何使用 CLGeocoder 的基于块的编码方法，执行正向和反向地理编码。用户的设备必须具有网络访问，以执行地理编码。

对于正向地理编码，可以提供地址字符串。对于反向地理编码，可以创建一个 CLLocation 实例并传递它。给完成块传递一个匹配地标的数组。如果该数组为空，则会显示一条错误解释原因。

Apple 列举了使用这个类的几个规则。第一，一次只执行一个地理编码请求。第二，只要有可能，就要重用现有的地理编码结果，而不是重新计算已知的细节。对于真实的更新，每分钟不要发送多个请求，并且直到用户已经移动了至少一段相当长的距离之后才能发送任何请求。最后，仅当周围有用户要查看结果时（也就是说，当应用程序没有挂起或者在后台运行时）才执行地理编码。

CLPlacemark 对象还在字典结构之外提供了各个属性以及相同的信息。这些属性包括如下这些。

- subThoroughfare 存储街道号码（例如，"1600"表示宾夕法尼亚林荫大道 1600 号）。
- thoroughfare 包含街道名称（例如，宾夕法尼亚林荫大道）。
- sublocality，当可用时，指本地邻近地区的名称或地标（例如，白宫）。
- subAdministrativeArea 通常指当地县、教区或其他行政区域。
- locality 存储城市（例如，华盛顿特区）。
- administrativeArea 对应于州，比如马里兰州或弗吉尼亚州。
- postalCode 是邮政编码或 ZIP 编码（例如，20500）。
- country 是自解释的，存储国家名称，比如美国。
- ISOcountryCode 提供简写的国家名称，比如 "US"。

addressDictionary 存储位置的 AddressBook 框架风格的版本。额外的属性包括 inlandWater 和 ocean，它们描述了与地标关联的任何本地的湖泊和海洋；还有一个迷人的属性 areasOf-Interest，指地标本地所有的，比如国家公园或者吸引人的事物。

秘诀 10-4　从坐标和描述中获取地址信息

```
- (void) reverseGeocode: (id) sender
{
    // Starting location
    CLLocation *location = [[CLLocation alloc]
        initWithLatitude:37.33168400
        longitude:-122.03075800];
    CLGeocoder *geocoder = [[CLGeocoder alloc] init];
    [geocoder reverseGeocodeLocation:location
        completionHandler:^(NSArray *placemarks, NSError *error)
    {
        if (!placemarks)
        {
            [self doLog:@"Error retrieving placemarks: %@",
                error.localizedFailureReason];
            return;
        }
```

```objectivec
        [self doLog:@"Placemarks from Location: %f, %f",
            location.coordinate.latitude,
            location.coordinate.longitude];

        for (CLPlacemark *placemark in placemarks)
        {
            [self doLog:@"%@", placemark.description];
        }
    }];
}
- (void) geocode: (id) sender
{
    // Retrieve coordinates from an address string
    CLGeocoder *geocoder = [[CLGeocoder alloc] init];
    NSString *address = @"1 Infinite Loop, Cupertino, CA 95014";
    [geocoder geocodeAddressString:address completionHandler:
        ^(NSArray *placemarks, NSError *error)
        {
            if (!placemarks)
            {
                [self doLog:@"Error retrieving placemarks: %@",
                    error.localizedFailureReason];
                return;
            }

            [self doLog:@"Placemarks from Description (%@):",
                address];

            for (CLPlacemark *placemark in placemarks)
            {
                [self doLog:@"%@", placemark.description];
            }
        }];
}
```

获取这个秘诀的代码

要查找这个秘诀的完整示例项目，可以浏览 https://github.com/erica/iOS-6-Advanced-Cookbook，并进入第 10 章的文件夹。

10.6 秘诀：查看位置

MKMapView 类可以给用户展示交互式地图，它们是在你提供的坐标和比例的基础上构建的。下面的代码段把地图的区域设置为 Core Location 坐标，显示"1 Infinite Loop"（无限循环圈 1 号）[1]周围 0.1° 的纬度和经度。在美国，具有这个范围的区域对应于相对较小的城市或者较大的城镇，大约是 5×7 英里。图 10-6（左图）在地图视图上显示 0.1°×0.1° 的范围：

```
// 1 Infinite Loop
CLLocation *location = [[CLLocation alloc]
    initWithLatitude:37.33168400 longitude:-122.03075800];
mapView.region = MKCoordinateRegionMake(
    location.coordinate, MKCoordinateSpanMake(0.1f, 0.1f));
```

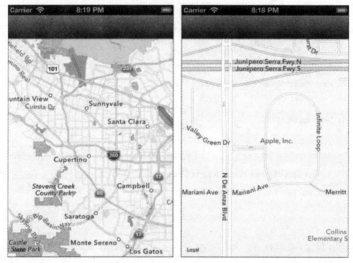

图 10-6　0.1° 的纬度×0.1° 的经度的坐标区域可以覆盖小城市或大城镇的区域，一边是 5~7 英里长（左图）。在一边把那个区域收缩到 0.005° 将产生街道级别的显示效果（右图）

由于地球的曲度，区域大小会发生变化。在赤道，1° 的经度对应于大约 69 英里（约等于 111 公里）。在极地，它将收缩到 0。纬度不受位置影响，1° 的纬度总是对应于大约 69 英里（约等于 111 公里）。

要查看邻近地区级别的地图数据，可以把坐标跨度缩小到 0.01°×0.01°。对于街道×街道级别，可以使用更小的跨度，比如说缩小一半，即 0.005° 的纬度×0.005° 的经度。图 10-6（右图）显示了这个范围的一个无限循环圈。

---

[1] Apple 公司的总部所在地——译者注。

可以避免处理纬度和经度度数，并通过以米为单位指定距离来创建区域。下面这个代码段把视图区域设置为中心坐标周围 500×500 平方米。这粗略近似于 0.005° ×0.005° 的纬度/经度的跨度，显示街道×街道的表示：

```
mapView.region = MKCoordinateRegionMakeWithDistance(
    location.coordinate, 500.0f, 500.0f);
```

10.6.1 查找最佳的位置匹配

秘诀 10-5 使用定时的方法执行所需的位置搜索。当用户点按 Find Me 按钮时，代码将启动一个 10 秒的计时器。在这个搜索期间，它将尝试找到可能最佳的位置。它使用每个点击位置返回的水平精度，来选择并保持最精确的地理定位。在时间结束时，视图控制器将放大它的地图视图，呈现检测到的位置。

秘诀 10-5 显示了在搜索期间及之后的当前用户位置。它通过把 showsUserLocation 属性设置为 YES 来执行该任务。当启用这个属性时，它将产生一个脉冲图钉，它最初出现在设备所在位置的地图视图的中心。这个位置是利用 Core Location 检测的，并将依据用户移动而更新。可以在模拟器中使用 Debug > Location 选项测试它。

无论何时启用这个属性，地图视图都会给 Core Location 分配一项查找设备的当前位置的任务。只要保持将这个属性设置为 YES，地图就会继续跟踪并定期更新用户位置。图钉周围的脉冲圆圈指示最新的搜索精度。记住：地图根本不会努力把用户位置放在地图视图的可见部分；你需要自己处理它。

在确定位置之后，秘诀 10-5 将允许用户开始与地图交互。启用 zoomEnabled 属性意味着用户可以挤压、拖动显示的地图，以及与之交互或者探索它。这个秘诀在允许进行这种交互前将等待，直到完成整个搜索工作，以确保直到控制返回给用户之后用户位置仍然保持在中心。

一旦完成了搜索，这个秘诀将通过调用 stopUpdatingLocation 来停止请求位置回调。与此同时，它将允许地图视图继续跟踪用户，并保持将 showsUserLocation 属性设置为 YES。

在取消预订更新之后，视图控制器实例将把它的位置管理器委托设置为 nil。在计时器结束之后，这种分配将阻止任何未决的回调到达控制器，否则，用户和未决的回调可能会争夺屏幕的控制权。

秘诀 10-5　在地图内展示用户位置

```
- (void)locationManager:(CLLocationManager *)manager
    didFailWithError:(NSError *)error
{
    NSLog(@"Location manager error: %@",
```

```objc
                error.localizedDescription);
}

- (void)locationManager:(CLLocationManager *)manager
    didUpdateLocations: (NSArray *) locations
{
    CLLocation *newLocation = [locations lastObject];

    // Keep track of the best location found
    if (!bestLocation)
        bestLocation = newLocation;
    else if (newLocation.horizontalAccuracy <
        bestLocation.horizontalAccuracy)
        bestLocation = newLocation;

    mapView.region = MKCoordinateRegionMake(
        bestLocation.coordinate,
        MKCoordinateSpanMake(0.1f, 0.1f));
    mapView.showsUserLocation = YES;
    mapView.zoomEnabled = NO;
}

// Search for n seconds to get the best location during that time
- (void) tick: (NSTimer *) timer
{
    if (++timespent == MAX_TIME)
    {
        // Invalidate the timer
        [timer invalidate];

        // Stop the location task
        [manager stopUpdatingLocation];
        manager.delegate = nil;

        // Restore the find me button
        self.navigationItem.rightBarButtonItem =
            BARBUTTON(@"Find Me", @selector(findme));

        if (!bestLocation)
        {
```

```objc
        // no location found
        self.title = @"";
        return;
    }

    // Note the accuracy in the title bar
    self.title = [NSString stringWithFormat:@"%0.1f meters",
        bestLocation.horizontalAccuracy];

    // Update the map and allow user interaction
    [mapView setRegion:
        MKCoordinateRegionMakeWithDistance(
            bestLocation.coordinate, 500.0f, 500.0f)
        animated:YES];
    mapView.showsUserLocation = YES;
    mapView.zoomEnabled = YES;
    }
    else
        self.title =
            [NSString stringWithFormat:@"%d secs remaining",
                MAX_TIME - timespent];
}
// Perform user-request for location
- (void) findme
{
    // disable right button
    self.navigationItem.rightBarButtonItem = nil;

    // Search for the best location
    timespent = 0;
    bestLocation = nil;
    manager.delegate = self;
    [manager startUpdatingLocation];
    [NSTimer scheduledTimerWithTimeInterval:1.0f
        target:self selector:@selector(tick:)
        userInfo:nil repeats:YES];
}

- (void) loadView
{
```

```
    self.view = [[UIView alloc] init];

    // Add a map view
    mapView = [[MKMapView alloc] initWithFrame:self.view.bounds];
    [self.view addSubview:mapView];

    if (!CLLocationManager.locationServicesEnabled)
    {
        NSLog(@"User has opted out of location services");
        return;
    }
    else
    {
        // User generally allows location calls
        manager = [[CLLocationManager alloc] init];
        manager.desiredAccuracy = kCLLocationAccuracyBest;
        self.navigationItem.rightBarButtonItem =
            BARBUTTON(@"Find Me", @selector(findme));
    }
}
```

获取这个秘诀的代码

要查找这个秘诀的完整示例项目,可以浏览 https://github.com/erica/iOS-6-Advanced-Cookbook,并进入第 10 章的文件夹。

10.7 秘诀:用户位置注释

秘诀 10-5 提供了一种方式,可以随着时间的推移可视化地跟踪一个被关注的位置事件。秘诀 10-6 把这种思想提升了一个档次,将随着时间的推移跟踪设备的移动。它不会随着时间的推移对位置进行抽样并选择最佳的效果,而是利用一种更容易的方法,同时实现类似的结果。秘诀 10-6 把与用户位置相关的任务交给地图视图及其 userLocation 属性处理。

为了支持这种做法,需要建立 showsUserLocation 属性,并把地图视图的跟踪模式设置为跟随用户:

```
mapView = [[MKMapView alloc] init];
mapView.showsUserLocation = YES;
mapView.userTrackingMode = MKUserTrackingModeFollow;
```

秘诀 10-6 每隔几秒钟就会检查用户位置。它将以多种方式更新地图视图,以反映那个位置。第一，它将使地图保持在用户的当前位置的中心。第二，它将给用户图钉添加自定义的注释，以显示当前的坐标。最后，它将尝试找到人类可读的地标以与之关联。如果它找到一个地标，就会在屏幕底部的文本视图中显示它，如图 10-7 所示。

注释是附加到地图中的位置上的弹出式视图。它们提供了标题和子标题，可以根据需要设置它们。图 10-7 显示了秘诀 10-6 构建的注释。MKUserLocation 类提供了对用户位置图钉及其关联的注释的直接访问。注释提供了两个可读写的属性：title 和 subtitle，可以根据需要设置这些属性。秘诀 10-6 把标题设置为 "Location Coordinates"，并把子标题设置为一个包含纬度和经度的字符串。

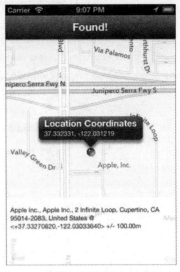

图 10-7 地图可以提供它们自己的用户位置信息，而无需直接给 Core Location 分配任务

这个类极大地简化了注释的编辑工作，但是在这种情况下，将限制于只能处理地图视图的用户位置属性。事实证明，针对注释的更一般的情况将更复杂。在本节后面的秘诀 10-7 中详细描述了它。

秘诀 10-6 通过 mapView 跟踪设备

```
// Search for n seconds to get the best location during that time
- (void) tick: (NSTimer *) timer
{
    self.title = @"Searching...";
    if (mapView.userLocation)
    {
        // Check for valid coordinate
        CLLocationCoordinate2D coord =
            mapView.userLocation.location.coordinate;
        if (!coord.latitude && !coord.longitude) return;

        // Update titles
        self.title = @"Found!";
        [mapView setRegion:MKCoordinateRegionMake(coord,
            MKCoordinateSpanMake(0.005f, 0.005f)) animated:NO];
        mapView.userLocation.title = @"Location Coordinates";
        mapView.userLocation.subtitle = [NSString stringWithFormat:
            @"%f, %f", coord.latitude, coord.longitude];
```

```objc
            // Attempt to retrieve placemarks
            CLGeocoder *geocoder = [[CLGeocoder alloc] init];
            [geocoder reverseGeocodeLocation:
                    mapView.userLocation.location
                    completionHandler:^(
                    NSArray *placemarks, NSError *error)
            {
                if (!placemarks)
                {
                    NSLog(@"Error retrieving placemarks: %@",
                        error.localizedFailureReason);
                    return;
                }

                NSMutableString *marks = [NSMutableString string];
                for (CLPlacemark *placemark in placemarks)
                {
                    [marks appendFormat:@"\n%@", placemark.description];
                    textView.alpha = 0.75f;
                    textView.text = marks;
                }
            }];
    }
}

- (void) loadView
{
    self.view = [[UIView alloc] init];

    // Add map
    mapView = [[MKMapView alloc] init];
    mapView.showsUserLocation = YES;
    mapView.userTrackingMode = MKUserTrackingModeFollow;
    [self.view addSubview:mapView];

    if (!CLLocationManager.locationServicesEnabled)
    {
        NSLog(@"User has opted out of location services");
        return;
    }
```

```
    else
    {
        [NSTimer scheduledTimerWithTimeInterval:5.0f target:self
            selector:@selector(tick:) userInfo:nil repeats:YES];
    }
}
```

获取这个秘诀的代码

要查找这个秘诀的完整示例项目,可以浏览 https://github.com/erica/iOS-6-Advanced-Cookbook,并进入第 10 章的文件夹。

10.8 创建地图注释

Cocoa Touch 定义了一个 MKAnnotation 协议。必须设计你自己的符合这个协议的类,它需要一个 coordinate 属性以及 title 和 subtitle 实例方法。程序清单 10-1 演示了如何执行该任务。它将构建一个简单的 MapAnnotation 类,提供协议所需的坐标、标题和子标题特性。额外的 tag 整数不是协议所需的,这里只是为了方便起见才添加它。

程序清单 10-1 构建地图注释对象

```
@interface MapAnnotation : NSObject <MKAnnotation>
- (id) initWithCoordinate: (CLLocationCoordinate2D) aCoordinate;
@property (nonatomic, readonly) CLLocationCoordinate2D coordinate;
@property (nonatomic, copy) NSString *title;
@property (nonatomic, copy) NSString *subtitle;
@property (nonatomic, assign) NSUInteger tag;
@end

@implementation MapAnnotation
- (id) initWithCoordinate: (CLLocationCoordinate2D) aCoordinate
{
    if (self = [super init])
        coordinate = aCoordinate;
    return self;
}
@end
```

10.8.1 创建、添加和删除注释

要使用注释，可以创建它们并把它们添加到地图视图中。一次可以添加单独一条注释：

```
anAnnotation = [[MapAnnotation alloc] initWithCoordinate:coord];
[mapView addAnnotation:anAnnotation];
```

此外，也可以构建一个注释的数组并同时添加它们：

```
[annotations addObject:annotation];
[mapView addAnnotations:annotations];
```

要从地图中删除注释，可以执行 removeAnnotation:，只删除一条注释；或者执行 removeAnnotations:，删除数组中的所有项目。

如果需要把地图视图返回到无注释状态，可以删除它的全部现有的注释。这个调用将通过 annotations 属性获取现有注释的数组，然后从地图中删除它们：

```
[mapView removeAnnotations:mapView.annotations];
```

这将删除注释过的用户位置和任何其他的程序注释，因此只会清除应用程序注释，你可能想测试每个注释，以确保它不属于 MKUserLocation 类：

```
- (void) clear
{
    NSArray *annotations =
        [NSArray arrayWithArray:mapView.annotations];
    for (id annotation in annotations)
        if (![annotation isKindOfClass:[MKUserLocation class]])
            [mapView removeAnnotation:annotation];
}
```

10.8.2 注释视图

注释对象不是视图。程序清单 10-1 中的 MapAnnotation 类不会创建任何屏幕上的元素。它是一个描述注释的抽象类。在需要时把注释描述转换为屏幕上的实际视图是地图视图的职责。那些视图属于 MKAnnotationView 类。可以通过查询地图，获取用于现有注释的注释视图，只需提供注释并请求匹配的视图即可。如果当前在地图视图上没有呈现注释，下面的调用将返回 nil：

```
annotationView = [mapView viewForAnnotation:annotation];
```

MKPinAnnotationView 是你可能想使用的一个方便的 MKAnnotationView 子类。它们是可以放到地图上的图钉。点按时，它们将显示一个标注视图。

10.8.3 自定义注释视图

在通过 addAnnotation: or addAnnotations:添加注释之后，地图视图将开始构建与那些注释对应的注释视图。当它完成时，它的委托（它必须声明 MKMapViewDelegate 协议）将接收到一个回调。当构建视图并添加到地图中之后，将利用 mapView:didAddAnnotationViews:通知委托。这个回调给应用程序提供了一个机会自定义那些注释视图。

注释视图的数组将作为第二个参数传递给那个回调。可以遍历这个数组，设置像视图的 image 这样的特性，或者自定义它的附属按钮。

程序清单 10-2 显示了如何基于注释来准备其中每个注释视图，以便使用它们。

程序清单 10-2　准备注释视图以便使用

```
- (void)mapView:(MKMapView *)mapView
    didAddAnnotationViews:(NSArray *)views
{
    // Initialize each view
    for (MKPinAnnotationView *mkaview in views)
    {
        // Only update custom items
        if (![mkaview isKindOfClass:
            [MKPinAnnotationView class]])
            continue;

        // Set the color to purple
        MapAnnotation *annotation = mkaview.annotation;
        mkaview.pinColor = MKPinAnnotationColorPurple;

        // Add buttons to each one
        UIButton *button = [UIButton
            buttonWithType:UIButtonTypeDetailDisclosure];
        mkaview.rightCalloutAccessoryView = button;
    }
}
```

这个示例设置一种图钉颜色并显示一个按钮，它允许应用程序利用注释视图响应用户交互。你将不会受限于内置的注释协议，利用程序清单 10-1 中定义的类最低限度地满足了它。可以利用你喜欢的任何实例变量和方法设计注释类，以便更多地控制如何查询注释，使注释视图做好准备。

每个注释视图都通过它的 annotation 属性提供了对其注释的直接访问。使用该注释数据构建所需的准确视图。下面介绍了你将希望在 MapKit 应用程序中自定义的一些注释视图属性。

每个 MKPinAnnotation 视图都会使用一种颜色，可以通过 pinColor 属性设置它。MapKit 提供了 3 种颜色选择：红色（MKPinAnnotationColorRed）、绿色（MKPinAnnotationColorGreen）和紫色（MKPinAnnotationColorPurple）。依据 Apple 的人机界面的指导原则，红色图钉指示目标点，这是用户可能想要探索或导航到的位置。绿色图钉是起点，用户可以从这个位置开始一段旅程。紫色图钉是用户指定的。当你鼓励用户把新数据添加到地图中时，可以使用紫色指示用户定义了它们。如你在前面的秘诀中所看到的，地图视图定义的淡蓝色图钉指示当前的用户位置。

每个注释视图都提供了两个槽，分别位于标注气泡的左右两边。rightCalloutAccessoryView 和 leftCalloutAccessoryView 属性允许向标注中添加按钮或者其他任何自定义的子视图。程序清单 10-2 添加了详细的用于展示信息的右侧标注。不过，不仅限于添加按钮，还可以根据需要添加图像视图或其他的标准 Cocoa Touch 视图。

canShowCallout 属性用于控制当点按一个按钮时是否会产生标注视图。这个属性默认是启用的，如果不希望用户点按以打开标注，可以把它设置为 NO。

可以通过把 calloutOffset 属性设置为一个新的 CGPoint 来偏移标注（通常情况下，它们将出现在所处理的图钉正上方）。也可以调整注释视图自身的 centerOffset 属性，来改变它的位置。视图的艺术默认是利用图钉注释设置的，但是可以通过把一个 UIImage 赋予视图的 image 属性，创建自定义的注释艺术。把自定义的艺术与中心偏移结合起来，可以产生你想要的准确的地图外观。

10.8.4 响应注释按钮的点按动作

MapKit 简化了按钮点按动作的管理。无论何时把标注附属的视图属性设置为一个控件，MapKit 都会接管控件回调。不需要添加目标和动作，MapKit 会为你处理这些。你只需实现 mapView:annotationView:calloutAccessoryControlTapped:委托回调，如秘诀 10-7 中所示。

秘诀 10-7 使用户能够通过点按 Tag 按钮，向当前地图中添加图钉，这将把图钉安置在地图的中心位置。无论何时用户调整地图，地图视图委托都会接收到一个 mapView:regionDidChange-

图 10-8　当用户放置图钉时，这些自定义的注释视图将报告时间和位置。点按附属视图将会计算从图钉到当前用户位置的距离，并在标题栏中显示它

Animated:回调。该回调通过它的 centerCoordinate 属性提取地图中心的坐标,并把它存储当前的坐标。

当用户点按附属视图上的展开按钮时,应用程序将会计算从那个图钉到当前用户位置的距离,并在屏幕顶部显示它(以米为单位)。图 10-8 显示了在 Apple 的 Infinite Loop(无限循环圈)总部使用它时的样子。

秘诀 10-7　创建加注释的交互式地图

```
// Test this using a single set location rather than zooming
// around highways

// Update current location when the user interacts with map
- (void)mapView:(MKMapView *)aMapView
    regionDidChangeAnimated:(BOOL)animated
{
    current = [[CLLocation alloc]
        initWithLatitude:mapView.centerCoordinate.latitude
        longitude:mapView.centerCoordinate.longitude];
}
- (void)mapView:(MKMapView *)aMapView
    annotationView:(MKAnnotationView *)view
    calloutAccessoryControlTapped:(UIControl *)control
{
    // Calculate coordinates and distance
    CLLocationCoordinate2D viewCoord = view.annotation.coordinate;
    CLLocation *annotationLocation =
        [[CLLocation alloc] initWithLatitude:viewCoord.latitude
            longitude:viewCoord.longitude];
    CLLocation *userLocation = mapView.userLocation.location;
    float distance = [userLocation
        distanceFromLocation:annotationLocation];

    // Set the title
    self.title = [NSString stringWithFormat:@"%0f meters", distance];
}

// Set colors and add buttons
- (void)mapView:(MKMapView *)mapView
    didAddAnnotationViews:(NSArray *)views
{
```

```objc
        // Initialize each view
        for (MKPinAnnotationView *mkaview in views)
        {
            if (![mkaview isKindOfClass:[MKPinAnnotationView class]])
                continue;

            // Set the color to purple
            mkaview.pinColor = MKPinAnnotationColorPurple;

            // Add buttons to each one
            UIButton *button = [UIButton buttonWithType:
                UIButtonTypeDetailDisclosure];
            mkaview.rightCalloutAccessoryView = button;
        }
}
- (void) tag
{
    // Create a new annotation
    MapAnnotation *annotation =
        [[MapAnnotation alloc] initWithCoordinate:current.coordinate];

    // Label it with time and place
    NSString *locString = [NSString stringWithFormat:@"%f, %f",
            current.coordinate.latitude, current.coordinate.longitude];
    NSDateFormatter *formatter = [[NSDateFormatter alloc] init];
    formatter.timeStyle = NSDateFormatterLongStyle;
    annotation.title = [formatter stringFromDate:[NSDate date]];
    annotation.subtitle = locString;

    // Add it
    [mapView addAnnotation:annotation];
}

// Clear all user annotations
- (void) clear
{
    NSArray *annotations = [NSArray arrayWithArray:mapView.annotations];
    for (id annotation in annotations)
        if (![annotation isKindOfClass:[MKUserLocation class]])
            [mapView removeAnnotation:annotation];
```

```objc
}

- (void) loadView
{
    self.view = [[UIView alloc] init];
    
    
    // Add map
    mapView = [[MKMapView alloc] init];
    mapView.showsUserLocation = YES;
    mapView.userTrackingMode = MKUserTrackingModeFollow;
    [self.view addSubview:mapView];
    
    if (!CLLocationManager.locationServicesEnabled)
    {
        NSLog(@"User has opted out of location services");
        return;
    }
    else
    {
        mapView.delegate = self;
        self.navigationItem.rightBarButtonItem = 
            BARBUTTON(@"Tag", @selector(tag));
        self.navigationItem.leftBarButtonItem = 
            BARBUTTON(@"Clear", @selector(clear));
    }
}
```

> **获取这个秘诀的代码**
>
> 要查找这个秘诀的完整示例项目，可以浏览 https://github.com/erica/iOS-6-Advanced-Cookbook，并进入第 10 章的文件夹。

10.9 小结

Core Location 与 MapKit 密切协作，提供了一些方式用于定位设备的位置，以及在一种基于地图的一致表示中展示相关的位置信息。在本章中，你发现了如何使用 Core Location 来获得实时的纬度和经度坐标，以及如何把这些坐标反向地理编码成实时的地址信息；获悉了如何以原始形式和计算的形式处理速度和行进方向；还学习了如何建立地图、调整它的原点，以及添加用户位置和自定义的注释。在结束本章的学习之前，还要思考以下几点。

- 在决定如何接近所需的位置之前，要了解你的用户以及他们将如何使用你的应用程序。一些 Core Location 特性更适合于驾驶汽车，另外一些则更适合于步行和骑自行车。
- 一定要进行测试。必须在现场以及在 Xcode 上对 Core Location 应用程序进行穷尽测试和调整，以便在 App Store 中获得最佳的结果。获取位置数据不是一门精密科学，要在应用程序中构建必需的富余空间。
- "噢，今天我没有在西经-104.28393° 看到你呢！"对于大多数人来说，地址比坐标更有意义。使用反向地理编码产生人类可读的信息，并使用前向地理编码把地址转换为坐标。
- 邮政编码/ZIP 编码对于 API 特别友好。即使你没有计划在应用程序中使用地图表示，ZIP 编码也会为传统的 GUI 集成做好准备。反向地理编码的 ZIP 编码有助于获取附近的零售信息，比如地址、电话号码以及关于附近的公园和吸引人的事物的信息。
- 良好设计的注释视图有助于把有意义的交互性引入地图视图中。不要害怕使用按钮、图像和其他自定义的元素，它们可以扩展地图的功用。

第 11 章 GameKit

本章介绍了可以通过 GameKit 创建相联系的游戏玩法的多种方式。GameKit 提供了一些特性，使应用程序能够超越单玩家/单设备场景，并过渡到使用 Game Center（游戏中心）和设备与设备之间的联网。

Apple 的 Game Center 添加了一种集中式服务，使游戏能够提供共享的排行榜和基于 Internet 的配对。GameKit 还为对等连接提供了即席的联网解决方案。这种即席的连接构建在一种称为 Bonjour 的技术的基础之上，它在设备之间提供了简单的、无配置的通信。

在本章中，你将发现如何使用 GameKit 构建相连接的应用程序，还将看到如何创建 Game Center 特性。你将学习把 GameKit Voice 添加到代码中，以便进行对讲机式的语音聊天。准备好了吗？现在就让应用程序连接起来吧！

11.1 启用 Game Center

Game Center 特性扩展了用户的体验。利用 Game Center，可以提供配对、排行榜及其他增强功能，使用户能够超越单个应用程序的界限，并获得一种共享式的团队体验。Game Center 不仅仅用于游戏，尽管它们显然非常适合于此目的。可以使用 Game Center 跟踪减肥和健身的进展、创建共享式设计空间，或者提供辅导支持。如果你可以想象一种方式，跨设备共享应用程序执行，那么 Game Center 将可用于提供你所需的基础设施。

在开始编码 Game Center 应用程序之前，需要执行下面几个步骤。

(1) 在 Apple 的 iOS 开发人员论坛（https://developer.apple.com/ios/manage/overview/index.action）上注册一个独特的应用程序标识符。

(2) 创建一个 1024 像素×1024 像素的应用程序图标和一幅 iPhone 大小的替换图像（640 像素×960 像素）。iTunes Connect 需要它们以构建新的应用程序清单。

(3) 访问 iTunes Connect（http://itunesconnect.apple.com）。使用你的标识符和艺术作品注册一个新应用程序，此时不需要最终确定信息，只需添加足够的细节，使应用程序清单能够正确保存即可。

(4) 访问 Manage Your Applications > Application Name > Manage Game Center，如图 11-1 所示。确保启用了 Game Center，默认就应该是这样。

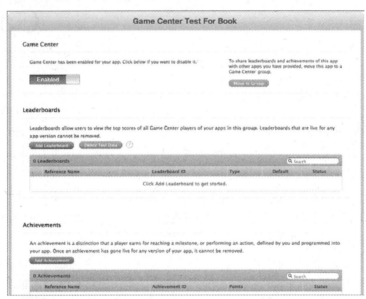

图 11-1　逐个应用程序地启用 Game Center。在建立应用程序时，可以设置新的排行榜和成绩，并且允许应用程序与其他 Game Center 特性交互

图 11-2　只能把 Game Center 特性用于在 iTunes Connect 上注册过的应用程序

直到创建了与项目的应用程序标识符匹配的 iTunes Connect 条目并且启用了 Game Center 之后，才能测试应用程序。如果无法提前设置 Game Center，将不能访问诸如记录用户和查看成绩之类的标准特性。一条说明原因的警告指出 Game Center 不识别你的游戏。当 Game Center 不识别当前运行的应用程序的标识符时，就会发生这种情况（参见图 11-2）。

对于本章中的秘诀，已经执行了上述的步骤。在使用这些 com.sadun.cookbook 应用程序时，无需执行任何进一步的设置，示例就应该会工作。

11.1.1　测试 Game Center 兼容性

如果计划部署 iOS 4.1 以前的版本，可能希望检查 Game Center 在当前设备上是否可用。可以轻松地检查类（例如，GKLocalPlayer）存在与否，以此推断是否提供了 GameKit Game Center 支持：

```
if (NSClassFromString(@"GKLocalPlayer")) ...
```

11.2 秘诀：登录到 Game Center

测试应用程序是否为 Game Center 进行了正确设置的最容易的方式是尝试签入用户。秘诀 11-1 获取共享的本地玩家单例对象，并为该玩家提供一个身份验证处理程序。

这种行为在 iOS 6 中改变了。以前用过的 authenticateWithCompletionHandler: 方法现在不建议使用了。如果计划部署到 iOS 5，就要确保包括特定于固件的测试，以运行正确的代码。

身份验证不一定意味着每次应用程序启动时用户都必须登录。会话可以持续相当长的时间。如果身份验证保持有效（player.authenticated），将利用欢迎回来的通知问候用户，而不是提示输入密码。

无论何时身份验证系统成功签入或者不能签入，都会触发 GKPlayerAuthenticationDidChange-NotificationName 通知。侦听它并更新 GUI，以反映当前的用户状态。仅当用户正确地进行了身份验证之后，才会展示 Game Center 特性：

```
[[NSNotificationCenter defaultCenter]
    addObserverForName:GKPlayerAuthenticationDidChangeNotificationName
    object:nil queue:[NSOperationQueue mainQueue]
    usingBlock:^(NSNotification *notification)
{
    [weakself updateUserGUI];
}];
```

可以通过重置然后退出 iOS 模拟器并重新运行应用程序，最好地测试身份验证状态。

注意：

在从事开发时，将在 GameKit 沙盒环境中执行测试。其中产生的所有分数和成绩都不会泄露给展示性的排行榜。iTunes Connect 允许从应用程序的 Game Center 管理页面中删除测试数据。在设备上执行测试时，可以访问 Game Center 应用程序，并在运行测试应用程序之前注销正常的账户。一旦运行应用程序，这应该允许登录到沙盒中。

秘诀 11-1　建立 Game Center 玩家

```
- (void) establishPlayer
{
    TestBedViewController __weak *weakself = self;
    [GKLocalPlayer localPlayer].authenticateHandler =
        ^(UIViewController *controller, NSError *error)
    {
        if (error)
        {
```

```objc
            NSLog(@"Error authenticating: %@",
                error.localizedDescription);
            alert(@"Restore game features by logging \
                in via the Game Center app.");
            return;
        }
        if (controller)
        {
            // User has not yet authenticated
            [weakself presentViewController:controller
                animated:YES completion:nil];
        }
    };
}
```

> **获取这个秘诀的代码**
> 要查找这个秘诀的完整示例项目，可以浏览 https://github.com/erica/iOS-6-Advanced-Cookbook，并进入第 11 章的文件夹。

11.3 设计排行榜和成绩

Game Center 提供了一对项目，有助于使用户参与游戏。排行榜使用户能够查看应用程序的共享的高分列表。成绩允许用户通过玩游戏解除锁定目标里程碑，例如，Scored First Win、Twenty Games Played 等。

这两个特性都是在 iTunes Connect 上建立的，可以在 iTunes Connect 应用程序信息页面中单击 Manage Game Center 按钮来创建它们（可以基于每个应用程序或者作为在多个应用程序当中共享的一组特性来创建）。下面几节介绍了这个过程中涉及的细节。

11.3.1 构建排行榜

选择 Leaderboard > Add Leaderboard > Single Leaderboard > Choose，为应用程序创建新的排行榜。iTunes Connect 将提示你指定以下信息。

- **Reference Name**（参考名称）：输入一个名称，创建它是为了你自己参考用的。这个名称将不会在 iTunes Connect 之外使用。例如，你可能想跟踪 Top Overall Point Scores （最高的总分）或 Fastest Reaction Time（最快的反应时间）。保持参考名称容易理解并且便于搜索。

- **Leaderboard ID**（排行榜 ID）：这是一个独特的字符串，应用程序使用它来标识特定的排行榜。出于简单起见，使用反向域命名。给应用程序标识符追加一个有特色的名称（例如，com.sadun.cookbook.topPoints）。
- **Score Format**（分数格式）：选择用于统计和显示分数的方式。可以选择整型（整型分数）、浮点型（浮点数，带有一位、两位或 3 位固定的小数）、时间间隔（分、秒或百分之一秒）或者货币（完整的单位货币或者单位货币以及两位小数）。还要选择是从高到低（高分胜出，比如分数）还是从低到高（低分胜出，比如流逝的时间最短）进行排序。
- **Score Range**（分数范围）（可选）：为应用程序输入可能最低和最高的分数。

在输入这些信息后，必须输入一条或多条本地化的描述，然后可以保存新的排行榜。图 11-3 显示了新创建的排行榜。

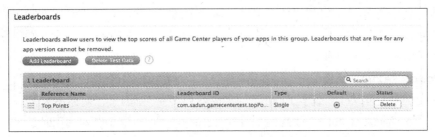

图 11-3　iTunes Connect 使你能够创建和管理排行榜。在排行榜投入使用后将不能删除它们。你创建的第一个排行榜是默认的排行榜，在添加多个排行榜后，如果你愿意，可以使用单选按钮选择一个不同的默认排行榜。也可以重新排列排行榜和成绩的顺序

1. 本地化描述

在把排行榜展示给用户时，将不会使用你输入的用于创建排行榜的参考名称。作为替代，将创建本地化描述，提供应用程序将使用的文本材料以及其他任何格式化选项。排行榜必须本地化成至少一种语言。单击 Add Language 并选择一种语言。

为你的排行榜输入一个现实的名称，这个名称将被应用程序使用并显示给用户。使之保持简短和贴切，比如 Top Scores、Best Shooters 或 Quickest Reactions。

选择一种分数格式，并输入用于项目的可选的后缀（小贴士：在后缀之前添加一个空格，以便用户可以赢得"50 coins"「50 个硬币」，而不是"50coins"）。这类信息可能因每种本地化方式而异，使你能够在美国得到"truckloads"（卡车负载），而在英国就会得到"lorry loads"（货车负载）。可以提供一幅可选的 512 像素×512 像素的图像，以补充每个排行榜。这幅图像也会进行本地化。可以为你支持的每个地区/文化创建不同的图像。

单击 Save，保存本地化设置。以后可以再次访问你刚才编辑的描述，并根据需要执行任何更改，但是在排行榜投入使用后将不能删除它。

11.3.2 创建成绩

通过在 iTunes Connect 中添加成绩来创建它们，其方式与创建排行榜非常相似。选择 Achievements > Set Up > Add New Achievement，并指定以下项目。

- **Reference Name**（参考名称）：这是一个人类可读的私有名称，不能在 iTunes Connect 外面使用（例如，"Great Start Achievement" 或 "100 Certified Kills"）。
- **Achievement ID**（成绩 ID）：提供一个独特的字符串，应用程序使用它来标识特定的成绩。与排行榜一样，你可能希望使用反向域命名，给应用程序标识符追加某个有意义的短语（例如，"com.sadun.cookbook.greatStart"）。
- **Hidden**（隐藏）：指定是否在 Game Center 中隐藏这些项目，直到用户实现了它们为止。这不会隐藏项目的数量；而只会隐藏名称和描述。隐藏项目允许利用你的用户意想不到的新成就使他们感到惊奇和快乐。如果没有隐藏，可以提供情感激励以揭示完整的项目集合以及预览这些项目。
- **Point Value**（点值）：对于每个成绩，可以分配最多 100 点；对于每个跨所有成绩的应用程序，可以分配最多 1000 点。这添加了一种限制因素，使你不会因为 "You Have Played This Game for 15 Minutes and 30 Seconds"（你玩这个游戏的用时是 15 分 30 秒）的成绩而压垮用户。

每个成绩都需要至少一条本地化描述。可以使用英语（如图 11-4 所示）添加基本的评论，以后可以再添加其他的语言。

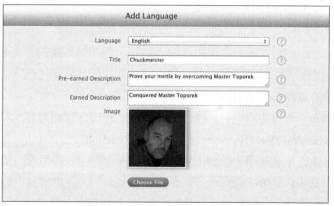

图 11-4 通过提供一个名称、赢得项目前后的两条描述以及一幅与成绩关联的图像来指定成绩元素。在测试时，要记住 iPhone 系列设备上可用于显示文本的空间小于 iPad

单击 Add Language，选择一种语言，输入一个将显示给读者的标题，并选择两条描述。Pre-earned Description 描述了开启它之前或者在开启它的过程中的成绩；Earned Description 则描述了开启它之后的成绩。例如，你可能具有 Prove Your Mettle by Overcoming Master Toporek 作为赢得项目前的版本，并将 Conquered Master Toporek 作为赢得项目后的版本。要使描述保持简短，否则将会自动裁剪它们。

未隐藏的成绩总会显示赢得项目前的描述，直到取得了它们之后。在赋予它一个（非 0 的）完成的百分比这样的值之后，隐藏的成绩也会把赢得项目前的描述显示给用户。如果用户朝着"征服主"的道路前进了 50%，就要确保赢得项目前的文本匹配那个概念的展示。

除了文本之外，每个成绩还需要一幅（本地化的）512 像素×512 像素或 1024 像素×1024 像素的艺术图像（参见图 11-4）。图像必须是 jpeg、jpg、tif、tiff 或 png 格式，至少 72 DPI，并且处于 RGB 颜色空间。

11.4 秘诀：访问排行榜

许多应用程序只有一个排行榜用于公布高分。其他应用程序则把排行榜用于多个类别，比如最准确的射手、最大的单词长度，或者每个星期执行的最多步骤。下面的方法用于为每个可用的排行榜获取类别（开发人员定义的排行榜标识符，比如 com.sadun.cookbook.topPoints）和标题（例如，Top Points）：

```
- (void) peekAtLeaderboards
{
    [GKLeaderboard loadLeaderboardsWithCompletionHandler:
        ^(NSArray *leaderboards, NSError *error)
    {
        if (error)
        {
            NSLog(@"Error retrieving leaderboards: %@",
                error.localizedFailureReason);
            return;
        }

        for (GKLeaderboard *leaderboard in leaderboards)
        {
            NSString *category = leaderboard.category;
            NSString *title = leaderboard.title;
            NSLog(@"%@ : %@", category, title);
        }
    }];
}
```

11.4.1 GKLeaderboard 类

每个 GKLeaderboard 实例都可以从 Game Center 获取分数。可以进行全局搜索以报告高分，或者使用范围来限制查询。例如，你可能指定一个 playerScope，把搜索限制于玩家和他的朋友，而不会返回那个组以外的所有高分。range 属性用于设置要显示哪些分数（顺便说一下，范围计数是从 1 开始，而不是从 0 开始）。通常，它默认为前 25 个高分，但是可以代之以选择前 10 个高分。也可以选择一个时间范围，只查看当天或本周等的高分。

秘诀 11-2 演示了如何加载用于特定类别的数据。它请求上一周内的前 10 个高分，然后尝试为每个玩家加载游戏中心显示名称。如果它失败了，就会退回使用 GKScore 的玩家 ID 属性。

秘诀 11-2　获取排行榜信息

```
- (void) peekAtLeaderboard: (GKLeaderboard *) leaderboard
{
    // top ten scores. Default range is 1,25
    leaderboard.range = NSMakeRange(1, 10);
    // Within last week
    leaderboard.timeScope = GKLeaderboardTimeScopeWeek;

    // Load in the scores
    [leaderboard loadScoresWithCompletionHandler:^(
        NSArray *scores, NSError *error)
    {
        if (error)
        {
            NSLog(@"Error retrieving scores: %@",
                error.localizedFailureReason);
            return;
        }

        // Retrieve player ids
        NSMutableArray *array = [NSMutableArray array];
        for (GKScore *score in scores)
            [array addObject:score.playerID];

        // Load the player names
        [GKPlayer loadPlayersForIdentifiers:array
            withCompletionHandler:
                ^(NSArray *players, NSError *error)
```

```
    {
        if (error)
        {
            // Report only with player ids
            for (GKScore *score in scores)
                NSLog(@"[%2d] %@: %@ (%@)",
                      score.rank, score.playerID,
                      score.formattedValue, score.date);
            return;
        }

        for (int i = 0; i < scores.count; i++)
        {
            // Report with actual player names
            GKPlayer *aPlayer = [players objectAtIndex:i];
            GKScore *score = [scores objectAtIndex:i];
            NSLog(@"[%2d] %@: %@ (%@)",
                  score.rank, aPlayer.displayName,
                  score.formattedValue, score.date);
        }
    }];
}];
}
```

获取这个秘诀的代码

要查找这个秘诀的完整示例项目，可以浏览 https://github.com/erica/iOS-6-Advanced-Cookbook，并进入第 11 章的文件夹。

11.5 秘诀：显示 Game Center 视图控制器

通常，不会直接访问排行榜数据。GameKit 提供了一个已经构建的友好得多的视图控制器，用于显示给你的用户（参见图 11-5）。秘诀 11-3 演示了如何创建、展示和关闭 Game Center 视图控制器。确保调用类声明 GKGameCenterControllerDelegate 协议。这将使代码能够捕获"已完成"委托回调，从而能够在用户完成了与视图控制器的交互时关闭它。

图 11-5　GameKit 提供了预先配置的 Game Center 视图控制器，可以从应用程序中访问它

对于那些熟悉 iOS 6 以前的 Game Center 的读者，这个新类复制了以前在单独的排行榜和成绩控制器中发现的特性。Apple 指出："GKGameCenterViewController 类把许多常见的 Game Center 特性聚集到单个用户界面中。它取代了 GKAchievementViewController 和 GKLeaderboardViewController，作为在游戏中显示 Game Center 内容的首选方式。"使用 viewState 属性设置由 Game Center 控制器显示的默认选项卡。

秘诀 11-3　展示 Game Center 视图控制器

```
- (void)gameCenterViewControllerDidFinish:
    (GKGameCenterViewController *)gameCenterViewController
{
    // You can save the user's selection here if desired
    // and re-use it later
    // e.g. _leaderboard_Category = 
    //    _gameCenterViewController.leaderboardCategory
    [self dismissViewControllerAnimated:YES completion:nil];
}

- (void) showGameCenterViewController
{
    GKGameCenterViewController *gvc = 
        [[GKGameCenterViewController alloc] init];
    gvc.gameCenterDelegate = self;
    [self presentViewController:gvc
        animated:YES completion:nil];
}
```

获取这个秘诀的代码

要查找这个秘诀的完整示例项目,可以浏览 https://github.com/erica/iOS-6-Advanced-Cookbook,并进入第 11 章的文件夹。

11.6 秘诀:提交分数

通过创建新的 GKScore 实例并设置它们的值来提交分数,如秘诀 11-4 中所示。可以像下面的方法所做的那样直接指定类别,或者可以只调用 init,并让 Game Center 代之以使用默认的排行榜。

每个用户都可以收集许多分数,通过你提交的类别来区分它们。不过,每个类别只存储一个值,并且只与一个排行榜相关联。你将设计希望排行榜怎样工作,并且带有多种不同的跟踪统计,比如准确性、杀死人数等。

秘诀 11-4 提交用户分数

```
- (void) createScore
{
    NSNumber *userScore = [self requestScore];
    if (!userScore) return;

    GKScore *score = [[GKScore alloc] initWithCategory:GKCATEGORY];
    score.value = userScore.intValue;
    [score reportScoreWithCompletionHandler:^(NSError *error){
        if (error)
        {
            NSLog(@"Error submitting score to game center: %@",
                error.localizedFailureReason);
            return;
        }

        NSLog(@"Success. Score submitted.");
    }];
}
```

获取这个秘诀的代码

要查找这个秘诀的完整示例项目,可以浏览 https://github.com/erica/iOS-6-Advanced-Cookbook,并进入第 11 章的文件夹。

11.7 秘诀：检查成绩

成绩是用户奋斗的目标。应用程序可以根据需要公开和重置每个成绩。由于成绩可以是部分获得的，可以给每一类成绩赋予一个 percentComplete 值。在 Game Center 视图控制器中将把它们显示为部分成绩（利用赢得项目前的文本描述）。

秘诀 11-5 以编程方式检查成绩。它将回到 Game Center，查看激活了哪些成绩以及它们的完成程度。如果已经公开了某个成绩，这将允许更新 GUI。例如，只能给那些经过了某种初始培训的用户提供高级武器。

> **注意：**
> 你可能想创建一个专用类，用于处理成绩语义，而不是把成绩代码散布在整个游戏中。很容易把 Game Center 代码与应用程序语义混合在一起，并且最终会导致巨大的混乱。在处理 Game Center 元素时，要牢记 MVC 设计模式，就像对其他应用程序开发任务那样。

秘诀 11-5　测试成绩

```
- (void) checkAchievement
{
    [GKAchievement loadAchievementsWithCompletionHandler:
        ^(NSArray *achievements, NSError *error)
    {
        if (error)
        {
            NSLog(@"Error loading achievements: %@",
                error.localizedFailureReason);
            return;
        }

        for (GKAchievement *achievement in achievements)
        {
            NSLog(@"Achievement: %@ : %f",
                achievement.identifier, achievement.percentComplete);
            if ([achievement.identifier isEqualToString:GKBEGINNER])
            {
                // unlock some GUI feature
            }
        }
    }];
}
```

> **获取这个秘诀的代码**
> 要查找这个秘诀的完整示例项目，可以浏览 https://github.com/erica/iOS-6-Advanced-Cookbook，
> 并进入第 11 章的文件夹。

11.8　秘诀：把成绩报告给 Game Center

要公开某个成绩，可以把它报告给 Game Center。如果报告出现错误，并且没有在 Game Center 中成功地更新，就要确保为将来的重复尝试设置某种备用措施，使得用户不会丢失所取得的进展。

为用户取得的成绩进行奖励。在秘诀 11-6 中，showsCompletionBanner 属性创建了图 11-6 中显示的可视化更新，可以让用户知道成绩已经公开。

图 11-6　完成报告可以让用户知道他们何时已经公开了成绩

每份成绩报告至少都代表一次网络访问。尽管报告成绩不是特别昂贵，但是你不会希望在应用程序中过度使用它。你不想请求之间相互"添堵"，并且在用户朝着某个目标取得微小的进展时不需要在一秒钟内把成绩更新许多次。可以代之以考虑在什么地方更新成绩是有意义的，而不要用户状态一有任何变化，就条件反射般地把它们插入到代码中。

秘诀 11-6　公开成绩

```
- (void) unlockAchievement
{
    GKAchievement *achievement =
        [[GKAchievement alloc] initWithIdentifier: GKBEGINNER];
    if (achievement)
    {
        achievement.percentComplete = 100.0f;
        achievement.showsCompletionBanner = YES;
        [achievement reportAchievementWithCompletionHandler:^(NSError *error)
         {
             if (error)
             {
                 NSLog(@"Error reporting achievement: %@",
                     error.localizedFailureReason);
```

```
            // Make sure to try again later in real
            // world deployment!!!

            return;
        }

        // Achievement is now unlocked
    }];
}
```

> **获取这个秘诀的代码**
> 要查找这个秘诀的完整示例项目，可以浏览 https://github.com/erica/iOS-6-Advanced-Cookbook，并进入第 11 章的文件夹。

11.8.1 重置成绩

你可能想要允许用户重置成绩，并从头重新开始游戏。用户可以重新赢得每个成绩，并再次体验其中的乐趣，甚至是利用相同的 Game Center ID。

下面的方法显示了一种实现它的典型方式。它将请求重置，并提供一个完成处理程序，用于响应用户成功或失败的请求。显然，你将希望与用户确认重置——并且可能希望提供一种方式，根据需要重新提交成绩来撤销此重置。不能逐个类别地重置成绩。重置将同时应用于所有赢得的项目：

```
- (void) resetAchievements
{
    [GKAchievement resetAchievementsWithCompletionHandler:
        ^(NSError *error)
    {
        if (error)
        {
            NSLog(@"Error resetting achievements: %@",
                error.localizedFailureReason);
            return;
        }

        // Achievements are now reset
    }];
}
```

11.9 秘诀：多玩家配对安排

GameKit 允许用户请求与其他玩家配对，无论那些玩家是熟悉的朋友还是其他匿名的人。配对安排者视图控制器将处理特定的邀请和一般的随机游戏配对。要开启一个配对安排会话，就创建一个新的配对请求。可以指定需要多少位玩家才能玩游戏，以及总共可以处理多少位玩家。

对于 Game Center 游戏玩法，可以为两位或 4 位玩家创建配对。你自己的服务器上托管的配对允许同时有最多 16 位玩家（本章没有介绍在 Game Center 之外托管的配对）。秘诀 11-7 为基本的双玩家游戏创建了一个请求，并且展示了利用该请求初始化的配对安排者控制器。

确保主类声明 GKMatchmakerViewControllerDelegate 协议，并且指定 matchmakerDelegate 协议。作为一个委托，控制器将响应常见的配对安排者状态更新，下面几节中将加以详细介绍。

在展示了控制器之后，将由用户选择是邀请朋友还是使用 Game Center 的自动配对机制。

秘诀 11-7　通过配对安排者请求配对

```
- (void) requestMatch
{
    // Clean up any previous game
    sendingView.text = @"";
    receivingView.text = @"";

    // This is not a hosted match, which allows up to 16 players
    GKMatchRequest *request = [[GKMatchRequest alloc] init];
    request.minPlayers = 2; // Between 2 and 4
    request.maxPlayers = 2; // Between 2 and 4

    GKMatchmakerViewController *mmvc =
        [[GKMatchmakerViewController alloc]
            initWithMatchRequest:request];
    mmvc.matchmakerDelegate = self;
    mmvc.hosted = NO;
    [self presentViewController:mmvc animated:YES completion:nil];
}
```

获取这个秘诀的代码

要查找这个秘诀的完整示例项目，可以浏览 https://github.com/erica/iOS-6-Advanced-Cookbook，并进入第 11 章的文件夹。

11.9.1 处理配对安排者失败

配对安排者委托协议使用下面两个（必需的）方法处理失败的配对安排，它们将处理用户取消控制器以及无法连接到 Game Center 等情形。下面显示了如何实现这两个方法，确保它们可以关闭模态视图控制器。下一节中将详细描述更多的委托方法：

```
- (void) matchmakerViewControllerWasCancelled:
    (GKMatchmakerViewController *)viewController
{
    [self dismissViewControllerAnimated:YES completion:nil];
}

- (void) matchmakerViewController:(GKMatchmakerViewController *)viewController
    didFailWithError:(NSError *)error
{
    [self dismissViewControllerAnimated:YES completion:nil];
    NSLog(@"Error creating match: %@",
        error.localizedFailureReason);
}
```

11.10 秘诀：响应配对安排者

实现（可选的）matchmakerViewController:didFindMatch:方法，开始 Game Center 托管的游戏（对于你自己的服务器托管的游戏，则代之以实现查找玩家回调）。在应用程序的当前玩家与另一个设备上的外部玩家之间成功地进行了配对之后，将调用查找配对委托方法。

一旦找到了配对，秘诀 11-8 将检查你是否已经在开始玩游戏，有时，比赛条件意味着你已经先设置了另一个配对；否则，它将关闭配对安排者控制器，把配对保存到一个局部实例变量中，并设置配对的委托。这个委托不同于配对安排者委托；配对委托将处理游戏中的状态更新，就像配对安排者委托用于处理玩游戏前的状态更新一样。

委托声明 GKMatchDelegate 协议，它从游戏中的其他参与者那里接收数据和状态更新，并在安排配对后处理实际的游戏玩法。

把配对保存到一个局部变量中。可以使用该对象把数据发送给其他玩家。它还允许发送查询、提供语音聊天，以及断开连接正在进行的游戏。

秘诀 11-8 响应找到的配对

```
- (void) matchmakerViewController:
    (GKMatchmakerViewController *)viewController
```

```objc
    didFindMatch:(GKMatch *)aMatch
{
    // Already playing. Ignore.
    if (matchStarted)
        return;

    if (viewController)
    {
        [self dismissModalViewControllerAnimated:YES];
        match = aMatch;
        match.delegate = self;
        self.navigationItem.rightBarButtonItem = nil;
    }

    // Normal matches now wait for player connection

    // Invited connections may be ready to go now. If so, begin
    if (!matchStarted && !match.expectedPlayerCount)
    {
        // Start game!
        [self activateGameGUI];
    }
}
```

获取这个秘诀的代码

要查找这个秘诀的完整示例项目，可以浏览 https://github.com/erica/iOS-6-Advanced-Cookbook，并进入第 11 章的文件夹。

11.10.1 开始游戏

在安排配对后，在开始玩游戏之前，要等待配对的 **expectedPlayerCount** 降至 0。每次玩家连接时，都把这个数字减 1。当它归 0 时，所有的玩家都会连接，并且游戏可以开始实际的配对。

可以在两个位置以编程方式使这个计数归 0，此时，就可以开始玩游戏了。它可能发生在配对安排者的查找配对委托中，或者可能发生在配对委托的 **match:player:didChangeState:** 回调中。检查这两个位置。

通常，邀请朋友玩的游戏出现在配对安排者回调中，而自动配对的游戏则出现在玩家状态回调中，此时玩家状态将变为已连接。当开发人员第一次在应用程序中实现配对时，这可能会使他们犯错误。与本章配套的示例代码演示了这两种情况。

11.11 秘诀：创建邀请处理程序

应用程序可以提前决定它将如何自动处理邀请请求。通过建立一个邀请处理程序，将创建一个代码块，无论何时共享的配对安排者对象接收到一个邀请都将执行它。应该在应用程序与 Game Center 的交互过程中及早设置共享式配对安排者的 inviteHandler 属性。

这个处理程序（如秘诀 11-9 中所示）接受两个参数：一个邀请和一个玩家数组。当游戏接收到一个直接邀请时，将使用 invitation 参数，并将其设置为某个非 nil 值。当检测到邀请时，就已经建立了配对；不应该创建新的配对请求。作为替代，只需展示配对安排者视图控制器，并预先加载邀请的详细信息。用户然后将等待主玩家开始游戏。视图控制器会提供一个 Cancel 选项，它允许用户关闭视图控制器并取消邀请。只有主玩家才能看到 Play Now 按钮。

> **注意：**
> 这里调用的 finishMatch 方法可能非常简单，就像给当前的配对对象发送 disconnect 并将游戏 GUI 重置为它的初始状态一样。

秘诀 11-9　实现邀请处理程序

```
- (void) addInvitationHandler
{
    [GKMatchmaker sharedMatchmaker].inviteHandler =
        ^(GKInvite *invitation, NSArray *playersToInvite)
    {
        // This cleans up any in-progress game and changes
        // the focus to handling the invitation. YMMV.
        [self finishMatch];

        if (invitation)
        {
            GKMatchmakerViewController *mmvc =
                [[GKMatchmakerViewController alloc]
                    initWithInvite:invitation];
            mmvc.matchmakerDelegate = self;
            [self presentViewController:mmvc
                animated:YES completion:nil];
        }
        else if (playersToInvite)
        {
            GKMatchRequest *request = [[GKMatchRequest alloc] init];
```

```
            request.minPlayers = 2;
            request.maxPlayers = 2; // 2-player matches for this example
            request.playersToInvite = playersToInvite;
            GKMatchmakerViewController *mmvc =
                [[GKMatchmakerViewController alloc]
                    initWithMatchRequest:request];
            mmvc.matchmakerDelegate = self;
            [self presentViewController:mmvc
                animated:YES completion:nil];
        }
    };
}
```

获取这个秘诀的代码

要查找这个秘诀的完整示例项目，可以浏览 https://github.com/erica/iOS-6-Advanced-Cookbook，并进入第 11 章的文件夹。

11.12 管理配对状态

在建立了配对之后，配对委托将管理配对状态中的变化。通常会导致事情崩溃的两种方式包括失败的配对（通往 Game Center 的连接终止）和失败的玩家连接（即从 Game Center 到另一位玩家的连接中断）。

通过提醒用户来响应失败的配对，清理正在进行的游戏，并回到配对前准备玩游戏的状态，如下面的方法中所示：

```
- (void)match:(GKMatch *) aMatch didFailWithError:(NSError *)error
{
    // Revert GUI
    [self setPrematchGUI];

    // Alert the user
    alert(@"Lost Game Center Connection: %@",
        error.localizedDescription);
}
```

对于丢失的玩家，可以视作游戏结束了，就像下面的方法中所做的那样，或者可能想提醒用户，并等待另一位玩家重新连接：

```
- (void)match:(GKMatch *) aMatch
```

```
        connectionWithPlayerFailed:(NSString *)playerID
        withError:(NSError *)error
{
        NSLog(@"Connection failed with player %@: %@",
            playerID, error.localizedFailureReason);
        [self setPrematchGUI];
}
```

为了简化重新连接，可以实现一个可选的委托方法。当这个方法返回 YES 时，Game Center 将自动重新邀请其连接丢失的玩家：

```
- (BOOL)match:(GKMatch *) aMatch
    shouldReinvitePlayer:(NSString *)playerID
{
        return YES;
}
```

11.13 秘诀：处理玩家状态改变

尽管当玩家的设备进入睡眠状态或者他们丢失了其 Internet 连接时，将会生成连接断开通知，但是玩家也可以用更有序的方式改变状态。例如，玩家可能点按 Quit 按钮，使应用程序把 disconnect 发送给配对对象。这将在断开连接前发送一种适当的状态改变，并且不会激活丢失的连接所进行的重新邀请查询。

在玩家第一次连接时，应用程序也会接收到状态改变。在这里应该检查当前配对期望的玩家计数，当计数归 0 时启动游戏。

秘诀 11-10 处理了连接和断开连接事件的状态回调，并且显示了一个示例，说明可能如何对这些状态改变做出反应：

秘诀 11-10　响应玩家状态

```
- (void)match:(GKMatch *) aMatch
    player: (NSString *) playerID
    didChangeState: (GKPlayerConnectionState) state
{
    if (state == GKPlayerStateDisconnected)
    {
        NSLog(@"Player %@ disconnected", playerID);
        [match disconnect];
        [self setPrematchGUI];
    }
```

```
        else if (state == GKPlayerStateConnected)
        {
            if (!matchStarted && !match.expectedPlayerCount)
            {
                [GKPlayer loadPlayersForIdentifiers:@[playerID]
                    withCompletionHandler:
                        ^(NSArray *players, NSError *error)
                    {
                        [self activateGameGUI];
                        NSString *opponentName = playerID;
                        if (!error)
                        {
                            GKPlayer *opponent = [players lastObject];
                            opponentName = opponent.displayName;
                        }
                        alert(@"Commencing Match with %@", opponentName);
                    }];
            }
        }
        else
        {
            NSLog(@"Player state changed to unknown");
        }
    }
```

获取这个秘诀的代码

要查找这个秘诀的完整示例项目，可以浏览 https://github.com/erica/iOS-6-Advanced-Cookbook，并进入第 11 章的文件夹。

11.14　秘诀：获取玩家名字

配对使用玩家标识符，而不是玩家名字。标识符（identifier）基本上是特定的 Game Center 玩家的字符串形式的数字指示符。你可以获取一个 **GKPlayer** 对象的数组，这些对象可以根据需要带有它们的人类可读的 **displayName** 名称属性。这个操作将异步运行，因此如果你想显示新对手的"名字"，就必须等待完成块执行。在本地存储对手（或多个对手），如秘诀 11-11 中所示，因此不需要把这个操作运行多次。秘诀 11-13 中及以后使用的 **PlayerHelper** 类演示了如何执行该任务。

秘诀 11-11　加载对手的名字

```objc
// Retrieve the player information for the opponent
[GKPlayer loadPlayersForIdentifiers:@[playerID]
    withCompletionHandler:^(NSArray *players, NSError *error)
{
    [self activateGameGUI];
    if (error) return;

    // Store the opponent object (2-player game)
    opponent = [players lastObject];

    // Announce the match
    GKPlayer *opponent = [players lastObject];
    alert(@"Commencing Match with %@", opponent.displayName);
}];
```

获取这个秘诀的代码

要查找这个秘诀的完整示例项目，可以浏览 https://github.com/erica/iOS-6-Advanced-Cookbook，并进入第 11 章的文件夹。

11.14.1　比较玩家

通常通过玩家的 ID 来比较它们。下面两个函数提供了用于这些检查的解决方案：

```objc
BOOL playerEqual(GKPlayer *p1, GKPlayer *p2)
{
    return [p1.playerID isEqualToString:p2.playerID];
}

BOOL playerIDCheck(GKPlayer *p1, NSString *playerID)
{
    return [p1.playerID isEqualToString:playerID];
}
```

11.14.2　获取本地玩家

GKLocalPlayer 类提供了对当前与应用程序交互的玩家的即时访问：

```objc
GKPlayer *me()
{
    return [GKLocalPlayer localPlayer];
}
```

11.15 游戏玩法

游戏主要涉及通过 **GKMatch** 对象来回发送数据,以及更新本地 GUI 来反映任何远程的改变。一般来讲,需要快速发送少量的数据以便进行实时更新。可以根据需要修改或打破这个规则,尤其是当不需要更新快速到达时。配对使用两类数据传输:可靠传输和不可靠传输。

如果数据必须连贯地到达并且不能丢失,就使用可靠传输。可靠传输是以它们发送的顺序递送的,保证会到达,并将会重发,直到完全传输了它们。在使用可靠的连接时,可以考虑 TCP(Transmission Control Protocol,传输控制协议)传输。

不可靠的数据只发送一次,并且可能会被接收方无序地接收。可以把这种方法用于在小分组中到达的数据,但是必须使用接近瞬时的传输,比如在实时射击动作游戏中。可以考虑为不可靠的连接使用 UDP(User Datagram Protocol,用户数据报协议)传输。

下面是一个简单的短信应用程序的示例,在把字符串数据输入到发送文本视图中时发送它。它使用可靠传输,确保每个字符都以输入它的顺序传输。当应用程序接收到数据时,它将把该信息转换回字符串,并在接收文本视图中显示它:

```
- (void)textViewDidChange:(UITextView *)textView
{
    NSError *error;
    NSData *dataToSend =
        [sendingView.text dataUsingEncoding:NSUTF8StringEncoding];
    BOOL success = [match sendDataToAllPlayers:dataToSend
        withDataMode:GKMatchSendDataReliable error:&error];
    if (!success)
        NSLog(@"Error sending match data: %@", error.localizedFailureReason);
}

- (void)match:(GKMatch *) aMatch didReceiveData:(NSData *)data
    fromPlayer:(NSString *)playerID
{
    NSString *received = [[NSString alloc]
        initWithData:data encoding:NSUTF8StringEncoding];
    receivingView.text = received;
}
```

11.16 序列化数据

在为复杂的游戏玩法发送数据时,你可能希望发送更加结构化的信息,而不是简单的字符串。

JSON 序列化为紧凑的结构化数据对象提供了一种易于使用的解决方案,这样的数据对象包括字典和数组,其中前者带有键/值对,后者则带有以一种不太关键的、预先规定的顺序发送的信息。例如,想象一下有一个应用程序,它将"掷"一粒本地骰子,然后传输它的值。掷出一个数字可能只是游戏可能使用的许多动作之一。创建键/值对字典将使游戏能够区分掷骰子、移动棋子、选择纸牌及其他特定于游戏的命令:

```
// Roll the [1d6] die
localRoll = @((random() % 6) + 1);

// Create an info dictionary for the roll
NSDictionary *dictionary = @{@"Roll" : localRoll};

// Transmit the roll as serialized JSON data
NSData *json = [NSJSONSerialization
    dataWithJSONObject:dictionary options:0 error:nil];
[match sendDataToAllPlayers:json
    withDataMode:GKMatchSendDataReliable error:nil];
```

在接收端,另一位玩家的应用程序可以反序列化数据,获取掷骰子的值,并在本地应用它。更复杂的动作(例如,如果玩家可以移动多枚棋子,而不仅仅是一枚棋子)则需要稍微复杂一点的结构,但是发送和接收组件将保持相同:把 JSON 可序列化的对象转换为数据,发送它,然后在接收端把它转换回对象形式:

```
- (void)match:(GKMatch *) aMatch
    didReceiveData:(NSData *)data
    fromPlayer:(NSString *)playerID
{
    NSDictionary *dict =
        [NSJSONSerialization JSONObjectWithData:data
            options:0 error:nil];
    NSString *key = [[dict allKeys] lastObject];
    if (!key) return;

    if ([key isEqualToString:@"Roll"])
    {
        // take action here, e.g. move n spaces
        // or extract which piece to move
    }
}
```

顶级 JSON 可序列化的对象必须是数组或字典。在此之下,它们可能包含更深一层的数组、字典、字符串、数字和 NSNull 实例。数字不能是 NaN（"不是一个数字"值）或无穷大。可以调用 isValidJSONObject:，测试对象是否是有效的 JSON 实例。

11.16.1 发送其他形式的数据

尽管 JSON 序列化实现起来很简单，但它可能不会提供足够的灵活性来满足你的需求。除了字典、数组、字符串或数字之外，如果还需要发送 NSData 或 NSDate 对象，则可能希望代之以调查标准的属性列表序列化。

属性列表提供了一种有用的抽象数据类型。属性列表对象可以指向数据（NSData）、字符串（NSString）、数组（NSArray）、字典（NSDictionary）、日期（NSDate）和数字（NSNumber）。在处理集合对象（即数组和字典）时，所有的成员和键也都必须是属性列表对象（即数据、字符串、数字和日期，以及嵌入式数组和字典）。

尽管这似乎有些限制，但是可以把大多数结构和对象转换为字符串，反之亦然。例如，可以使用内置的 NSStringFromCGPoint()或 NSStringFromClass()函数，或者可以创建自己的函数。下面两个方法扩展了 UIColor 类，提供跨 GameKit 连接将颜色信息作为字符串进行发送所需的功能：

```
@implementation UIColor (utilities)
- (NSString *) stringFromColor
{
    // Recover the color space and store RGB or monochrome color
    const CGFloat *c = CGColorGetComponents(self.CGColor);
    CGColorSpaceModel csm =
        CGColorSpaceGetModel(CGColorGetColorSpace(self.CGColor));
    return (csm == kCGColorSpaceModelRGB) ?
        [NSString stringWithFormat:@"%0.2f %0.2f %0.2f %0.2f",
            c[0], c[1], c[2], c[3]] :
        [NSString stringWithFormat:@"%0.2f %0.2f %0.2f %0.2f",
            c[0], c[0], c[0], c[1]];
}

+ (UIColor *) colorWithString: (NSString *) colorString
{
    // Read a color back from a string
    const CGFloat c[4];
    sscanf([colorString cStringUsingEncoding:NSUTF8StringEncoding],
        "%f %f %f %f", &c[0], &c[1], &c[2], &c[3]);
    return [UIColor colorWithRed:c[0] green:c[1] blue:c[2] alpha:c[3]];
}
@end
```

当采用属性列表形式时，可以序列化数据，并将其作为单个块进行发送。在接收方，将准备好使用反序列化的数据。程序清单 11-1 显示了示范性的 transmit 和 receivedData:方法，用于处理序列化和反序列化。这段代码来自于一个示例，它在一个 NSDictionary 对象中存储各种不同的游戏状态。

> **注意：**
> 可以使用 NSKeyedArchiver 和 NSKeyedUnarchiver 类，以及这里演示的 NSPropertyListSerialization 类和 CFPropertyListCreateWithData 函数。

程序清单 11-1　序列化和反序列化属性列表

```
- (void) transmit
{
    NSString *errorString;

    // Send a copy of the local points to the peer
    // by serializing the property list into data
    NSData *plistdata = [NSPropertyListSerialization
        dataFromPropertyList:pointData
        format:NSPropertyListXMLFormat_v1_0
        errorDescription:&errorString];

    if (plistdata)
        [match sendDataToAllPlayers: plistdata
            withDataMode:GKMatchSendDataReliable error:nil];
    else
        NSLog(@"Error serializing property list: %@", errorString);
}

- (void)match:(GKMatch *) aMatch
    didReceiveData:(NSData *)data
    fromPlayer:(NSString *)playerID
{

    // Deserialize the data back into a property list
    CFStringRef errorString;
    CFPropertyListRef plist =
        CFPropertyListCreateWithData (kCFAllocatorDefault,
        (CFDataRef)data, kCFPropertyListMutableContainers,
        kCFPropertyListXMLFormat_v1_0, &errorString);
```

```
    if (!plist)
    {
        NSLog(@"Error deserializating data: %@", errorString);
        return;
    }

    // Do something with the received data
}
```

11.17 秘诀：同步数据

尽管 Apple 通过 Game Center 提供了一个逐回合玩游戏的选项（参见下一节），但是你可能想开发一些游戏，以一种更直接的风格执行类似的特性，其中玩家将不能自由地离开和返回，开始他们的那一个回合。你可以通过同步每位玩家之间的数据，为直接配对实现一款逐回合风格的游戏。例如，回合制的游戏的最常见的任务之一是选择谁先开始玩游戏。可以生成一个随机数并把它与对手的随机数做比较，创建你自己的通过掷骰子确定谁先开始的系统。

由于每个应用程序可能与另一个应用程序稍微有些不同步，永远也不要假定你的掷骰子动作将会在另一位玩家之前或之后发生。作为替代，可以在本地存储两个值，并在检查赢家之前等待两个项目可用。这意味着在接收数据的方法和在你一方执行掷骰子动作的方法中都应该调用启动检查。

秘诀 11-12 演示了如何执行该任务。它总会检查它是否可以决定一位赢家。如果是，该方法就会返回，因为在选择了赢家之后可能会把它调用多次。接下来，它将检查本地和远程掷骰子动作是否存在。仅当它们都可用时，它才会执行检查。

如果出现平局，它将重置两个值，开始新一轮掷骰子并执行检查；否则，它将选择赢家（在这里更高的值将会胜出），并且遵从另一位玩家（opponentGoesFirst）的意见来决定是否开始玩游戏。

秘诀 11-12　通过掷骰子确定谁先开始

```
- (void) checkStartupWinner
{
    // Already resolved the startup winner?
    if (startupResolved)
        return;

    // Need data from both remote and local rolls in order to
    // determine a winner
```

```objc
        if (!remoteRoll || !localRoll)
        {
            [self performSelector:@selector(checkStartupWinner)
                withObject:nil afterDelay:1.0f];
            return;
        }

        // Both rolls are in. Decide a winner
        NSLog(@"Remote roll: %@, local roll: %@", remoteRoll, localRoll);
        unsigned int local = localRoll.unsignedIntValue;
        unsigned int remote = remoteRoll.unsignedIntValue;

        // Handle a tie conflict by resetting and trying again
        if (local == remote)
        {
            NSLog(@"TIE!");
            remoteRoll = nil;
            localRoll = nil;
            [self sendRoll:GKROLLFORFIRST];
            return;
        }

        startupResolved = YES;

        if (remote > local)
        {
            // they go first.
            opponentGoesFirst = YES;
        }
        else
        {
            // you go first
            opponentGoesFirst = NO;
        }

        // reset both rolls
        localRoll = nil;
        remoteRoll = nil;
    }
```

> **获取这个秘诀的代码**
> 要查找这个秘诀的完整示例项目，可以浏览 https://github.com/erica/iOS-6-Advanced-Cookbook，并进入第 11 章的文件夹。

11.18 秘诀：逐回合地安排配对

基于回合的配对允许应用程序深入检查一种 Game Center 特性，它基于一次一步的玩法来调节游戏，比如西洋跳棋、拉米牌戏、纵横拼字游戏或桥牌。它引入了存储和转发游戏状态，允许每位玩家走一步，然后把更新的状态传回 Game Center 存储。

基于回合的配对安排者允许用户异步连接以及通过邀请来连接，就像标准的游戏配对安排者所做的那样。它们之间的区别在于如何建立游戏，以及如何通过传递一个虚拟的"接力棒"将游戏控制从一个人转移给下一个人，再转移给另外一个人。持有"接力棒"的人将具有游戏的控制权，直到她走一步为止。

从这里将"接力棒"传递给下一位玩家——它可能是按顺序排列的下一位玩家，或者是基于另一个规则选出的某位玩家。例如，在每一局结束之后，它可能传递给得到最高分或最低分的人。或者，玩家可能选择将某个项目丢弃给谁，同时传递游戏控制权。

如果实现这种控制取决于你以及游戏的需求。如何建立配对并实现逐回合的机制遵循几种基本的模式。启动逐回合配对的过程借鉴了标准游戏的方法。你将创建相同类型的标准配对请求，指定最小和最大的玩家数量。

使用该请求初始化 GKTurnBasedMatchmakerViewController 并展示它。处理程序类将实现 GKTurnBasedMatchmakerView-ControllerDelegate 协议，委托将响应成功的和不成功的配对尝试。把 showExistingMatches 属性设置为 YES 将使用户能够选择新的配对，以及展示现有的配对。秘诀 11-13 显示了请求配对过程中涉及的步骤，这种实现可能返回一种全新的配对或者已经在进行的配对。

逐回合地安排配对不会总是获得成功。与标准的游戏中心控制器一样，要响应用户点按 Cancel 按钮以及任何连接失败情况。实现对这些情况的基本响应，以关闭控制器，然后把界面恢复到用户可以再次尝试建立配对的状态。

秘诀 11-13 开始配对

```
// Build a 2-player match request
- (GKMatchRequest *) matchRequest
{
    GKMatchRequest *request = [[GKMatchRequest alloc] init];
```

```objc
        request.minPlayers = 2; // Between 2 and 4
        request.maxPlayers = 2; // Betseen 2 and 4
        return request;
}


// Request a match from the turn-based matchmaker
- (void) requestMatch
{
    GKMatchRequest *request = [self buildMatchRequest];
    GKTurnBasedMatchmakerViewController *viewController =
        [[GKTurnBasedMatchmakerViewController alloc]
            initWithMatchRequest:request];
    viewController.turnBasedMatchmakerDelegate = self;
    viewController.showExistingMatches = YES;
    [self presentViewController:viewController
        animated:YES completion:nil];
}


// User selected match
- (void)turnBasedMatchmakerViewController:
        (GKTurnBasedMatchmakerViewController *)viewController
    didFindMatch:(GKTurnBasedMatch *)aMatch
{

    // Handle dismissal
    if (viewController)
        [_delegate dismissViewControllerAnimated:YES completion:nil];

    // Add match to matches
    MatchHelper *helper = [MatchHelper helperForMatch:match];
    _matchDictionary[match.matchID] = helper;
    [helper loadData];
    [helper loadParticipants];

    // Set this match to the current match
    [_delegate chooseMatch:helper];
}

// Game Center Fail
- (void)turnBasedMatchmakerViewController:
```

```
            (GKTurnBasedMatchmakerViewController *)viewController
    didFailWithError:(NSError *)error
{
    [self dismissViewControllerAnimated:YES completion:^(){
        alert("Error creating match: %@",
            error.localizedDescription);
    }];
}

// User cancel
- (void)turnBasedMatchmakerViewControllerWasCancelled:
            (GKTurnBasedMatchmakerViewController *)viewController
{
    [self dismissViewControllerAnimated:YES completion:nil];
}
```

获取这个秘诀的代码

要查找这个秘诀的完整示例项目，可以浏览 https://github.com/erica/iOS-6-Advanced-Cookbook，并进入第 11 章的文件夹。

11.19 秘诀：响应基于回合的邀请

与许多 Game Center 特性一样，你的代码应该处理游戏中的配对请求和外部邀请请求。通过一个处理程序委托来处理游戏更新，比如邀请，如秘诀 11-14 中所示。声明并实现 GKTurnBasedEventHandlerDelegate。这个协议为配对安排提供了邀请入口点，并且允许应用程序响应逐回合的更新。

要进行预订，可以像下面这样设置处理程序的委托。这是一个共享的处理程序，它将为游戏中当前涉及的用户更新所有的配对：

```
[GKTurnBasedEventHandler sharedTurnBasedEventHandler].delegate
    = turnByTurnHelper;
```

在建立了委托之后，可能接收到邀请回调。这些邀请来源于其他玩家，并且明确地针对他们的朋友。实现 handleInviteFromGameCenter:方法来响应他们。

与以前用户指示的配对安排一样，可以设置配对安排控制器的配对请求。不过，对于邀请，可以把请求的 playersToInvite 属性设置为传递给你的数组。此外，还可以把视图控制器的 showExistingMatches 属性设置为 NO。在处理邀请时，只应该处理将你邀请到的配对。

秘诀 11-14 处理邀请

```
- (void)handleInviteFromGameCenter:
    (NSArray *)playersToInvite
{
    GKMatchRequest *request = [self matchRequest];
    request.playersToInvite = playersToInvite;
    GKTurnBasedMatchmakerViewController *viewController =
        [[GKTurnBasedMatchmakerViewController alloc]
            initWithMatchRequest:request];
    viewController.showExistingMatches = NO;
    viewController.turnBasedMatchmakerDelegate = self;
    [_delegate presentViewController:viewController
        animated:YES completion:nil];
}
```

获取这个秘诀的代码

要查找这个秘诀的完整示例项目，可以浏览 https://github.com/erica/iOS-6-Advanced-Cookbook，并进入第 11 章的文件夹。

11.20 秘诀：加载配对

每个游戏启动并且玩家进行身份验证时，你都应该利用 Game Center 签入。把 GKTurnBased-EventHandler 委托设置为捕获回合和状态改变，并加载进你的玩家可能参与的任何现有的配对。逐回合的配对可能经历几天、几周或者几个月的时间，而你的玩家可能有许多配对在进行过程中：

```
[[NSNotificationCenter defaultCenter] addObserverForName:
        GKPlayerAuthenticationDidChangeNotificationName
    object:nil queue:[NSOperationQueue mainQueue]
    usingBlock:^(NSNotification *notification)
{
    BOOL authenticated =
        [GKLocalPlayer localPlayer].isAuthenticated;
    if (authenticated)
    {
        turnByTurnHelper = [[TurnByTurnHelper alloc] init];
        turnByTurnHelper.delegate = self;
        [GKTurnBasedEventHandler
```

```
            sharedTurnBasedEventHandler].delegate = turnByTurnHelper;
        [turnByTurnHelper loadMatchesWithCompletion:^()
        {
            [weakself nextMatch];
        }];
    }
}];
```

秘诀 11-15 演示了如何请求当前的配对集。这个过程将异步运行，使得这种实现将包括在主线程上运行的完成处理程序，使你能够在配对数据完全到达后设置游戏界面。

这个方法将存储在本地字典中获取的配对，并使用自定义的 MatchHelper 类管理每个配对。可以看到，助手类的第一个任务是加载配对数据和参与者，完成下载也要花一些时间。

尽管一旦从 Game Center 加载完配对就能立即使用配对参与者 ID，还是不能使用实际的玩家名字（参见秘诀 11-11）、照片，以及表示当前配对状态的数据。

秘诀 11–15　从 Game Center 加载配对

```
- (void) loadMatchesWithCompletion: (CompletionBlock) completion
{
    if (!_matchDictionary)
        _matchDictionary = [NSMutableDictionary dictionary];

    NSLog(@"Loading matches from Game Center");
    [GKTurnBasedMatch loadMatchesWithCompletionHandler:
        ^(NSArray *theMatches, NSError *error)
    {
        if (error)
        {
            NSLog(@"Error retrieving matches: %@",
                error.localizedDescription);
            return;
        }

        NSLog(@"Number of matches: %d", theMatches.count);
        for (GKTurnBasedMatch *match in theMatches)
        {
            MatchHelper *helper = [MatchHelper helperForMatch:match];
            _matchDictionary[match.matchID] = helper;
            [helper loadData];
            [helper loadParticipants];
```

```
            }

            [[NSOperationQueue mainQueue] addOperationWithBlock:^()
            {
                if (completion)
                    completion();
            }];
        }];
    }
```

获取这个秘诀的代码

要查找这个秘诀的完整示例项目，可以浏览 https://github.com/erica/iOS-6-Advanced-Cookbook，并进入第 11 章的文件夹。

11.21 秘诀：响应玩法

你已经知道，GKTurnBasedEventHandlerDelegate 提供了一种方式来响应外部邀请。当回合发生时，相同的协议可以让你知道这一点。当用户完成了一个回合并更新了游戏数据之后，以及当用户退出时，这些就会发生。你的游戏一接收到这种回调就会确定实际的场景是什么，并且相应地做出响应。

秘诀 11-16 提供了几个方法，它们源于示例代码中的几个类。像这样把它们组合在一起，以使你能够查看流程是如何进行的。该秘诀开始于回合回调，它将在主线程上调用游戏的 takeTurn: 方法。

这个方法将继续检查更新的回合是否用于所显示的游戏。如果用户参与了多个游戏，更新可能来自于他当前没有查看的游戏。

它还会检查更新的配对是否已经完成。当配对的状态更新为"已完成"时，配对就完成了。对于简单的双玩家游戏，当任何配对参与者的状态更新为"已完成"时，也可以把配对视作完成了。一个配对"完成了"，也就意味着所有的配对都"完成了"。并非所有的游戏都是这样。可以想象一下多玩家配对，其中一个角色死亡了，但是其他的角色继续在战斗。可以为特定的游戏玩法需求修改这种方法。

takeTurn:方法在工作时将检查每种可能性。如果另一位玩家退出，但是游戏仍在继续进行，它将终止配对。如果配对在继续进行，它将检查这是哪位玩家的回合，并且更新 GUI 以反映这种状态。在更新 GUI 时，不要忘记请求最新的玩法数据，以使用户准备好仅当数据到达后才开始玩游戏。

在允许本地用户继续玩游戏之后，通过调用配对的结束回合方法来完成回合。你将按顺序传递更新的游戏状态数据以及其后的参与者的数组。对于双玩家游戏，该数组仅仅只是另一位参与者。

与游戏结束请求不同，另一位参与者是否还未配对是无关紧要的。要使另一位玩家加入你的新游戏，在可以找到另一个玩家之前，必须开始第一个回合。在测试时，在尝试加入第二次配对之前，总是要在一个设备上开始第一个回合。

秘诀 11-16　处理回合事件

```
// This is a Turn Event Delegate Method
- (void)handleTurnEventForMatch:(GKTurnBasedMatch *) match
            didBecomeActive:(BOOL)didBecomeActive
{
    MatchHelper *helper = [self matchForID:match.matchID];
    helper.didBecomeActive = didBecomeActive;

    [helper loadDataWithCompletion:^(BOOL success) {
        if (!success) return;
        [[NSOperationQueue mainQueue] addOperationWithBlock:^()
         {
             [_delegate takeTurn:helper];
         }];
    }];
}
// This game method is called by the delegate
- (void) takeTurn: (MatchHelper *) match
{
    BOOL isCurrentMatch =
        matchEqual(currentMatch.match, match.match);
    BOOL matchEnded = match.matchIsDone;

    // Should I quit?
    if (matchEnded && match.amActive)
    {
        [match winMatch];
    }

    // Update ended match?
    if (matchEnded)
    {
        if (isCurrentMatch)
```

```objc
            [self chooseMatch:currentMatch];
        else
            alert(@"Match %@ has ended", match.matchID);
        return;
    }

    // Match has not ended. It is someone's turn
    if (!isCurrentMatch && match.isMyTurn)
    {
        alert(@"Your turn for match %@", match.matchID);
        return;
    }

    if (!isCurrentMatch)
    {
        NSLog(@"Non-turn activity on match %@", match.matchID);
        return;
    }

    // It is the current match and it is your turn
    [self chooseMatch:currentMatch];
}

// The following helper methods are implemented by the
// match helper class to support gameplay
- (BOOL) isMyTurn
{
    if (!_match.currentParticipant.playerID)
        return NO;
    return playerIDCheck(me(),
        _match.currentParticipant.playerID);
}

- (BOOL) amActive
{
    GKTurnBasedParticipant *me = myParticipantForMatch(_match);
    return (me.status == GKTurnBasedParticipantStatusActive);
}

- (BOOL) matchIsDone
```

```objc
{
    if (_match.status == GKTurnBasedMatchStatusEnded)
        return YES;
    
    // Note -- this assumes "done for one is done for all"
    for (GKTurnBasedParticipant *participant in _match.participants)
    {
        if (participant.status == GKTurnBasedParticipantStatusDone)
            return YES;
    }
    
    return NO;
}

// End turn at this player's end
- (void) endTurnWithTimeout: (NSTimeInterval) timeout
    withCompletion: (SuccessBlock) completion
{
    if (!myCurrentPlayerForMatch(_match))
    {
        NSLog(@"Error: You are not current player for match %@",
            _match.matchID);
        return;
    }
    
    NSMutableArray *participants = [NSMutableArray
        arrayWithArray:_match.participants];
    GKTurnBasedParticipant *me = myParticipantForMatch(_match);
    [participants removeObject:me];
    
    // Unlike ending games, it's okay to pass unmatched participants
    // to the following method
    [_match endTurnWithNextParticipants:participants
                     turnTimeout:timeout
                       matchData:_data
                completionHandler:^(NSError *error)
    {
        if (error)
            NSLog(@"Error completing turn: %@", error.localizedDescription);
```

```
            [[NSOperationQueue mainQueue] addOperationWithBlock:^()
            {
                if (completion)
                    completion(error == nil);
            }];
        }];
    }
```

获取这个秘诀的代码

要查找这个秘诀的完整示例项目,可以浏览 https://github.com/erica/iOS-6-Advanced-Cookbook,并进入第 11 章的文件夹。

11.22 秘诀:结束游戏玩法

大多数游戏最终都会结束。玩家可能退出、获胜、失利、打成平手,或者取得某个名次,比如第二名、第三名或第四名。你的游戏可能产生这些场景之一,或者你的用户可能直接从 Game Center 视图控制器中退出。下面的方法用于捕获 GUI 回调,并把退出请求重定向到我的配对助手类:

```
// Quit through GUI
- (void)turnBasedMatchmakerViewController:
        (GKTurnBasedMatchmakerViewController *)viewController
    playerQuitForMatch:(GKTurnBasedMatch *) match
{
    if (viewController)
        [_delegate dismissViewControllerAnimated:YES completion:nil];
    MatchHelper *helper = _matchDictionary[match.matchID];
    [helper quitMatch];
}
```

秘诀 11-17 显示了如何终止配对,而不管它是否是玩家的回合("轮到他"或者"未轮到他")。每种配对都有一个结果。这是一个枚举,在玩游戏结束时指定玩家的状态。Game Center 使用这些状态来显示尚未从系统中删除的完成的配对。受到 Game Center 支持的合法结果如下:

- **GKTurnBasedMatchOutcomeNone**
- **GKTurnBasedMatchOutcomeQuit**
- **GKTurnBasedMatchOutcomeWon**
- **GKTurnBasedMatchOutcomeLost**
- **GKTurnBasedMatchOutcomeTied**

- **`GKTurnBasedMatchOutcomeTimeExpired`**
- **`GKTurnBasedMatchOutcomeFirst`**
- **`GKTurnBasedMatchOutcomeSecond`**
- **`GKTurnBasedMatchOutcomeThird`**
- **`GKTurnBasedMatchOutcomeFourth`**
- **`GKTurnBasedMatchOutcomeCustomRange`**

在从回合中退出时，必须把参与者的数组传递给 nextParticipants:参数。永远不要对没有配对的玩家（即 Game Center 尚未为你找到的玩家）执行该操作。如果还没有其他合法的玩家（检查参与者的 PlayerID 以进行确认，如果它们是 nil，那么玩家就没有配对），就代之以退出回合。

秘诀 11-17　结束游戏

```
// Quit immediately
- (void) quitOutOfTurnWithOutcome:
    (GKTurnBasedMatchOutcome) outcome
{
    GKTurnBasedParticipant *participant =
        myParticipantForMatch(_match);
    participant.matchOutcome = outcome;

    [_match participantQuitOutOfTurnWithOutcome:outcome
                withCompletionHandler:^(NSError *error)
    {
        if (error)
            NSLog(@"Error while quitting match out of turn %@: %@",
                _match.matchID, error.localizedDescription);
        else
            NSLog(@"Participant quit match out of turn: %@",
                _match.matchID);
    }];
}

// Finish the game
- (void) finishMatchWithOutcome:
    (GKTurnBasedMatchOutcome) outcome
{
    GKTurnBasedParticipant *participant =
        myParticipantForMatch(_match);
    if (!participant)
```

```objc
    {
        NSLog(@"Error: Cannot finish game. \
            You are not playing in match %@.", _match.matchID);
        return;
    }
    
    participant.matchOutcome = outcome;
    
    NSArray *participants = self.otherCurrentParticipants;
    BOOL isCurrent = myCurrentPlayerForMatch(_match);
    
    if ((participants.count == 0) || !isCurrent)
    {
        // no other valid players or out of turn
        [self quitOutOfTurnWithOutcome:outcome];
        return;
    }
    
    [_match participantQuitInTurnWithOutcome:outcome
                            nextParticipants:participants
                                  turnTimeout:GKTurnTimeoutNone
                                    matchData:_data
                            completionHandler:^(NSError *error)
    {
        if (error)
            NSLog(@"Error while quitting match %@ in turn: %@",
                  _match.matchID, error.localizedDescription);
        else
            NSLog(@"Participant did quit in turn");
    }];
}

// You quit the game
- (void) quitMatch
{
    [self finishMatchWithOutcome:GKTurnBasedMatchOutcomeQuit];
}

// You win the game
- (void) winMatch
```

```
{
    [self finishMatchWithOutcome:GKTurnBasedMatchOutcomeWon];
}

// You lose the game
- (void) loseMatch
{
    [self finishMatchWithOutcome:GKTurnBasedMatchOutcomeLost];
}
```

获取这个秘诀的代码

要查找这个秘诀的完整示例项目,可以浏览 https://github.com/erica/iOS-6-Advanced-Cookbook,并进入第 11 章的文件夹。

11.23　删除配对

在游戏结束之后,可以请求从 Game Center 中删除它。确保玩家已经结束了玩游戏。当玩家仍处于活动状态时,将不能删除游戏。下面的方法用于请求 Game Center 执行删除操作。如果任何其他的玩家在游戏中仍然处于活动状态时,该方法将不会出错,但它也不会成功。因此,在请求删除配对之后,要特别小心地记录下它:

```
- (void) removeFromGameCenter
{
    [_match removeWithCompletionHandler:^(NSError * error)
    {
        if (error)
        {
            NSLog(@"Error removing match %@: %@",
                _match.matchID, error.localizedDescription);
            return;
        }
        
        NSLog(@"Match %@ removed from Game Center",
            _match.matchID);
        _terminated = YES;
    }];
}
```

在调试期间，有时只想为测试用户清除所有正在进行的配对。秘诀 11-18 采用了一种笨方法来退出和删除所有当前的配对。这种方法显然不打算用于部署到 App Store 最终用户，但它在调试工具箱中提供了一个方便的工具。

秘诀 11-18 清除当前玩家的 Game Center 配对

```
- (void) removeAllMatches
{
    // This is nuclear armageddon. Prepare!
    [GKTurnBasedMatch loadMatchesWithCompletionHandler:
    ^(NSArray *matches, NSError *error)
    {
        if (error)
        {
            NSLog(@"Error loading matches: %@",
            error.localizedFailureReason);
            return;
        }

        NSLog(@"Attempting to remove %d matches", matches.count);
        for (MatchHelper *helper in _matchDictionary.allValues)
        {

            GKTurnBasedMatch *aMatch = helper.match;
            GKTurnBasedParticipant *me =
                myParticipantForMatch(aMatch);
            if (me && (me.status ==
                GKTurnBasedParticipantStatusActive))
            {
                NSLog(@"Quitting match %@", aMatch.matchID);
                [helper quitOutOfTurnWithOutcome:
                    GKTurnBasedMatchOutcomeQuit];
                sleep(1);
                NSLog(@"Removing Match %@", aMatch.matchID);
                [_matchDictionary removeObjectForKey:aMatch.matchID];
                [helper removeFromGameCenter];
            }
            else
            {
                NSLog(@"Removing Match %@", aMatch.matchID);
```

```
            [_matchDictionary removeObjectForKey:aMatch.matchID];
            [helper removeFromGameCenter];
        }
    }
}];
}
```

> **获取这个秘诀的代码**
> 要查找这个秘诀的完整示例项目，可以浏览 https://github.com/erica/iOS-6-Advanced-Cookbook，并进入第 11 章的文件夹。

11.24　秘诀：Game Center 语音

每个 Game Center 配对都可以建立语音聊天，它们最适合于在玩家之间进行对讲式的通话。你的应用程序可以为每一个人创建一个群聊，把聊天限制在团队成员之间，或者在两位玩家之间创建单独的一对一通道。这是由于所有的聊天都是命名的。当玩家连接到相同名称的聊天中时，他们可以倾听并参与聊天。你的应用程序将控制名称，并且 API 提供的钩子将使你能够构建高级的 GUI 控件来管理聊天。

> **注意：**
> 确保对连接到 Wi-Fi 的 Voice Chat（语音聊天）进行测试，而不是测试蜂窝网络数据，因为你可能在尝试连接到聊天时遇到错误。

11.24.1　测试聊天可用性

在给用户提供聊天之前，要测试该特性是否可用。GKVoiceChat 类提供了一个简单的检查，用于确定当前在设备上是否允许 IP 语音（Voice over IP，VoIP）。如同 Apple 在其文档中所描述的："一些国家或电话运营商可能限制通过 IP 服务提供语音。"

```
if (![GKVoiceChat isVoIPAllowed])
    return;
```

11.24.2　建立播放和录制音频会话

要使用 Game Center 语音，需要把项目链接到 AVFoundation 框架。然后，在程序启动时，把共享的 AVFoundation 音频会话设置成使用播放和录制类别。秘诀 11-19 把共享的会话更新为播放和录制音频。

秘诀 11-19　为语音聊天建立音频会话

```objc
- (BOOL) establishPlayAndRecordAudioSession
{
    NSLog(@"Establishing Audio Session");
    NSError *error;
    AVAudioSession *audioSession =
        [AVAudioSession sharedInstance];
    BOOL success = [audioSession setCategory:
        AVAudioSessionCategoryPlayAndRecord error:&error];
    if (!success)
    {
        NSLog(@"Error setting session category: %@",
            error.localizedDescription);
        return NO;
    }
    else
    {
        success = [audioSession setActive: YES error: &error];
        if (success)
        {
            NSLog(@"Audio session is active");
            return YES;
        }
        else
        {
            NSLog(@"Error activating audio session: %@",
                error.localizedDescription);
            return NO;
        }
    }

    return NO;
}
```

获取这个秘诀的代码

要查找这个秘诀的完整示例项目，可以浏览 https://github.com/erica/iOS-6-Advanced-Cookbook，并进入第 11 章的文件夹。

11.24.3 创建语音聊天

为你想在应用程序内使用的每个聊天创建并保存一个 GKVoiceChat 实例，通常把它赋予一个 strong 实例变量或者一个数组。每个语音聊天通道都使用一个简单的字符串来标识它。例如，你可能为配对创建一个"GeneralChat"：

```
chat = [match voiceChatWithName:@"GeneralChat"];
```

目前没有更多的安全元素可用。所有的聊天都限制于当前的配对，并受到应用程序控制。你可以根据需要以编程方式提供对聊天的访问。

11.24.4 开始和停止聊天

要开始聊天，可以创建它，把它的 active 属性设置为 NO，可以选择初始化它的 volume 属性，并给它发送 start 消息。聊天将立即准备好使用，但是还不会开始记录或发送数据。直到你明确使之处于活动状态时，它才会做这些事情：

```
chat = [match voiceChatWithName:@"GeneralChat"];
chat.active = NO;
chat.volume = 1.0f;
[chat start];
```

要结束聊天，就给它发送 stop。以后可以再次开始聊天，如果给保存它的变量重新赋值，将允许释放它：

```
[chat stop];
chat = nil;
```

11.24.5 聊天状态监测

当对聊天调用 start 或 stop 时，Game Center 将发出通知，指示本地玩家加入了聊天。有 4 种可能的语音聊天状态：连接、断开连接、讲话和沉默。对于其中每种状态都会发出通知。因此，如果你具有一个活动的聊天并且你的用户暂停下来深吸一口气，那么当她再次开始说话时，GKVoiceChat 将传播一个 GKVoiceChatPlayerSilent 事件，其后接着 GKVoiceChatPlayerSpeaking 事件。

可以使用这些事件来更新本地图标，当用户正在讲话时将突出显示它们，当他们没有讲话时则使这些图标变暗淡，或者当用户不再是聊天的一份子时就从视图中删除它们。给每个聊天添加一个状态更新处理程序，当玩家的聊天状态改变时用于控制 GUI 展示：

```
chat.playerStateUpdateHandler =
    ^(NSString *playerID, GKVoiceChatPlayerState state)
```

```
{
    switch (state)
    {
        case GKVoiceChatPlayerSpeaking:
            // Highlight player's picture
            break;
        case GKVoiceChatPlayerSilent:
            // Dim player's picture
            break;
        case GKVoiceChatPlayerConnected:
            // Show player name/picture
            break;
        case GKVoiceChatPlayerDisconnected:
            // Hide player name/picture
            break;
    }
};
```

11.24.6　实现聊天按钮

它助于把聊天视作按键通话式的项目，仅当用户决定参与讲话时才激活它们。例如，可能创建一个通话按钮，当按下它时就激活本地麦克风，释放它时就停用麦克风。下面显示了如何定义该按钮：

```
talkButton = [UIButton buttonWithType:UIButtonTypeRoundedRect];
[talkButton setTitle:@"Speak" forState:UIControlStateNormal];
talkButton.frame = CGRectMake(0.0f, 0.0f, 200.0f, 40.0f);
[self.view addSubview:talkButton];

[talkButton addTarget:self action:@selector(startSpeaking)
    forControlEvents:UIControlEventTouchDown];
[talkButton addTarget:self action:@selector(stopSpeaking)
    forControlEvents:UIControlEventTouchUpInside |
    UIControlEventTouchUpOutside];
```

两个回调只用于把聊天的活动属性设置为 YES（启用麦克风并使之处于活动状态）或 NO（禁用麦克风）。如您已经看到的，在创建聊天时，最初总是把该属性设置为 NO：

```
- (void) stopSpeaking
{
    chat.active = NO;
}
```

```
- (void) startSpeaking
{
    chat.active = YES;
}
```

11.24.7 控制音量

聊天的 volume 属性的范围在 0.0（静音）~1.0（最大音量）之间。这可以控制聊天在本地设备的扬声器上的播放声音有多大。此外，还可以控制是否逐个玩家地侦听其他玩家。通过调用 setMute:forPlayer:，可以使个别玩家静音。像玩家状态更新处理程序一样，这个特性打算集成到逐个玩家的图标 GUI 中，允许用户逐个使其他人静音：

```
[chat setMute:YES forPlayer: playerID];
```

可以通过遍历聊天的玩家标识符列表，使所有玩家都静音。这只会影响所选聊天的音频。如果玩家在参与多个聊天，就必须使每个聊天都静音，或者把每个聊天的 volume 属性都设置为 0.0：

```
- (void) muteChat: (GKVoiceChat *) aChat
{
    NSLog(@"Muting chat %@", aChat.name);
    for (NSString *playerID in aChat.playerIDs)
        [aChat setMute:YES forPlayer:playerID];
}

- (void) unmuteChat: (GKVoiceChat *) aChat
{
    NSLog(@"Unmuting chat %@", aChat.name);
    for (NSString *playerID in aChat.playerIDs)
        [aChat setMute:NO forPlayer:playerID];
}
```

11.25 GameKit 对等服务

GameKit 在 Game Center 环境之外提供了 iOS 设备之间的对等连通性。该框架有助于创建互连的应用程序，实时地交换真实的数据。在其默认实现中，GameKit 在工作时将创建和管理即席的 Bluetooth（蓝牙）网络，允许设备相互寻找对方，建立一条连接，并通过该连接传输数据。GameKit 还允许查找、连接以及传输数据给相同 Wi-Fi 网络上的设备。

使用 Bluetooth 和 Wi-Fi 提供了一种快速、可靠的方法进行设备间通信。不幸的是，使用

GameKit 的对等通信的速度不快，尽管在 iOS 6 下它比旧固件要可靠得多。事实证明，使用 GameKit 对等连接可能会使最终用户受挫。如果你决定在应用程序中包括这个 GameKit 特性，就应该期望相应地测试你的支持承诺。

标准的使用场景要求用户拥有多个设备，并在封闭的住所使用它们。无论对等的思想有多好以及企业有多英明，它都是一次投资败笔。使用 Game Center 更简单、健壮，并为那些特性提供了更大的潜在用户群。

话虽如此，下面还是快速介绍 GameKit 的对等方面。

11.25.1　GameKit Bluetooth 的局限性

对于 Bluetooth，需要地理位置比较接近。连接被限制于大约 10 米（30 英尺）。考虑你的用户在会堂的会议室里或者在同一间办公室里工作，包括一起乘坐火车或汽车的人。在这种范围内，应用程序可以建立对等连接。

GameKit 提供了极佳的性能用于短小、瞬时的信息传输。Apple 建议 GameKit 传输限制 1000 字节以下。尽管 GameKit 可以处理大块的信息传输，一次最多传输 95 KB，但它并不打算用于普通的设备间的数据传输。如果尝试同时发送太多的数据，将会接收到传输错误。

如果必须传输大文件，将需要把那些文件分解成容易管理的块。确保使用标准的握手协议和分组校验和，以确保数据的可靠性。

GameKit 的 Bluetooth 联网可以在所有现代的 iOS 设备上工作。Info.plist 文件的 UIRequired-DeviceCapabilities 条目中的 peer-peer 键可让你明确要求通过 Bluetooth 提供对等连通性。

11.25.2　Bonjour 会话

GameKit 的对等连接是使用 Bonjour 联网构建的。Bonjour 是 Apple 用于零配置联网的商标，使设备能够宣传和发现网络服务。从 Mac OS X 版本 10.2 起就构建了 Bonjour 支持，它提供了这些特性，而不会引起人们对它本身的注意。例如，Bonjour 加强了一些特性，可以让用户为 iTunes 查找共享的音乐，或者连接到无线打印机，而无需进行自定义的配置。当这些服务变得可用时，它们将自动出现；当它们不可用时，则会自动消失。它是一个强大的 OS 特性。

GameKit 提供了与 Bonjour 相同的能力，但是不会构建复杂的 Bonjour 回调，用于注册和检测设备。利用 GameKit，将使用"对等选择"控制器请求一条连接，然后在连接建立之后管理"会话"。

GameKit 的会话对象提供了单个焦点用于数据传输管理。每个会话都使用一个独特的名称（你选择的）来宣传自身。当应用程序寻找另一个设备进行连接时，它将使用这个名称来标识兼容的设备。

如果使用 Bonjour 浏览设备寻找那个名称，那么将会失败。Apple 使用 SHA-1 散列编码服务名称。例如，一个名为"MacBTClient Sample"的服务将变为"_11d7n7p5tob54j._udp."Bonjour 服务。GameKit 将自动转换你提供的名称，以使它知道如何查找匹配的服务。

11.25.3 服务器、客户和对等方

GameKit 提供了 3 种会话模式：应用程序可以充当服务器、客户和对等方。服务器宣传服务并初始化会话，允许客户搜索并连接到它们。这是一类智能打印机使用的行为，可以让客户查找并使用它的能力。它对于提供固定功能的设备很方便，但是对于大多数 iOS 应用程序（尤其是游戏）并不是最佳的选择。

对等方是作为服务器和客户工作的。它们同时执行宣传和搜索。当某个对等方选择一种服务时，将对用户和开发人员隐藏它的客户/服务器角色。这使得很容易为 iOS 开发对等方法。你不必构建单独的客户和服务器应用程序，一个基于对等的应用程序将会做所有这些工作。

11.25.4 对等连接过程

对等方选择过程是由一个名为 GKPeerPickerController 的类处理的。它提供了一系列内置的交互式提醒对话框，用于自动执行宣传设备可用性和选择对等方的任务。不强求使用这个类，也可以绕过它，并创建一个自定义的类来搜索并连接到对等方。

不过，对于简单的连接，GKPeerPickerController 类提供了一个准备好使用的接口，它回避了检测并与对等方协商的需要。要使用对等方选择器，可以分配它，设置一个委托（它必须实现 GKPeerPickerControllerDelegate 协议），并显示它。

1. 显示对等方选择器

下面的代码用于分配并显示一个新的对等方选择控制器，并把它的连接风格设置为附近。这跳过了一个可选的交互步骤，其中用户将在 Online（联机）和 Nearby（附近）模式之间做出选择。在展示时，它将显示图 11-7 中的界面。不必使用对等方选择器来建立 GameKit 会话。iOS SDK 允许创建你自己的自定义接口，来处理底层的 GameKit 连接：

```
// Create and present a new peer picker
GKPeerPickerController *picker = [[GKPeerPickerController alloc]
    init];
picker.delegate = self;
picker.connectionTypesMask = GKPeerPickerConnectionTypeNearby;
[picker show];
```

图 11-7 这是展示给用户的用于对等 Bluetooth/Wi-Fi 连接的第一个屏幕

当掩码还包括联机类型（GKPeerPickerConnectionTypeOnline）时，选择器首先将询问用户要使用哪种类型的连接，然后再转向附近的连接界面或者你构建的自定义的联机界面。

2. 按下 Cancel 按钮

用户可能取消对等方选择器提醒。当他们这样做时，委托将接收到一个 peerPickerController-DidCancel:回调。如果在应用程序中显示一个 Connect 按钮，就要确保在此时恢复显示它，以使用户可以再次尝试：

```
- (void) peerPickerControllerDidCancel: (GKPeerPickerController *)picker
{
    [self setupConnectButton];
}
```

3. 创建会话对象

作为选择器委托，必须提供一个关于请求的会话对象。会话将提供一个抽象类，用于创建和管理设备之间的数据套接字，它属于 GKSession 类，必须利用会话标识符进行初始化。这是一个独特的字符串，用于创建 Bonjour 服务，并把两个宣传相同服务的 iOS 设备（对等方）链接在一起。通过把显示名称设置为 nil，会话将使用内置的设备名称：

```
- (GKSession *)peerPickerController:(GKPeerPickerController *)picker
      sessionForConnectionType:(GKPeerPickerConnectionType)type
{
    // Create a new session if one does not already exist
```

```
    if (!self.session) {
        self.session = [[GKSession alloc] initWithSessionID:
            (self.sessionID ? self.sessionID : @"Sample Session")
            displayName:nil sessionMode:GKSessionModePeer];

        self.session.delegate = self;
    }
    return self.session;
}
```

尽管这是一个可选的方法，但你通常都希望实现它，以便可以设置会话 ID 和模式。一旦检测到另一个具有相同的宣传服务 ID 的 iOS 设备，对等方选择器将把对等方显示为兼容的匹配，如图 11-8 中所示。

等待对等方选择器列表可能要花几秒钟的时间（典型情况），或者最多几分钟的时间（在现代固件上极少见）。Apple 建议总是通过干净的重新启动来进行调试。如果调试延迟的时间比较长，足以令人感到沮丧，就一定要重新启动。

在正常使用情况下，连接延迟通常最多会持续 45 秒。提醒用户要耐心等待。在图 11-9 中，Bear 是用于另一个运行相同应用程序的 iOS 设备的设备名称。当用户点按名称 Bear 时，这个设备将自动进入客户模式，并且 Bear 将进入服务器模式。

图 11-8　对等方选择器列出了可以充当对等方的所有设备

4. 客户和服务器模式

当设备变成客户模式时，它将停止宣传它的服务。以前在图 11-8 中显示的设备选择对话框将在服务器单元上发生变化。客户的对等方名称将变暗淡，以深灰色显示，并会在下方显示"is not available"（不可用）这些文字。几秒钟后（这实际上可能长达 1 分钟，以便再次提醒用户关于延迟的情况），两个单元都将更新它们的对等方选择器的显示内容。

图 11-9 显示了这个过程中的服务器和客户对等方选择器。在服务器接收到连接请求时客户将等待（左图）。在服务器上，主机用户必须接受或拒绝连接（中图）。如果他们拒绝连接，更新的对等方选择器将通知客户（右图）。如果他们接受连接，两个委托都会接收到一个新的回调。

图 11-9　一旦选择了合作伙伴,在服务器决定是接受还是拒绝连接时(中图)客户将进入等待模式(左图)。如果服务器拒绝连接,客户将接收到一个针对它的通知(右图)

委托回调允许新的对等方关闭对等方选择器,并设置他们的数据接收处理程序:

```
- (void)peerPickerController:(GKPeerPickerController *)picker
    didConnectPeer:(NSString *)peerID
    toSession: (GKSession *) session
{
    // Dismiss the picker, then set the data handler
    [picker dismiss];
    [self.session setDataReceiveHandler:self withContext:nil];
}
```

11.25.5　发送和接收数据

数据处理程序(在这里是 self)必须实现 receiveData:fromPeer:inSession:context:方法。发送给该方法的数据将使用一个 NSData 对象,没有钩子或句柄用于部分数据接收和处理。对于交互式程度较高的应用程序,当数据作为单个块到达时,要保持数据脉冲比较短(1000 字节以下)。

```
- (void) receiveData:(NSData *)data fromPeer:(NSString *)peer
    inSession: (GKSession *)session context:(void *)context
{
    // handle data here
}
```

通过会话对象发送数据。可以以可靠模式或不可靠模式发送它。可靠模式将使用错误检查以及重试,直到数据正确发送为止。使用 TCP 传输,保证所有的数据项都按发送它们的顺序到达。如果利用不可靠模式,将使用 UDP 传输一次发送数据,并且不会进行重试,数据可能无序到达。

在必须保证正确递送时，要使用可靠模式（GKSendDataReliable）；对于必须在几乎一瞬间到达的短数据脉冲，则使用不可靠模式：

```
- (void) sendDataToPeers: (NSData *) data
{
    // Send the data, checking for success or failure
    NSError *error;
    BOOL didSend = [self.session sendDataToAllPeers:data
        withDataMode:GKSendDataReliable error:&error];
    if (!didSend)
        NSLog(@"Error sending data to peers: %@",
            Error.localizedDescription);
}
```

你将在这里遇到的一个错误源于以可靠模式对大量数据进行排队，这将产生一个缓冲区填满错误。

11.25.6 状态改变

下面的会话委托回调可让你知道何时对等方的状态改变了。你希望寻找的两种状态是连接和断开连接，其中在对等方选择器关闭之后连接最终发生时，将出现连接状态；而当另一个用户退出应用程序、手动断开连接或者移出范围之外时，将出现断开连接状态：

```
- (void)session:(GKSession *)session peer:(NSString *)peerID
    didChangeState:(GKPeerConnectionState)state
{
    /* STATES: GKPeerStateAvailable, GKPeerStateUnavailable,
    GKPeerStateConnected, GKPeerStateDisconnected,
    GKPeerStateConnecting */

    if (state == GKPeerStateConnected)
    {
        // handle connected state
    }

    if (state == GKPeerStateDisconnected)
    {
        // handle disconnection
    }
}
```

要强制会话断开连接，可以使用 disconnectFromAllPeers 方法：

```
- (void) disconnect
{
    [session disconnectFromAllPeers];
}
```

11.25.7 创建 GameKit 助手

在与本章配套的示例代码中可以找到一个简化的对等助手类。这个类隐藏了大部分的 GameKit 连接细节和数据传输细节，同时演示了如何使用这些特性。更重要的是，它把 GameKit 处理过程分解成它的两个关键细节：连接和数据。

11.25.8 对等语音聊天

GameKit 的 In-Game Voice 服务可以让应用程序创建一条把两个设备连接在一起的对讲机式的语音通道。GKVoiceChatService 类添加的语音功能位于正确的 GameKit 之外。聊天服务连接到 iPhone 的音频播放和录制系统中，因此语音聊天可以倾听和播放音频。这样，语音聊天可以通过 GameKit 发送它的数据，以及播放它从 GameKit 接收到的数据。

GKVoice 期望利用 GKSession 和 GKPeers 传输它的数据。如果需要把语音传输用于另一种连接风格，那么你必须自己编写那一层代码。

11.25.9 实现语音聊天

在处理语音时，开始时的方式没有什么不同。显示一个对等方选择器并协商连接，就像通常利用 GameKit 所做的那样。在对等方连接之后，区别就随之而来了。你需要建立语音聊天，并把数据重定向到该服务，或者重定向来自该服务的数据。

一旦连接到新的对等方，就可以设置基本的语音聊天细节。激活播放和录制音频会话（如本章前面所示），设置默认的聊天服务客户，并与该对等方开始一个新的语音聊天。通过设置 client 属性，确保类接收到协商数据所需的语音聊天回调：

```
// Set the voice chat client and start voice chat
[GKVoiceChatService defaultVoiceChatService].client = self;
if (![[GKVoiceChatService defaultVoiceChatService]
    startVoiceChatWithParticipantID: peerID error: &error])
{
    NSLog(@"Error starting voice chat");
    return;
}
```

你的主类必须声明 GKVoiceChatClient 协议来执行该任务。当聊天服务通过麦克风收集数据时，它将触发 voiceChat-Service:sendData:toParticipantID:回调。在这里，可以把语音数据重定向到正常的 GameKit 会话。对于仅用于语音的连接，连同数据一起发送即可。当应用程序同时处理语音和其他数据时，可以构建一个字典，并利用键标记数据（比如@"voice"），或者当类通过正常的 receiveData:fromPeer:inSession:context:回调接收数据时，可以应用相同的方法。对于只传输语音的情况，可以使用 receivedData:fromParticipantID:把数据发送给聊天服务。语音聊天使你能够把游戏音频与游戏中的语音混合在一起。对于混合有语音和数据的应用程序，可以反序列化数据，确定分组包含的是语音还是普通的数据，并把该数据重定向到合适的接收方：

```
- (void)voiceChatService:(GKVoiceChatService *)voiceChatService
    sendData:(NSData *)data
    toParticipantID:(NSString *)participantID
{
    // Send the next burst of data to peers
    [self.session sendData: data toPeers:[NSArray arrayWithObject:
        participantID] withDataMode: GKSendDataReliable error: nil];
}


- (void) receiveData:(NSData *)data
    fromPeer:(NSString *)peer
    inSession: (GKSession *)session context:(void *)context
{
    // Redirect any voice data to the voice chat service
    [[GKVoiceChatService defaultVoiceChatService]
        receivedData:data fromParticipantID:peer];
}
```

11.25.10 创建"联机"GameKit 连接

在 GameKit 对等世界中，"联机"目前意味着除 Game Center 或 Bluetooth 之外的任何有效的连接风格。你可能使用本地 WLAN 网络连接相同网络上的另一个设备，或者通过 WWAN（即蜂窝服务）或 Wi-Fi 连接到基于 Internet 的远程主机。GameKit 只能把你带到这么远，如图 11-10 所示。通过选择 "Online"，你的用户将依靠你来创建一条通往另一个设备或服务的自定义连接。

通过给对等方选择器掩码提供联机选项，创建这个包含两个

图 11-10 "Online" GameKit 连接意味着自行联网

项目的对话框。在所有其他的方式中,在创建和展示标准的 GameKit 对等方选择控制器方面没有什么变化:

```
- (void) startConnection
{
    if (!self.isConnected)
    {
    GKPeerPickerController *picker = [[GKPeerPickerController
        alloc] init];
    picker.delegate = self;
    picker.connectionTypesMask =
        GKPeerPickerConnectionTypeNearby |
        GKPeerPickerConnectionTypeOnline;
    [picker show];
    if (self.viewController)
        self.viewController.navigationItem.rightBarButtonItem =
            nil;
    }
}
```

在 peerPickerController:didSelectConnectionType:回调中捕获用户选择。你可以假定如果用户选择的是"Nearby",那么握手对话框将为你处理好一切。不过,如果用户选择"Online",就由你自己决定如何把事情转到下一步。你将需要关闭选择器,并显示下一阶段的连接任务。在这里,控制将从对等方选择器传递给自定义的类,由你自己决定怎样创建这个类。下面的示例方法使用 BonjourHelper,它是在本书的上一版中引入的。无论使用哪个类,它的作用都是开启一条联机连接,选择要与之共享数据的对等方,并把该数据直接提供给那个对等方:

```
- (void)peerPickerController:(GKPeerPickerController *)picker
    didSelectConnectionType:(GKPeerPickerConnectionType)type
{
    if(type == GKPeerPickerConnectionTypeOnline)
    {
        [picker dismiss];

        // Establish your own custom connection class here
        [BonjourHelper sharedInstance].sessionID = self.sessionID;
        [BonjourHelper sharedInstance].viewController =
            self.viewController;
        [BonjourHelper sharedInstance].dataDelegate =
            self.dataDelegate;
```

```
    [BonjourHelper connect];
  }
}
```

11.26 小结

GameKit 在 iPhone 开发工具箱中提供了一个令人兴奋的新工具。它的易于使用的 Game Center 连接使得能够很轻松地交付在远程 iOS 设备之间通信的应用程序,以满足信息传输和玩游戏的需要。在本章中,你学习了如何构建那些连接以及产生实时的数据传输,允许游戏及其他应用程序在不同的设备之间协调信息。下面列出了关于这些技术的最后几点考虑。

- GameKit 一开始只推出了 OS X 版本。你的用户现在可以跨平台使用它,考虑深入研究为 OS X 和 iOS 开发客户。

- 仅仅由于 GameKit 这个名称中含有单词"Game"(游戏)并不意味着不能使用 GameKit 和 Game Center 在设备之间传输其他类型的信息和数据。不要由于 Apple 的游戏品牌宣传服务而限制你的应用程序。GameKit 提供了一种极好的数据传输基础设施,特定于游戏的特性位于它们的顶部,许多优秀的实用程序都使用这些技术。

- 在本地(尤其是在测试期间)使用 Voice Chat(语音聊天)时,记住附近的用户可能会产生将引起反馈失真的声音循环,除非他们使用耳机。对于对等使用,考虑彼此相距 10 英尺的人可以在不使用技术的情况下轻松地交谈。对于远程 Game Center 语音聊天,要知道该技术多么容易被滥用,并且它为你的用户提供了本地静音选项。

- 不要忘记激励你的用户。排行榜和成绩提供了极佳的方式,它们能超越单个设备或单个会话激励玩家参与玩游戏。成绩和排行榜显示现在能使用户快速、容易地给你的应用程序评级。在 iOS 6 中,用户可以点按他们获得的成绩,在社交场合夸耀他们自己。

- Game Center 配对设计用于不带计时器的回合制游戏(每次移动或者每个游戏)。尽管可以在 Game Center 顶部构建你自己的计时器,但它不是一个受 Apple 直接支持的特性,并且可能潜在地影响应用程序评审。

- 逐回合的游戏介绍在这一版中进行了大规模更新。不过,仍然有一些细枝末节需要处理,还有一些错误需要清除。如果你有关于改进代码的任何建议,请让我知道。可以把你的意见提交到 github 信息库,或者给我发送电子邮件即可。

第 12 章 StoreKit

StoreKit 提供了应用程序中的采购功能，可以把它集成进你的软件中。利用 StoreKit，最终用户可以使用他们的 iTunes 凭证从应用程序内购买可以解锁的特性、媒体订阅或可消费的资产，比如鱼食或阳光。他们将在最初从 App Store 获得并安装应用程序之后执行这些购买活动。本章介绍了 StoreKit，并且说明了如何使用 StoreKit API 为用户创建采购选项。在本章中，你将初步了解 StoreKit，并且学习如何在 iTunes Connect 上设置产品以及本地化它们的描述。你将看到它将采用什么来创建测试用户，以及如何顺利地通过多种不同的开发/部署障碍。本章将介绍如何恳请用户提出购买请求，以及如何把那些请求交给商店以便付款。在学完本章后，你将了解关于 StoreKit 的从产品创建到销售的基本情况。

12.1 初识 StoreKit

当应用程序需要比"购买一次，一直使用"更复杂的购买模型时，可以考虑 StoreKit。StoreKit 给开发人员提供了一种方式，使他们能够从应用程序内销售产品，以创造额外的收入来源。有许多原因使用 StoreKit。你可能支持一种预订模型，按需提供额外的游戏关卡，或者引入其他未锁定的特性。

利用 StoreKit，可以选择你想销售的项目，并设置它们的价格。StoreKit 和 iTunes 将负责处理细节，它们提供了基础设施，通过一系列 API 调用和委托回调给应用程序引入了店面。

12.1.1 履约

在购买时，用户不会下载新代码。所有基于 StoreKit 的应用程序在交付时都将带有它们已经构建的特性。例如，StoreKit 购买可能使用户能够访问应用程序中某些最初设置为禁用限制的部分。他们也可以下载或者解锁新数据集（它们现在可以由 Apple 托管），或者授权访问基于订阅的 Web 新闻馈送，还能够升级游戏中的项目以获得更多的能力或者更长的玩游戏时间。StoreKit 提供了让用户付费访问这些特性的方式，使它们能够在购买后仍然存在。

1. StoreKit 的局限性

不能使用应用程序中的购买功能来销售"硬"资产（比如 T 恤衫）或中间货币（比如用于 Web 站点的商店信用积分），并且还会禁止真实的赌博。通过应用程序中的购买功能销售的任何货物都必须以数字方式递送给应用程序。可购买的项目绝对不能包括色情物品、仇恨言论或诽谤。

12.1.2　StoreKit 开发悖论

不幸的是，StoreKit 提出了一条悖论：尽管可以轻松地把 IAP 特性添加到已经在 App Store 上销售的应用程序，但是直到已经把应用程序提交给 iTunes 之后，才能为新应用程序完全开发和测试应用程序中的购买功能。并且，在知道你还没有完成开发的情况下，将不能把应用程序完全提交给 iTunes。那么，开发人员要做什么呢？你怎样为 StoreKit 正确地开发新应用程序？

幸运的是，有一个解决方案，图 12-1 中显示了这个方法。为了解决 StoreKit 悖论，可以给 iTunes Connect 上传一个完全可以工作但是功能没有完全充实的应用程序。你在执行该操作时，完全清楚在将来某个时刻将会替换你的二进制文件。

Apple 把这种占位符描述为可批准的应用程序，可以把它提交给 App Store 评审过程并批准销售。如果应用程序被拒绝，应用程序中的项目将停止工作。必须具有一个已被批准或者正处于评审过程中的应用程序，才能为它开发 IAP 项目。

Apple 在其技术说明 TN2259 中这样解释了它：直到应用程序准备好进行应用程序评审以获得批准

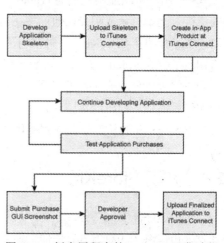

图 12-1　新应用程序的 StoreKit 开发过程

之后，才能把开发二进制文件上传到 iTunes Connect。如果在 iTunes Connect 中展示二进制文件并且它不是全功能的，那么应用程序评审将评审二进制文件，并且可能会拒绝开发二进制文件。如果应用程序评审拒绝 iTunes Connect 中最新的二进制文件，在应用程序购买功能中执行的测试将会失败。在这种情况下，解决办法是上传不带有应用程序中的购买特性的二进制文件，它可以被应用程序评审批准。一旦二进制文件获得批准，就要利用应用程序中的购买特性对二进制文件进行测试。

> **注意：**
> 在提交应用程序以便进行测试时，要在 iTunes Connect 价格标签中缩短可用性日期。在你做好准备之前，这可以阻止没有为黄金时间做好准备的应用程序因为疏忽而出现在 App Store 上进行销售。在做好准备时，可以重置那个日期。

12.1.3 开发和测试

对于新应用程序，在提交了应用程序并在 iTunes Connect 上定义了至少一个应用程序中的购买项目之后，就可以开始完全开发并测试应用程序及其购买功能（对于现有的应用程序，可以在不提交可批准骨架的情况下创建应用程序中的购买项目）。

无需收取信用卡，即可使用 StoreKit 的沙盒版本以及测试用户账户购买新项目。沙盒 StoreKit 使你能够在付费前后或期间测试应用程序特性。

12.1.4 提交

在完成开发并准备把最终版本提交给 App Store 时，将在 iTunes Connect 上完成 StoreKit 开发过程。

- 上传一个截屏图，显示用于应用程序购买的 GUI。
- 如果上传了一个占位符应用程序，并且它还没有通过评审流程，你可以自己拒绝它（如果它被批准，就代之以上传新版本即可）。
- 你上传应用程序的完全可工作的新版本以进行评审，其中包含应用程序中的购买支持。
- 在 iTunes Connect 上，提交应用程序中的购买以进行评审。

下面几节介绍了这个过程的许多细节。你将阅读到 StoreKit 过程的特定细节，并将学习如何把 StoreKit 添加到你的应用程序中。

12.2 创建测试账户

测试账户在 StoreKit 开发场景中扮演着关键角色。在开始开发新的基于 StoreKit 的应用程序之前，要创建一个或多个新的用户账户。这些账户将使你能够登录到 iTunes, 在不收费的情况下测试应用程序的付款功能。

下面说明了如何添加新用户。登录到 iTunes Connect, 并选择 Manage Users > Test User, 然后单击 Add New User。iTunes Connect 将展示图 12-2 中所示的表单。在填写这个表单时，要记住以下几点。

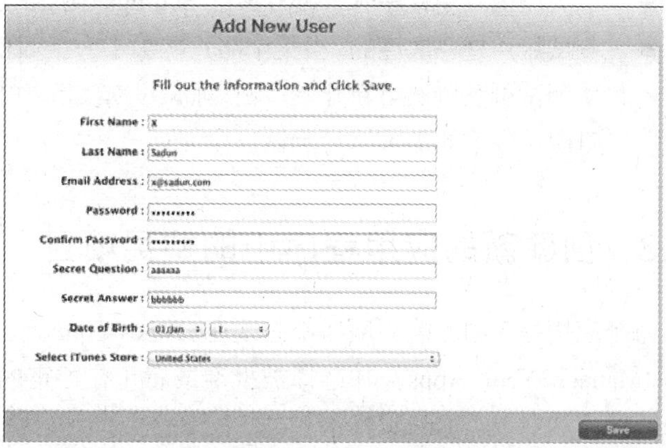

图 12-2　通过填写这个表单在 iTunes Connect 中添加新的测试用户

- 每个电子邮件地址都必须是唯一的，但它不必是真实的。只要该电子邮件地址不与系统中任何其他的电子邮件地址相冲突，那就没有什么问题。如你可能猜到的，其他开发人员已经采用了容易输入的地址，比如 abc.com、abcd.com 等。
- 名字不必是真实的，生日也是一样。考虑使用基本的按字母顺序排列的命名系统，比如 "a Sadun"、"b Sadun"、"c Sadun" 等。每一个人都出生在 1 月 1 日。
- Apple 总在不断改变它的密码规则。一般来讲，密码至少必须包含 8 个字符，并且包括一个大写字符、一个小写字符和一个数字。密码不能包含 3 个连续的相同字符，它不能如此简单，以至于 Apple 会将其标记为需要变得"复杂"。如你所想象的，当需要一遍又一遍地输入项目时，这可能是一种痛苦。单个容易记住的一次性密码可用于所有的测试账户（例如，AlphaBeta1）。你可能想投资一种输入实用程序，比如 Type2Phone（houdah.com）。这个应用程序使你能够通过从计算机到手机的蓝牙连接输入密码，从而简化了重复性的文本输入。
- 在这种环境下，秘诀问题/答案文本框是无意义的，但是不能把它们保持为空。不能为这两个文本框输入相同的字符串，并且问题文本框至少必须包含 6 个字符。可以考虑使用问题/答案对，比如 "aaaaaa" 和 "bbbbbb"，以简化账户创建。
- 选择一个 iTunes 商店是必需的，这个商店设置了测试的区域。如果你计划为多个商店使用多种语言支持，就要确保在每个受影响的区域都创建一个测试账户。
- 可以自由地删除用户账户并添加新的用户账户。如果用完了尚未购买任何项目的用户账户，只要根据需要创建新用户账户即可。
- 你永远都不希望在 Settings 应用程序中签入虚假的用户"账户"。如果尝试这样做，iOS 设备将强制你同意它的标准用户协议，然后尝试从你那里提取一张有效的信用卡。可以使用 Settings 注销一个账户，但要避免使用它登录一个账户。
- 每次测试应用程序时，要养成在 Settings 中注销账户的习惯。
- 测试账户是一次性的。当某个账户购买了一个项目后，可以自由地转移到下一个项目上，使测试可以再次重新开始。测试账户没有实际的上限，可以根据需要创建许多测试账户，对应用程序的需求执行测试。

12.3 创建新的应用程序中的购买项目

每个应用程序中的购买项目都必须在 iTunes Connect 上注册。要创建新的购买功能，可以导航到 Manage Your Apps，并选择被批准或者正在评审的任何应用程序。单击 Manage In-App Purchases，然后单击 Create New。

iTunes Connect 将提示你选择一种应用程序中的购买类型，可以从以下选项中做出选择：

- 可消费品——创建将被用户在正常使用应用程序期间用完的所购物品，比如升压电源、额外的子弹或鱼食。可消费品可能会过期，并且它们在使用期间将随着时间的推移而自然减少。
- 不可消费品——创建一次性购买项目，比如专业的步枪、访问额外的层次或全功能的界面，或者其他解锁物品。不可消费品不会过期，也不会随着使用而减少。
- 可自动延期的预订——这些使用户能够在固定的时间段购买内容，它将在每个时间段结束时自动对用户收费，直到用户选择退出服务为止。可自动延期的预订将自动传播给注册到相同 Apple ID 账户的所有设备。
- 免费预订——用于把免费内容放到 Newstand 应用程序中。
- 非延期预订——提供限时访问，比如对电影存档的 1 年期访问。

选择你想创建的应用程序中购买的类型，然后填写新购买项目的细节。首先可以输入关于购买的基本细节，设置它的价格和可用性，并输入一幅截屏图以便进行评审。

12.3.1 填写细节区域

细节区域包含参考名称、产品 ID 和语言列表。图 12-3 显示了这个屏幕的样子。参考名称应该以一种容易理解的方式描述购买；产品 ID 提供了独特的标识符，用于在 App Store 数据库中查找产品。

图 12-3　通过创建独特的产品 ID 来设置购买细节

1. 参考名称

参考名称用于为 iTunes Connect 的搜索结果以及在 App Store 中的应用程序的 Top In-App Purchase 区域中提供一个名称。因此，要输入一个有意义的名称（例如，"Unlock Sky Master Level 3 Purchase"），以便帮助你和其他人知道项目是什么，以及将如何在应用程序中使用它。

2. 产品ID

产品ID是一个独特的标识符，与用于应用程序的应用程序标识符相似。通常，我会使用我的应用程序ID并给它追加一个购买名称，比如com.sadun.scanner.optionalDisclosure。你需要这个标识符来查找商店，并获取关于这个购买项目的详细信息。对于产品ID可以应用与应用程序ID相同的规则。不能把一个标识符使用多次，也不能从App Store中"删除"它。一旦注册了它，就会永久注册它。如果在批准前删除它，将不能重新创建它。

还需要对购买添加一个或多个本地化语言描述。

12.3.2 添加本地化描述

每个可购买项目都必须向应用程序描述它自身。记住：应用程序是这种信息的主要消费者。在本地化描述中，需要指定一个显示名称（要购买的产品的名称）和描述（向用户解释购买的产品是什么以及它有什么作用）。

其中后两个元素将本地化成特定的语言。可以为任何或所有语言创建数据，只要定义了至少一种语言即可。在没有创建一个或多个名称/描述对的情况下，将不能提交一个新的购买项目。对于主要面向美国商店的开发人员，单独一个英语条目应该就能涵盖他们的需要。尽管如此，你还是可能考虑在以后某个时间借助专业翻译人员的帮助来修改它。

如果应用程序是在全世界销售的，你将可能希望镜像用于应用程序描述和应用程序中的特性的现有本地化。例如，如果你的iTunes商店宣传材料提供了日本的本地化，并且应用程序提供了一个日本的语言版本，还将需要创建日本的本地化的应用程序中的购买描述。如果不这样做，仍然可以使用应用程序中的购买，但是语言将默认为你所提供的任何本地化语言。

> **注意：**
> 总是要使用说本族语的人来本地化、编辑和校对文本。"Google 翻译"不能代替正确的本地化。像Traducto（traductoapp.com）这样的翻译社将在不同技能级别提供翻译和校对，包括说本族语的人和专业翻译人员。

1. 输入产品信息

在输入本地化产品信息数据时（参见图12-4），要记住以下几点。

- 你的应用程序是该信息的消费者。在iTunes Connect中输入的文本有助于创建应用程序将展示给用户的购买GUI。
- 用户的语言设置将选择本地化选项。如果本地化不可用，它将默认为所提供的其他描述之一。
- 如果计划使用带有Buy/Cancel选项的简单提醒表，要使措词保持严密，不要废话太多。

- 如果使用更复杂的视图，也要考虑这一点。

无论你怎样创建GUI，都要记得描述必须传达购买的动作以及所购买项目的描述，例如，"在购买时，这个选项将展示这个应用程序的细节屏幕。这些屏幕甚至会显示关于所扫描的MDNS服务的更多数据。"诸如"额外的细节屏幕"或"展示更多的细节"之类较短的描述不会向用户解释购买将如何工作，以及他们可以期望接收到什么。

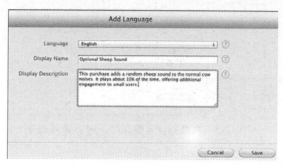

图12-4 单独添加每种语言

注意：
在iTunes Connect上进行评审期间，可以编辑项目显示细节。在购买的项目被批准之后，必须提交新的改变以便进行评审。

12.3.3 填写定价区域

定价区域用于指定购买将如果定价。保持选中Cleared for Sale项目，这个复选框可以确保应用程序（包括开发和分发）都能够以编程方式访问购买项目。

注意：
在评审期间，可以随时更改定价层和Cleared for Sale复选框。在购买的项目被批准后，必须提交新的改变以便进行评审。不能编辑标识符或者重用现有的标识符，也不能在创建购买项目之后更改产品的类型。

12.3.4 提供购买GUI截屏图

For Review区域出现在项目单的底部。直到完成了开发并且调试了应用程序之后，才会使用这个区域。在测试完产品并准备上传它以进行评审之前，可以把产品的状态保持为"Waiting for Screenshot"。当处于那个位置时，可以把截屏图上传到所提供的字段中。截屏图必须显示正在进行的应用程序中的购买活动，演示你构建的自定义GUI。

图12-5显示了提交区域。有效图片的大小至少需要有640像素×960像素。截屏图将突出显示你如何开发购买特性，提交一幅突出显示购买的图像。

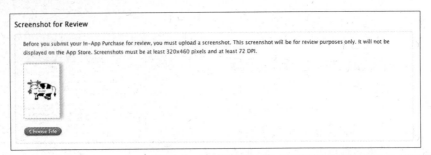

图 12-5 在准备评审购买产品时，必须提交一幅截屏图，把应用程序中的购买 GUI 显示给 Apple

12.3.5 提交应用程序中的购买产品以进行评审

在完成了沙盒测试并且确信应用程序和应用程序中的购买活动准备好让 Apple 进行评审之后，可以提交购买项目，以执行它自己的评审流程。上传新应用程序，返回到应用程序的 Manage In-App Purchases，选择任何购买项目，并单击 Submit for Review。

12.4 构建店面 GUI

Apple 的 StoreKit 框架没有提供一个内置的 GUI 用于请求用户购买，必须创建你自己的 GUI。通过创建 SKProductsRequest 实例，从 App Store 获取本地化价格和描述。这个类将基于你提供的标识符集合向商店询问该信息。必须在 iTunes Connect 上把每个标识符都注册为一个应用程序中的购买项目。

分配一个新的产品请求实例，并利用那个集合初始化它。可以为你已经建立的项目以及计划在将来添加的项目添加标识符。由于每个标识符实质上都是一个字符串，因此可以创建一个循环，依据某种命名模式构建标识符（例如，com.sadun.app.sword1、com.sadun.app.sword2），以便将来进行扩展。下面这个代码段用于搜索单个项目：

```
#define PRODUCT_ID          @"com.sadun.moo.baaa"

// Create the product request and start it
SKProductsRequest *productRequest =
    [[SKProductsRequest alloc] initWithProductIdentifiers:
        [NSSet setWithObject:PRODUCT_ID]];
productRequest.delegate = self;
[productRequest start];
```

在使用产品请求时，委托必须声明并实现 SKProductsRequestDelegate 协议，这包括 3 个简单的回调。程序清单 12-1 显示了这些回调方法，它们用于一个简单的应用程序。当接收到响应时，这段代码将寻找一个产品（通过上面的代码段只能请求一个产品），并获取它的本地化价格和描述。

然后，它将使用描述作为提醒文本以及两个按钮（价格和"No Thanks"）构建简单的提醒。这个提醒将充当基本的购买 GUI，图 12-6 显示了它的样子。

图 12-6　这个提醒是利用从 App Store 获取的本地化描述构建的

> **注意：**
> 如果没有以某种方式连接到网络，StoreKit 将不会工作。参阅第 5 章"联网"，查找用于帮助检查网络访问的秘诀。

程序清单 12-1　产品请求回调方法

```
- (void)request:(SKRequest *)request
    didFailWithError:(NSError *)error
{
    NSLog(@"Error: Could not contact App Store properly: %@",
        error.localizedFailureReason);
}


- (void)requestDidFinish:(SKRequest *)request
{
    NSLog(@"Request finished");
}


- (void) alertView: (UIAlertView *) alertView
    clickedButtonAtIndex: (NSInteger)answer
{
    NSLog(@"User %@ buy", answer ? @"will" : @"will not");
    if (!answer) return;
```

```objc
    // Ready to purchase - make purchase here
    // (see next section)
}


- (void)productsRequest:(SKProductsRequest *)request
    didReceiveResponse:(SKProductsResponse *)response
{
    // Find a product
    SKProduct *product = [[response products] lastObject];
    if (!product)
    {
        NSLog(@"Error: Could not find matching products");
        return;
    }

    // Retrieve the localized price
    NSNumberFormatter *numberFormatter =
        [[NSNumberFormatter alloc] init];
    [numberFormatter setFormatterBehavior:
        NSNumberFormatterBehavior10_4];
    [numberFormatter setNumberStyle:NSNumberFormatterCurrencyStyle];
    [numberFormatter setLocale:product.priceLocale];
    NSString *priceString =
        [numberFormatter stringFromNumber:product.price];

    // Show the information
    UIAlertView *alert = [[UIAlertView alloc]
        initWithTitle:product.localizedTitle
        message:product.localizedDescription
        delegate:self
        cancelButtonTitle:@"No Thanks"
        otherButtonTitles: priceString, nil];
    [alert show];
}
```

12.5 购买项目

要从应用程序中购买项目，首先可以添加一个交易观察者。例如，可以在应用程序委托的带选项完成启动方法或者在主视图控制器的加载视图或视图加载方法中执行该操作。无论你使用哪种观察者，都要确保类描述并实现 SKPaymentTransactionObserver 协议：

```
[[SKPaymentQueue defaultQueue] addTransactionObserver: someClassInstance ];
```

具有合适的观察者之后，就可以使用程序清单 12-1 创建的 GUI 来开始实际的购买：

```
// Ready to purchase the item
SKPayment *payment =
    [SKPayment paymentWithProduct:product];
[[SKPaymentQueue defaultQueue] addPayment:payment];
```

StoreKit 提示用户确认应用程序中的购买项目，如图 12-7 所示，然后接管购买过程。用户在可以继续下面的操作之前，可能需要登录到某个账户。

图 12-7　在从用户界面转移到实际的 App Store/StoreKit 购买系统之后，用户必须确认购买项目。这些截屏图是在执行沙盒测试期间拍摄的

12.5.1　签出 iTunes 账户以进行测试

要使用在 iTunes Connect 上建立的测试账户，确保签出当前的真实账户。启动 Settings 应用程序，选择 Store 参数设置，并单击 Sign Out。

如前所述，不要尝试利用测试账户凭证再次签入。只需退出 Settings，并返回应用程序。在单击 Buy 之后，将提示你签入 iTunes。在该提示之后，可以选择 Use Existing Account，并输入账户的详细信息。

12.5.2　在购买后重新获得程序控制

付款交易观察者将基于付款过程的成功或失败接收回调。程序清单 12-2 显示了一个骨架，用于响应完成的和未完成的付款。在用户完成了购买过程之后，交易将会成功或失败。如果成功，就会执行用户为其付款的任何动作，可能是通过下载数据或者解除锁定特性。

程序清单12-2 响应付款

```
- (void)paymentQueue:(SKPaymentQueue *)queue
    removedTransactions:(NSArray *)transactions
{
}


- (void) completedPurchaseTransaction:
    (SKPaymentTransaction *) transaction
{

    // PERFORM THE SUCCESS ACTION THAT UNLOCKS THE FEATURE HERE

    // For example, update user defaults, (preferably) the keychain,
    // or local variables. User defaults is *not* secure
    [[NSUserDefaults standardUserDefaults] setBool:YES forKey:@"baa"];
    [[NSUserDefaults standardUserDefaults] synchronize];
    hasBaa = YES;

    // Update GUI accordingly
    if (purchaseButton)
    {
        [purchaseButton removeFromSuperview];
        purchaseButton = nil;
    }
    // Provide some feedback to the user that the transaction
    // was successful, e.g.
    AudioServicesPlaySystemSound(baasound);

    // Finish transaction
    [[SKPaymentQueue defaultQueue] finishTransaction: transaction];

    // Always say thank you.
    UIAlertView *okay = [[UIAlertView alloc]
        initWithTitle:@"Thank you for your purchase!"
        message:nil delegate:nil cancelButtonTitle:nil
        otherButtonTitles:@"OK", nil];
    [okay show];
}


- (void) handleFailedTransaction: (SKPaymentTransaction *) transaction
```

```objectivec
{
    // Process any transaction error
    if (transaction.error.code != SKErrorPaymentCancelled)
    {
        UIAlertView *okay = [[UIAlertView alloc]
            initWithTitle:@"Transaction Error. Try again later."
            message:nil delegate:nil cancelButtonTitle:nil
            otherButtonTitles:@"OK", nil];
        [okay show];
    }

    // Complete the pending transaction
    [[SKPaymentQueue defaultQueue] finishTransaction: transaction];

    // Restore the GUI
}

- (void)paymentQueue:(SKPaymentQueue *)queue
    updatedTransactions:(NSArray *)transactions
{
    for (SKPaymentTransaction *transaction in transactions)
    {
        switch (transaction.transactionState)
        {
            case SKPaymentTransactionStatePurchased:
            case SKPaymentTransactionStateRestored:
                [self completedPurchaseTransaction:transaction];
                break;
            case SKPaymentTransactionStateFailed:
                [self handleFailedTransaction:transaction];
                break;
            case SKPaymentTransactionStatePurchasing:
                [self repurchase];
                break;
            default: break;
        }
    }
}
```

12.5.3 注册购买

可以使用许多方法中的任何一种来注册购买。可以同步 Web 服务器、创建本地文件、设置用户默认值（对于严肃的业务工作不建议这样做，但它非常合适于了解 IAP 如何工作），或者添加密钥链条目。你选择的解决方案留给你自己完成，只要不失去对购买的跟踪即可。在用户购买了解锁的特性、描述或数据之后，必须保证应用程序提供承诺的元素。

1. 安全和侵权

通常，你将希望在密钥链中存储购买项目。使用密钥链可以提供额外的好处：在删除并在以后重新安装应用程序之后，在这里存储的数据可以继续存在。

当使用不在现场的服务器注册购买并进行身份验证之后，确保在设备上重演这些设置。无论用户是否具有网络访问，都必须使用他们的应用程序。本地设置（例如，"在 2011 年 6 月 6 日前启用的服务"）将使应用程序能够运行并提供适当的反馈，甚至在预订的服务不可访问时也能如此。

永远不要假定用户会绑定到任何特定的服务。可以使用 iOS 6 及更高版本中提供的 identifierForVendor 跨安装的应用程序跟踪设备。

12.5.4 恢复购买

在卸载然后重新安装应用程序的设备上或者在与相同的 iTunes/Apple ID 账户关联的另一台安装应用程序的设备上，可能恢复购买活动。如果在多个设备（比如总共 5 款 iPhone、iPad 和 iPod 家族）上使用了顾客的账户，利用这些账户凭证通过任何设备进行的购买活动将使所有的设备都能够下载该购买项目，而不会另外收费。

StoreKit 使你能够恢复购买，这对于可消费的项目和预订项目特别重要，其中你将不希望允许用户重新购买已经有效的项目。对于不可消费的项目，用户可以重新购买，而不会无止境地花钱。对于这些不可消费的项目，可以简单地提交购买请求。App Store 界面将展示一个窗口，通知用户他已经购买了这个项目，并且他可以免费再次下载它。

要恢复与 iTunes 账户关联的购买活动，可以调用 restoreCompletedTransactions。它就像是添加一笔付款，并且涉及相同的回调。要把重新购买与购买分开来捕获，可以将 SKPayment-TransactionStateRestored 作为付款交易状态进行检查（参见程序清单 12-2）：

```
- (void) repurchase
{
    // Repurchase an already purchased item
    [[SKPaymentQueue defaultQueue] restoreCompletedTransactions];
}
```

这是由于购买事件提供的不是一个而是两个可能的成功结果。第一个是完成的购买，用户购买了项目，并且付款处理完成。第二个是这里描述的恢复购买。确保付款队列处理程序将寻找这两种状态。

这里有一个漏洞。考虑提供一个可消费的购买项目，比如发送传真的信用。如果用户卸载了应用程序然后重新安装它，任何重新购买功能都可能恢复一件已经在使用的资产。在设计具有可消费产品的应用程序时，必须更多地考虑安全基础设施，并且需要服务器端记账，以记录用户余额和消费的资产。

继续前进并恢复购买，但是要确保这些购买将适当地与服务器数据库协调一致。如下一节中将介绍的，Apple 利用购物收据为每个购买事件提供了一个独特的标识符。重新购买的项目将保留那个原始的标识符，使你能够区分新的购买事件和恢复的购买事件。

12.5.5 购买多个项目

用户可以购买多份可消费的项目和预订。设置付款的 quantity 属性，请求多次购买。下面这个代码段将添加对 3 份产品的请求，也许会增加 3 个月的预订、对某个字符的 3000 次点击，等等：

```
SKPayment *payment = [SKPayment paymentWithProduct:product];
payment.quantity = 3;
[[SKPaymentQueue defaultQueue] addPayment:payment];
```

12.5.6 处理注册购买中的延迟

如果购买连接有服务器，并且不能完成购买注册过程，那么不要最后确定交易。直到保证为顾客完成了所有的编制工作之后，才能调用 finishTransaction:。

万一在应用程序终止前没有利用用户新购买的项目设置他们，也没有问题。交易仍将保持在购买队列中，直到下一次启动应用程序为止。可以给你另外一次机会来尝试完成工作。

12.6 验证收据

成功的购买交易将包含一个收据。这个收据是以原始的 NSData 格式发送的，对应于编码的 JSON 字符串。它包含一个签名和购买信息。

Apple 强烈建议利用它的服务器验证所有收据，以阻止欺骗并且确保顾客实际地购买了他们所请求的项目。程序清单 12-3 显示了如何执行该操作。

必须利用 POST 方法把请求发送给 Apple 的两台服务器之一。使用的 URL 依赖于应用程序的部署，可以为生产软件和沙盒使用 buy.itunes.apple.com，为开发则使用 itunes.apple.com。

请求主体包括一个 JSON 字典。这个字典由一个键（"receipt-data"）和一个值（交易收据数据的 Base64 编码的版本）组成。我通常使用 CocoaDev NSData Base 64 扩展（来自 www.cocoadev.com/index.pl?BaseSixtyFour），把 NSData 对象转换为 Base64 编码的字符串。CocoaDev 为 Mac 和 iOS 开发人员提供了许多优秀的资源。

有效的收据将返回如下所示的 JSON 字典。收据包括交易标识符、用于所购买项目的产品 ID、用于主机应用程序的软件包 ID 和购买日期。最重要的是，它将返回一种状态：

```
{"receipt":
    {
        "item_id":"467440745",
        "original_transaction_id":"1000000008472082",
        "bvrs":"1.0",
        "product_id":"com.sadun.moo.baa",
        "purchase_date":"2011-09-23 15:18:22 Etc/GMT",
        "quantity":"1", "bid":"com.sadun.Moo",
        "original_purchase_date":"2011-09-22 20:48:46 Etc/GMT",
        "transaction_id":"1000000008535652"},
"status":0}
```

有效的收据总是具有一种 0 状态。任何非 0 数字都指示收据是无效的。

仅仅检查状态对于验证来说是不够的。通常，应该通过服务器而不是通过设备执行这种检查。可以建立一台代理服务器，截获对验证服务器的呼叫，并给所有的请求返回 JSON {"status":0}，这不是非常困难。更重要的是，连同验证请求一起发送的收据数据可以轻松地进行反序列化。

诸如 Urban Airship 和 Beeblex 之类的第三方服务为 IAP 提供了安全的收据验证。寻找限时的令牌和强大的加密。加密可以阻止中间人攻击；限时令牌则可以阻止重放攻击。它们一起使得简单的代理非常不可能成功地哄骗 IAP 收据，并欺骗应用程序提供某些信息。

程序清单 12-3　检查收据

```
- (void) checkReceipt: (SKPaymentTransaction *) transaction
{
    // Retrieve the receipt data and encode it with Base 64
    NSString *receiptData =
        [transaction.transactionReceipt base64Encoding];

    // Construct a dictionary for the receipt-data
    NSDictionary *dictionary =
        [NSDictionary dictionaryWithObject:receiptData
            forKey:@"receipt-data"];

    // Translate to JSON data
```

```objc
    NSData *json = [NSJSONSerialization
        dataWithJSONObject:dictionary options:0 error:nil];
    if (!json)
    {
        NSLog(@"Error creating JSON receipt representation");
        return;
    }

    // Select target
    NSString *urlsting = SANDBOX ?
        @"https://sandbox.itunes.apple.com/verifyReceipt" :
        @"https://buy.itunes.apple.com/verifyReceipt";

    // Create the request
    NSMutableURLRequest *urlRequest = [NSMutableURLRequest
        requestWithURL:[NSURL URLWithString: urlsting]];
    if (!urlRequest)
    {
        NSLog(@"Error creating the URL request");
        return;
    }
    [urlRequest setHTTPMethod: @"POST"];
    [urlRequest setHTTPBody:json];

    // Post the request and retrieve the result
    NSError *error;
    NSURLResponse *response;
    NSData *result = [NSURLConnection
        sendSynchronousRequest:urlRequest
        returningResponse:&response error:&error];
    // Check any errors here, look for status information, store
    // transaction data, etc.

    // Just demonstrate by showing the result data
    NSString *resultString = [[NSString alloc]
        initWithData:result encoding:NSUTF8StringEncoding];
    NSLog(@"Receipt Validation: %@", resultString);
}
```

12.7 小结

StoreKit 框架提供了一种极佳的方式使应用程序赚钱。如你在本章中所读到的，你可以建立自己的店面，通过应用程序销售服务和特性。下面列出了最后几点考虑。

- 尽管新应用程序的整个设置和测试流程似乎有一点"到底先有鸡还是先有蛋"的意思，但是利用最小的代价开发和部署一个基于 StoreKit 的应用程序显然是可能的。下一次为已经批准的应用程序添加应用程序中的购买功能将变得更容易。
- 总是要记住 IAP 有两个提交步骤。确保应用程序和 IAP 项目准备好让 Apple 评审。
- 在新的购买设置完全、绝对、100%完成前，要避免结束交易，即使这意味着等待应用程序重新启动。与此同时，要通知用户购买过程正在经历意料之外的延迟。
- 你的方法只能从注册到当前运行的应用程序的项目那里请求产品信息，而不能跨应用程序共享请求。
- 只要有可能，就要使用脱离设备的收据验证。过去一年间众多的事件显示绕过安全性有多容易。
- 开发人员大量使用 IAP，并且通常不考虑用户体验。在我看来，在应用程序中需要 IAP 对年轻人来说是一种罪恶，Apple 政策应该强烈反对这样做。此外，如果你的应用程序需要 IAP 来绕过玩游戏环节，就需要严肃地重新考虑游戏设计。
- 要重点关注为付费用户提供良好的体验，而不是把精力放在与侵权作斗争上。如果反侵权保护甚至激怒了一位付费顾客，你就输掉了一场战争。
- 不要忘记建立购买观察者。三个臭皮匠，顶个诸葛亮，这个步骤比其他任何 StoreKit 问题都重要。

第 13 章 推送通知

当脱离设备的服务需要直接与用户通信时，推送通知提供了一种解决方案。就像本地通知允许应用程序在预定的时间联系用户一样，推送通知可以从基于 Web 的系统递送消息。推送通知可以让设备显示一条提醒、播放一段自定义的声音，或者更新应用程序商标。脱离设备的服务可以用这种方式联系基于 iOS 的客户，使他们能够知道新的数据或更新。

与大多数其他的 iOS 开发领域不同的是，几乎所有的推送故事都是在设备之外发生的。开发人员创建基于 Web 的服务，来管理和部署信息更新。推送通知把那些更新传送给特定的设备。在本章中，你将学习推送通知的工作原理，并深入研究创建自己的基于推送的系统所需了解的细节。

13.1 推送通知简介

推送通知（push notification）也称为远程通知（remote notification），指一种通过外部设备发送给 iOS 设备的消息。这些基于推送的服务可以与通常会检查信息更新的任何类型的应用程序协同工作。例如，一种服务可能在 Facebook、Twitter 或 Google Plus 上轮询新的更新；扫描附近危险的天气系统；为家中的安全系统响应传感器；或者邀请你参加共享的会议。当新信息可以为客户所用时，服务将通过 Apple 的远程通知系统推送该更新。通知直接传送给设备，后者则注册成接收那些更新。

推送的关键是：这些消息来源于设备外部。它们是客户—服务器范型的一部分，使基于 Web 的服务器组件能够通过一种 Apple 提供的服务与 iOS 客户通信。利用推送，开发人员可以给 iOS 设备发送几乎即时的更新，而不依赖于用户启动特定的应用程序。作为替代，处理过程则发生在服务器端。当推送消息到达时，iOS 客户可以做出相应的响应，比如显示一个商标、播放一段声音或显示一个提醒框。

把应用程序逻辑转移到服务器上限制了客户端的复杂性，允许许多客户共享单个更新系统。考虑刚才提到的危险天气服务。一个专用的站点可以监测 NOAA 及其他大气数据馈送，并根据需要给每个预订服务的客户发送远程通知。脱离现场的处理还可以为基于 iOS 的应用程序节约能量，它们现在可以依赖于推送，而不需要使用 iOS 设备的本地 CPU 资源来监测重要的信息改变并对其做出反应。

推送不仅绑定到本地资源，它还提供了一种有价值的解决方案，用于同超越轮询和更新应用程序之外的基于 Web 的服务通信。例如，推送可能使你能够挂钩到一种训练服务，传送有益的证词，甚至当应用程序没有运行时也可如此；或者挂钩到一种玩游戏服务，给你发送关于即将到来的比赛的提醒通知。

从社交网络到监测 RSS 馈送，推送使 iOS 用户能够保持在异步数据馈送的顶部。它提供了一种强大的解决方法，用于把 iOS 客户连接到各类基于 Web 的系统。利用推送，你编写的服务可以连接到你所安装的 iOS 库，并以一种干净、实用的方式传达更新。

13.1.1 推送的工作原理

推送通知绑定到特定的应用程序，需要进行多项安全检查。推送提供者（生成推送请求并把它们传送给 Apple 的服务器组件）只能与那些宿主其应用程序中的 iOS 设备通信，这些设备是联机的，并且选择了接收远程消息。用户在推送更新中具有最终的发言权，他们可以允许或禁止这类通信，并且编写良好的应用程序可以让用户根据意愿决定加入或者决定退出服务。

服务器与客户之间的通信链的工作方式如下。推送提供者通过一台中心 Apple 服务器递送消息请求，并经由该服务器传送给它们的 iOS 客户。在正常的使用情况下，服务器将触发某个事件（比如新的邮件或者即将到来的约会），并生成针对特定 iOS 设备的通知数据。它把这个消息请求发送给 APNS（Apple Push Notification Service，Apple 推送通知服务）。这个通知将使用 JSON 格式化，并且每个通知受限于 256 字节，因此可以通过该消息推送的信息相当有限。这种格式化和大小确保 APNS 把带宽限制于可能最严格的配置。

APNS 提供了一种集中式系统，可以与现实世界里的 iOS 设备协商通信。它把消息传送给指定的设备，iOS 设备上的处理程序决定如何处理消息。如图 13-1 所示，推送提供者与 APNS 交流，发送它们的消息请求，APNS 再与 iOS 设备交流，把那些消息转发给设备上的处理程序。

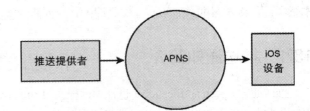

图 13-1　提供者通过 Apple 的集中式推送通知服务发送消息，与 iOS 设备通信

13.1.2 多个提供者支持

APNS 被构建成支持多个提供者连接，使许多服务能够同时与它通信。它提供了多个进入服务的网关，使得每个推送服务在发送它的消息前都不必等待可用性。图 13-2 演示了提供者与 iOS 设备之间的多对多关系。APNS 使提供者能够通过多个网关同时连接，每个提供者都可以把消息推送给许多不同的设备。

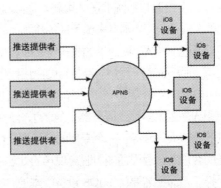

图 13-2　Apple 的 APNS 在其面向提供者的一端提供了许多网关，使许多提供者能并行连接。每个推送提供者都可能连接到任意数量的 iOS 设备

13.1.3 安全

安全是远程通知的主要组件。推送提供者必须为它使用的每个应用程序签署 SSL（Secure Sockets Layer，安全套接字层）证书。服务将不能与 APNS 通信，除非它们利用该证书对自己进行了身份验证。它们还必须提供一个称为令牌（token）的独特的键，用于标识单个设备上的特定应用程序。

在获取经过验证的消息和设备令牌之后，APNS 将联系所涉及的设备。每个 iOS 设备都必须以某种方式联机以接收通知。可以把它们连接到蜂窝数据网络或者 Wi-Fi 热点。APNS 将建立一条与设备的连接，并转发通知请求。如果设备是脱机的并且 APNS 服务器不能建立连接，就会对通知进行排队处理，便于以后递送。

一旦接收到请求，iOS 就会执行许多检查。当用户为给定的应用程序禁用推送更新时，推送请求将会被忽略；用户可以在其设备上的 Settings 应用程序中执行该操作。当允许更新时，只有在那个时候，iOS 设备才会确定客户应用程序当前是否在运行。如果是，它将通过应用程序委托直接发送一条消息给运行的应用程序。如果不是，它将执行某种提醒，比如显示文本、播放声音或者更新商标。

当提醒显示时，用户通常可以选择关闭提醒或者点按 View。如果他们选择 View，iOS 设备将启动所涉及的应用程序，并给它发送通知消息，当它运行时将会接收到该消息。如果用户点按 Close，通知将被忽略，并且应用程序不会启动。

从服务器到 APNS 到设备再到应用程序的路径构成了推送通知的核心流程。每个阶段都会沿着该路径移动消息。尽管多个步骤在现实生活中听起来可能比较广泛，但是通知几乎是即时到达的。在设置了证书、标识符和连接之后，实际的信息递送将变得微不足道。几乎所有的工作都在于首先设置那条传送链，然后产生你想要递送的信息。

确保将所有的证书和设备令牌都视作敏感信息。在服务器上存储这些项目时，必须确保它们一般是不可访问的。如果这类信息不受控制，它们就可能会被第三方利用。这很可能会导致 Apple 撤销你的 SSL 推送证书。这将会禁用你销售的任何应用程序的所有远程通知，并且可能会强制你从商店里取走应用程序。

13.1.4 推送限制

在现实场景中，推送通知不像你可能想要的那样可靠，它们可能相当脆弱。Apple 不保证每个通知都会递送，也不保证通知会按顺序到达。永远不要通过推送来发送至关重要的信息。把这个特性预留用于有益的通知，用它们来更新和通知用户，但是用户可能会错失它们，不过不会导致严重的后果。

推送递送队列中的项目可能会被新的通知所取代，这意味着通知可能会竞争，并且可能会在沿途丢失。尽管 Apple 的反馈服务可以报告失败的递送（即不能通过推送服务正确发送的消息，特别是对于已经从设备上删除的应用程序），但是不能获取关于相冲突的通知的信息。从 APN 服务的角度讲，丢失的消息仍然会被成功地"递送"。

13.1.5 推送通知与本地通知

iOS 提供了远程推送通知和本地通知支持。本地通知（在 Core iOS Developer's Cookbook 一书中讨论了它）是在设备上创建和调度的；推送通知则是从远程服务器上发出的，并将通过 APNS 发送给设备。本章只介绍了远程通知。

13.2 配置推送

要开始推送开发，必须访问 Apple 的 iOS 开发人员配置门户网站（Provisioning Portal），这个门户网站位于 http://developer.apple.com/ios/manage/overview/index.action。利用你的 iOS 开发人员凭证登录，以获得该站点的访问权限。在这个门户网站上，可以执行所需的步骤，创建可以与推送服务关联的新应用程序标识符。

下面几节将引领你通过这个过程。你将看到如何创建新的标识符、生成证书，以及请求特殊的配置文件，以便可以构建支持推送的应用程序。如果没有支持推送的配置文件，应用程序将不会接收到远程通知。

13.2.1 生成新的应用程序标识符

在 iOS 配置门户网站上，单击 App IDs，在 Web 页面左边的列中可以找到这个选项。这将打开一个页面，允许创建新的应用程序标识符。每个推送服务都基于单个标识符，必须创建并设置它，以允许远程通知。不能对推送应用程序使用通配符标识符；每个支持推送的应用程序都需要独特的标识符。

在 App IDs 区域中，单击 New App ID；这个按钮出现在 Web 页面的右上方。单击该按钮时，站点将打开一个新的 Create App ID 页面，如图 13-3 所示。输入一个名称，描述新的标识符（比如"Tutorial Push Application"）和新的软件包标识符。

这些 ID 通常使用反向域名模式，比如 com.domainname.appname 和 com.sadun.pushtutorial。标识符必须是独特的，不能与 Apple 的系统中注册的其他任何应用程序标识符发生冲突。保持将软件包种子（bundle seed）设置为群组标识符，如果不想这样做，除非有具有某个非常有说服力的理由。

单击 Submit，添加新的标识符。这将

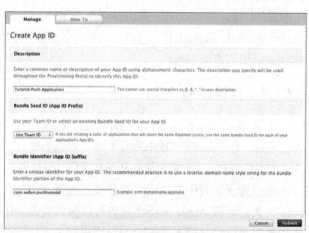

图 13-3　在 iOS 配置门户网站上创建新的应用程序标识符

不可撤销地把应用程序 ID 添加到 Apple 的系统中，其中现在将把它注册给你。此时，将返回到 App ID 页面，其中显示了它的标识符列表，并且现在将准备把该标识符建立为遵从推送的。

> **注意：**
> 在创建了应用程序标识符之后，Apple 没有提供任何方式从程序门户网站上删除它。

13.2.2 生成 SSL 证书

在 App ID 页面上，可以看到哪些标识符用于推送，哪些则没有。Apple Push Notification 列显示了是否为每个应用程序 ID 启用了推送。这个列的 3 种状态如下。

- Unavailable（灰色），用于不再可用的 ID。
- Configurable（黄色），用于能够使用推送但是尚未设置成这样做的应用程序。
- Enabled（绿色），用于为推送做好准备的应用程序。

在推送列中，可以找到用于每个应用程序标识符的两个圆点：一个用于 Development（开发），另一个用于 Production（生产）。这些选项是单独配置的。定位新的应用程序 ID，确保显示了用于 Development 的黄色 Configurable，并单击 Configure。这个选项出现在最右边的列中，单击该选项时，浏览器将打开一个新的 Configure App ID 页面，允许把标识符与推送通知服务关联起来。

在页面向下大约一半的位置将出现一个 Enable Push Notification Services 复选框。选中这个复选框，开始证书创建过程。当选中时，页面右边的两个 Configure 按钮将变成启用状态。单击一个按钮，加载指导页面，说明如何执行下面的操作。它将指导你创建一个安全的证书，服务器将使用它来签署发送给 APNS 的消息。

如所指示的那样，启动 Keychain Access 应用程序。这个应用程序位于 Macintosh 上的 /Applications/Utilities 文件夹中。在启动时，可以选择 Keychain Access > Certificate Assistant > Request a Certificate From a Certificate Authority。需要再次执行这个步骤，即使已经为开发人员和分发证书创建了以前的请求也是如此。新的请求将会添加可以唯一地标识 SSL 证书的信息。

在 Certificate Assistant 打开后，输入电子邮件地址并添加一个可识别的公共名称，比如 "Push Tutorial App"。这个公共名称很重要，它可以为将来的操作带来方便，因此要选择一个容易识别并且准确地描述了项目的名称。公共名称可让你把 OS X Keychain Access 实用程序中看上去比较相似的密钥链项目彼此区分开。

在指定了公共名称之后，选择 Saved to Disk，并单击 Continue。Certificate Assistant（证书助手）将提示你选择一个保存的位置（"桌面" 就比较方便）。单击 Save，等待证书生成，然后单击 Done，返回到 Web 浏览器，并单击 Continue。现在将准备好提交证书签署的请求。

单击 Choose File，并导航到刚才生成的请求。选择它，并单击 Choose。然后单击 Generate，构建新的 SSL 推送服务证书。这可能要花一两分钟的时间，因此要有耐心，并且不要关闭 Web 页面。生成证书后，单击 Continue。可以单击 Download 获取新的证书。最后，单击 Done，返回到 App ID 页面，其中在应用程序 ID 旁边应该会显示一个新的绿色 Enabled 指示器。Apple 还会给你发送一封电子邮件，确认你的证书请求已被批准。

> **注意：**
> 如果你需要再次下载 SSL 证书的公钥部分，可以单击 Configure，返回到 Configure App ID 页面。在该页面中，可以单击 Download，请求另一个副本。通常，应该把证书私钥和公钥部分导出到一个 p12 文件中，以便安全保存。如果丢失了私钥，将不得不重新生成整个证书。

如果计划从 Macintosh 运行 Push Server（便于在部署服务器前或者在完全开发了服务器部分之前从事开发工作），就要向密钥链中添加新项目。它将出现在证书中，可以单击证书旁边的小三角形，展示在创建证书请求时使用的公共名称来确认证书。

13.2.3 特定于推送的配置

不能为支持推送的应用程序使用通配符群组配置文件。作为替代，必须只为那个应用程序创建单个配置文件。这意味着如果你打算创建应用程序的开发、临时和分发版本，那么除了你已经为其他工作创建的任何配置文件之外，还必须请求 3 个新的移动配置文件。

在 Xcode 的最新版本中，可以在 Organizer 中构建配置文件（按下 Command+Shift+2 组合键 > Devices > Provisioning Profiles > New）。输入配置文件名称，选择已经配置的应用程序 ID，并选择你想使用的设备。有一个 Select All 按钮，可以使这个操作变得更容易。单击 Next 按钮，可以让 Xcode 为你做所有余下的工作。

此外，还可以转到开发人员门户网站的 Provisioning 区域。通过单击相应的标签，选择 Development 或 Distribution 配置文件。然后单击 New Profile，开始创建新的配置，此时将打开 Create Provisioning Profile 页面。

- `Development Provision`：对于开发，可以输入一个配置文件名称，比如"Push Tutorial Development"。检查你将使用的证书，并从弹出式列表中选择应用程序标识符。然后选择你将使用的设备，并单击 Submit。
- `Distribution Provision`：对于分发，可以选择 App Store 或 Ad Hoc。为新配置输入一个名称，比如"Push Tutorial Distribution"或"Push Tutorial Ad Hoc"。从弹出式列表中选择应用程序标识符。对于仅临时性的分发，可以选择要包括在配置中的设备，然后单击 Submit，完成配置。

生成配置文件可能花一会儿功夫。稍等片刻，并重新加载页面。配置状态应该会从 Pending 变为 Active。下载新的配置，并通过 Organizer > Devices > Provisioning Profiles 把它添加到 Xcode 中。单击 Import，并从下载文件夹中选择配置，然后单击 Open。

13.2.4 创建推送兼容的应用程序

在把支持推送的配置文件安装到 Xcode 中后,就准备创建推送兼容的应用程序。首先确保新的软件包标识符匹配刚才在 TARGETS > Info > Bundle identifier 中创建的配置。然后,在 Build Settings > Code Signing Identity 中,为新的配置文件选择开发人员身份。你将为调试和发布版本使用单独的身份和配置文件。

13.3 注册应用程序

利用推送兼容的移动配置文件签署应用程序只是处理推送通知的第一步。应用程序必须请求利用 iOS 设备的远程通知系统注册它自身。可以利用单个 UIApplication 调用执行该操作,如下所示。application: didFinishLaunchingWithOptions:委托方法为调用它提供了特别方便的位置:

```
[[UIApplication sharedApplication]
    registerForRemoteNotificationTypes:types];
```

这个调用告诉 iOS 应用程序希望接受推送消息。你传递的类型指定了应用程序将接收什么类型的提醒,iOS 提供了 3 类通知。

- **UIRemoteNotificationTypeBadge**:这个通知在 SpringBoard 上给应用程序图标添加一个红色商标。
- **UIRemoteNotificationTypeSound**:声音通知可让你从应用程序软件包中播放声音文件。
- **UIRemoteNotificationTypeAlert**:这种风格将使用提醒通知利用自定义的消息在 SpringBoard 或者任何其他的应用程序中显示文本提醒框。

选择你想使用的类型或者它们的组合。它们都是位标志,结合起来可以告诉通知注册过程你想怎样执行下面的操作。例如,下面的标志允许提醒和商标,但是不允许有声音:

```
types = UIRemoteNotificationTypeBadge | UIRemoteNotificationTypeAlert;
```

执行注册,更新用户设置,并且允许用户通过 Settings > Notifications 自定义他们的推送参数设置。其中,用户可以选择如何展示通知提醒风格(横幅、提醒或者什么也不展示)、允许哪些类型(商标、提醒和声音),以及是否在锁定屏幕上显示提醒等。

要从这类通知的主动参与中删除你的应用程序,可以发送 unregisterForRemoteNotifications。这将针对所有的通知类型取消注册你的应用程序,并且不接受任何参数:

```
[[UIApplication sharedApplication] unregisterForRemoteNotifications];
```

13.3.1 获取设备令牌

直到应用程序生成设备令牌并递送给服务器之后,它才能接收推送消息。它必须把该设备令牌发送给不在现场的服务，以推送实际的通知。本节后面的秘诀 13-1 没有实现服务器功能，它只提供了客户软件。

令牌绑定到一个设备上。它与 SSL 证书相结合，唯一地标识了 iOS 设备，并且可用于把消息发送回使用的设备。要知道的是，在恢复设备固件之后，设备令牌可以改变。

设备令牌是作为注册的副产品创建的。一旦接收到注册请求，iOS 将会联系 APNS。它使用 SSL 请求。似乎有些显而易见的是，设备必须连接到 Internet。如果没有，请求将会失败。利用活动的连接，iOS 把请求转发给 APNS，并等待它利用设备令牌做出响应。

APNS 将构建设备令牌并把它返回给 iOS，后者反过来又通过应用程序委托回调（即 application:didRegisterForRemoteNotificationsWithDeviceToken:）把它传递回应用程序。应用程序必须获取设备令牌，并把它传递给服务的提供者组件，在这里将需要安全地存储它。任何可以访问设备令牌的人和应用程序的推送凭证都可以把消息发送给设备。必须把该信息视作敏感信息，并相应地加以保护。

> **注意：**
> 有时，生成令牌可能要花一些时间。通过在每次应用程序运行时注册，考虑在应用程序中设计可能的延迟。直到创建了令牌并上传到站点之后，才能给用户提供远程通知。

13.3.2　处理令牌请求错误

有时，APNS 不能创建令牌或者设备不能发送请求。例如，不能通过模拟器生成令牌。UIApplicationDelegate 方法 application:didFailToRegisterForRemoteNotificationsWithError:可让你处理这些令牌请求错误。一般来讲，需要获取错误并将其显示给用户：

```
// Provide a user explanation for when the registration fails
- (void)application:(UIApplication *)application
    didFailToRegisterForRemoteNotificationsWithError:(NSError *)error
{
    NSString *status = [NSString stringWithFormat:
        @"%@\nRegistration failed.\n\nError: %@", pushStatus(),
        error.localizedFailureReason];
    tbvc.textView.text = status;
}
```

13.3.3　响应通知

iOS 使用一组操作链（见图 13-4）来响应推送通知。当应用程序运行时，将直接把通知发送给 UIApplicationDelegate 方法 application:didReceiveRemoteNotification:。以 JSON 格式发送的有效载荷将自动转换为 NSDictionary，并且应用程序可以根据需要自由地使用该有效载荷中的信息。当应用程序已经在运行时，不会调用更多的声音、商标或提醒。

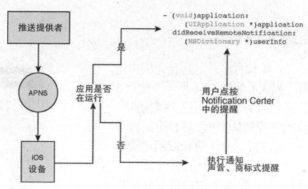

图 13-4　仅当应用程序没有运行时,才会展示可视的和能够听见的通知。如果用户点按 Notification Center（通知中心）中的提醒,应用程序将会启动,并将有效载荷作为通知发送给应用程序委托

```
// Handle an actual notification
- (void)application:(UIApplication *)application
    didReceiveRemoteNotification:(NSDictionary *)userInfo
{
    NSString *status = [NSString stringWithFormat:
        @"Notification received:\n%@",userInfo.description];
    tbvc.textView.text = status;
    NSLog(@"%@", userInfo);
}
```

图 13-5　远程提醒可以出现在锁定屏幕上、SpringBoard 中、应用程序中和 Notification Center（通知中心）中。用户可以在锁定屏幕上滑动提醒或者在 Notification Center（通知中心）中点按它,以切换到通知应用程序。在这里,应用程序是 HelloWorld,其名称清楚地显示在提醒上

当应用程序没有运行时,iOS 将执行注册和用户设置允许的所有请求的通知。这些通知可能包括播放声音、标记应用程序和显示提醒。以接收到通知时,播放声音也可能会触发 iPhone 振动。

默认情况下,通知将添加到 Notification Center（通知中心）中（参见图 13-5）（用户可以通过 Settings > Notification > Application Name > Notification Center 禁用它）。当用户点按列表中的通知时,应用程序将启动,并把通知作为启动选项。

一旦启动,应用程序委托就会接收到已经运行的应用程序将会看到的相同的远程通知回调。当设备锁定时,提醒将出现在锁定屏幕上和 Notification Center（通知中心）中,除非用户覆盖了默认的设置。

13.4 秘诀：推送客户骨架

秘诀 13-1 介绍了一个基本的客户，它允许用户注册和取消注册推送通知。界面（如图 13-6 所示）使用 3 个选项开关来控制要注册的服务。当应用程序启动时，它将查询应用程序支持的远程通知类型，并且更新选项开关以与之匹配。此后，客户将记录注册和取消注册动作，并调整选项开关，以与设置的实际情况保持同步。

位于界面左上方和右上方的两个按钮可以让用户取消注册和注册他们的应用程序。取消注册将禁用与应用程序关联的所有服务，它将提供干净的清除。与之相比，注册应用程序需要标志来指示请求的是哪些服务。

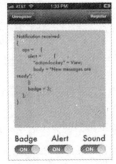

图 13-6 秘诀 13-1 中介绍的推送客户骨架可以让用户指定他们想注册哪些服务

在请求新的服务时,总会提示用户批准它们。图 13-7 显示了出现的对话框。用户必须通过明确授予应用程序权限来进行确认。如果用户不这样做，即点按 Don't Allow 按钮，标志将保留它们以前的设置。

不幸的是，无论用户是否同意，确认对话框在关闭时都不会生成一个回调。要捕获这个事件，可以侦听在对话框把控制返回给应用程序时生成的普通通知（UIApplicationDidBecome- ActiveNotification）。它是一种入侵方式，但是它可以让你知道用户何时做出响应以及用户是怎样响应的。在秘诀 13-1 中，confirmationWasHidden:方法捕获这个通知并更新选项开关，以匹配任何新的注册设置。

作为一种骨架系统，这个推送客户除了显示递送的用户信息有效载荷的内容之外，不会实际地响应推送通知。图 13-8 演示了在图 13-6 中发送的实际有效载荷，这种显示是在应用程序委托中的 application:didReceiveRemoteNotification:方法中执行的。

图 13-7 用户必须明确授权应用程序接收远程通知

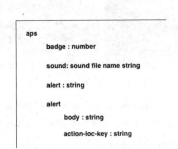

图 13-8 aps 字典可能包括商标请求、声音文件名称和/或提醒字符串。由于 Notification Center（通知中心），动作锁定键现在不那么重要。如果用户禁用 Notification Center（通知中心）（这很少见），他们将看到旧式的提醒，这将显示动作按钮

> **注意：**
> 在线示例项目中包括的 3 个声音文件（ping1.caf、ping2.caf 和 ping3.caf）可让你利用真实的音频来测试声音通知。

秘诀 13-1　推送客户骨架

```
// Important: Set up your own app id and provisioning profile
// before attempting to play with this code

@implementation TestBedViewController
@synthesize textView;

// Basic status
NSString *pushStatus ()
{
    return [[UIApplication sharedApplication]
        enabledRemoteNotificationTypes] ?
        @"Notifications were active for this application" :
        @"Remote notifications were not active for this application";
}

// Fetch the current switch settings
- (NSUInteger) switchSettings
{
    NSUInteger settings = 0;
    if (badgeSwitch.isOn) settings =
        settings | UIRemoteNotificationTypeBadge;
    if (alertSwitch.isOn) settings =
        settings | UIRemoteNotificationTypeAlert;
    if (soundSwitch.isOn) settings =
        settings | UIRemoteNotificationTypeSound;
    return settings;
}

// Change the switches to match reality
- (void) updateSwitches
{
    NSUInteger rntypes = [[UIApplication sharedApplication]
        enabledRemoteNotificationTypes];
    badgeSwitch.on = (rntypes & UIRemoteNotificationTypeBadge);
```

```objc
        alertSwitch.on = (rntypes & UIRemoteNotificationTypeAlert);
        soundSwitch.on = (rntypes & UIRemoteNotificationTypeSound);
    }
}

// Register application for the services set out by the switches
- (void) registerServices
{
    if (![self switchSettings])
    {
        textView.text = [NSString stringWithFormat:
            @"%@\nNothing to register. Skipping.", pushStatus()];
        [self updateSwitches];
        return;
    }

    NSString *status = [NSString stringWithFormat:
        @"%@\nAttempting registration", pushStatus()];
    textView.text = status;
    [[UIApplication sharedApplication]
        registerForRemoteNotificationTypes:[self switchSettings]];
}

// Unregister application for all push notifications
- (void) unregisterServices
{
    NSString *status = [NSString stringWithFormat:@"%@\nUnregistering.",
        pushStatus()];
    textView.text = status;

    [[UIApplication sharedApplication] unregisterForRemoteNotifications];
    [self updateSwitches];
}

- (void) loadView
{
    self.view = [[UIView alloc] init];

    // Workaround for catching the end of user choice
    TestBedViewController __weak *weakself = self;
    [[NSNotificationCenter defaultCenter]
```

```objc
        addObserverForName:UIApplicationDidBecomeActiveNotification
        object:nil queue:[NSOperationQueue mainQueue]
        usingBlock:^(NSNotification *notification)
     {
         [[UIApplication sharedApplication]
             registerForRemoteNotificationTypes:
                 [weakself switchSettings]];
         [weakself updateSwitches];
     }];

    self.navigationItem.rightBarButtonItem =
        BARBUTTON(@"Register", @selector(registerServices));
    self.navigationItem.leftBarButtonItem =
        BARBUTTON(@"Unregister", @selector(unregisterServices));
    [self updateSwitches];
}
@end


#pragma mark - Application Delegate
@implementation TestBedAppDelegate
// Retrieve the device token
- (void)application:(UIApplication *)application
    didRegisterForRemoteNotificationsWithDeviceToken:(NSData *)deviceToken
{
    NSUInteger rntypes = [[UIApplication sharedApplication]
        enabledRemoteNotificationTypes];
    NSString *results = [NSString stringWithFormat:
        @"Badge: %@, Alert:%@, Sound: %@",
        (rntypes & UIRemoteNotificationTypeBadge) ? @"Yes" : @"No",
        (rntypes & UIRemoteNotificationTypeAlert) ? @"Yes" : @"No",
        (rntypes & UIRemoteNotificationTypeSound) ? @"Yes" : @"No"];

    // Show the retrieved Device Token
    NSString *status = [NSString stringWithFormat:
        @"%@\nRegistration succeeded.\n\nDevice Token: %@\n%@",
        pushStatus(), deviceToken, results];
    tbvc.textView.text = status;
    NSLog(@"deviceToken: %@", deviceToken);

    // Write token to file for easy retrieval in iTunes.
```

```objc
    // Handy for learning push dev - but not a great idea for
    // App Store deployment. Set UIFileSharingEnabled to YES
    [deviceToken.description writeToFile:[NSHomeDirectory()
        stringByAppendingPathComponent:@"Documents/DeviceToken.txt"]
        atomically:YES encoding:NSUTF8StringEncoding error:nil];
}


// Respond to failed registration
- (void)application:(UIApplication *)application
    didFailToRegisterForRemoteNotificationsWithError:(NSError *)error
{
    NSLog(@"Error registering for remote notifications: %@",
        error.localizedFailureReason);
    NSString *status = [NSString stringWithFormat:
        @"%@\nRegistration failed.\n\nError: %@",
        pushStatus(), error.localizedFailureReason];
    tbvc.textView.text = status;
}


// Handle an actual notification
- (void)application:(UIApplication *)application
    didReceiveRemoteNotification:(NSDictionary *)userInfo
{
    NSString *status = [NSString stringWithFormat:
        @"Notification received:\n%@", userInfo.description];
    tbvc.textView.text = status;
    NSLog(@"%@", userInfo);
}


// Little work-around for showing text
- (void) showString: (NSString *) aString
{
    tbvc.textView.text = aString;
}


// Report the notification payload when launched by alert
- (void) launchNotification: (NSNotification *) notification
{
    // Workaround allows the text view to be created if needed first
    [self performSelector:@selector(showString:)
```

```
            withObject:[[notification userInfo] description]
            afterDelay:1.0f];
}

- (BOOL)application:(UIApplication *)application
    didFinishLaunchingWithOptions:(NSDictionary *)launchOptions
{
    window = [[UIWindow alloc] initWithFrame:[[UIScreen mainScreen] bounds]];
    tbvc = [[TestBedViewController alloc] init];
    UINavigationController *nav =
        [[UINavigationController alloc] initWithRootViewController:tbvc];
    window.rootViewController = nav;
    [window makeKeyAndVisible];

    // Listen for remote notification launches
    [[NSNotificationCenter defaultCenter]
        addObserver:self selector:@selector(launchNotification:)
        name:@"UIApplicationDidFinishLaunchingNotification" object:nil];

    NSLog(@"Launch options: %@", launchOptions);
    return YES;
}
@end
```

获取这个秘诀的代码

要查找这个秘诀的完整示例项目，可以浏览 https://github.com/erica/iOS-6-Advanced-Cookbook，并进入第 13 章的文件夹。

13.5　构建通知有效载荷

通过 APNS 递送推送通知需要 3 样东西：SSL 证书、设备 ID 以及带有你想要发送的通知的自定义有效载荷，其中有效载荷使用 JSON 格式化。你已经知道了如何生成证书和产生设备标识符，需要把它们传递给服务器。构建 JSON 有效载荷实质上涉及把良好定义的小字典转换成 JSON 格式。

JSON（JavaScript Object Notation，JavaScript 对象表示法）是一种基于键—值对的简单数据互换格式。JSON Web 站点（www.json.org）提供了这种格式的完整的语法细目列表，它允许表示字符串、数字和数组值。APNS 有效载荷包含最多 256 字节，其中必须包含完整的通知信息。

通知有效载荷必须包括一个 aps 字典。这个字典定义了一些属性，可以产生发送给用户的声音、商标和/或提醒。此外，还可以添加自定义的字典，其中带有需要发送给应用程序的任何数据，只要保持在 256 字节的限制内即可。图 13-8 显示了基本的（非本地化的）提醒的层次结构。

aps 字典包含一种或多种通知类型。其中包括你已经学过的标准类型：商标、声音和提醒。商标和声音通知都接受一个参数，其中商标是通过一个数字设置的，声音则通过一个字符串设置，它引用一个已经存储在应用程序软件包中的文件。如果找不到那个文件（或者开发人员传递 default 作为参数），就会为带有声音请求的任何通知播放默认的声音。当未包括商标请求时，iOS 将从应用程序图标中删除任何现有的商标。

13.5.1 本地化的提醒

在处理本地化的应用程序时，可以利用两个额外的键构造 aps > alert 字典。使用 loc-key 传递一个在应用程序的 Localizable.strings 文件中定义的键，iOS 将查找这个键，并利用所发现的用于当前本地化的字符串替换它。

有时，本地化字符串使用像 "%@" 和 "%n$@" 这样的参数。如果对于正在使用的本地化也是如此，则可以通过 loc-args 把那些参数作为字符串的数组进行传递。通常，Apple 建议不要使用复杂的本地化，因为它们可能会消耗 256 字节的带宽中的绝大部分带宽。

13.5.2 从字典转换为 JSON

在设计字典之后，必须把它转换为 JSON。JSON 格式比较简单，但是很精确。NSJSONSerialization 类用于创建一个 JSON 字符串。下面显示了基本的调用：

```
NSData *jsonData = [NSJSONSerialization
    dataWithJSONObject:mainDict options:0 error:nil];
```

13.5.3 自定义的数据

只要有效载荷还有剩余的空间，就要记得严格的字节预算，可以以键-值对的形式发送额外的信息。这些自定义的项可以包括数组和字典，以及字符串、数字和常量。你将定义如何使用和解释这种额外的信息。将整个有效载荷字典发送给应用程序，使得你一起传递的任何信息都可以通过用户字典供 application:didReceiveRemoteNotification: 方法使用。

含自定义的键—值对的字典不必展示一个提醒，以便于进行最终用户交互；如果应用程序没有运行，这样做将允许用户选择打开应用程序。如果应用程序已经启动，键—值对将作为有效载荷字典的一部分到达。

13.5.4 在启动时接收数据

当客户接受到通知时，点按它将会启动你的应用程序。在启动后，iOS 将会给应用程序委托发送一个可选的回调。该委托将通过实现一个名为 application:didFinishLaunchingWithOptions: 的方法来获取它的通知字典。

iOS通过启动选项参数把通知字典传递给委托方法。对于远程通知，这是正式的回调，用于在提醒框启动时获取数据。当 iOS 接收到通知并且应用程序没有运行时，不会调用didReceiveRemoteNotification:方法。

这个"完成启动"方法实际上被设计成处理许多不同的启动情况，推送通知只是其中之一。其他情况包括通过本地通知、通过 URL 模式处理等打开。在任何情况下，该方法必须返回一个布尔值。通常，如果处理了请求，就返回 YES；如果没有处理，就返回 NO。在远程通知启动时，实际上将忽略这个值，但是仍然必须返回一个值。

注意：
用户点按 Close 并在以后打开应用程序时，不会在启动时发送通知。必须手动签入服务器，以获取任何新的用户信息。不保证应用程序会接收到提醒。除了点按 Close 之外，提醒还可能会简单地丢失。总是要设计应用程序，以使之不仅仅依赖于接收推送通知来更新它自身及其数据。

13.6 秘诀：发送通知

通知处理涉及多个步骤（参见图 13-9）。首先，构建刚才讨论过的 JSON 有效载荷。接下来，为你想发送到的设备获取 SSL 证书和设备令牌。怎么存储它们留给你自己决定，但是必须记住它们是敏感信息。然后打开一条通往 APNS 服务器的安全连接。最后，与服务器握手，发送通知包，并关闭连接。

这是最基本的通信方式，并且假定你只有一个有效载荷需要发送。事实上，可以建立一个会话，并且一次发送许多分组；不过，本书把它作为一个练习留给读者完成，因为它将以不同于 Objective-C 的语言创建服务。Apple 开发人员论坛（devforums.apple.com）上正在举办关于推送提供者的讨论，并且提供了一个极佳的起点，用于查找 PHP、Python、Ruby、Perl 及其他语言的示例代码。

要知道的是，APNS 可能对一系列反复建立又断开的快速连接做出糟糕的反应。如果要同时发送多个通知，可以继续前进并在单个会话期间发送它们。否则，APNS 可能会利用一个拒绝服务式攻击来搅乱推送通知的递送。

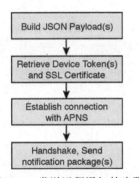

图 13-9 发送远程通知的步骤

秘诀 13-2 演示了如何将单个有效载荷发送给 APNS，并显示了实现图 13-9 中的第四个即最后一个方框所需的步骤。这个秘诀是围绕 Stefan Hafeneger 开发的代码构建的，并且使用了 Apple 的 ioSock 示例源代码。从本书的上一版起，对它做了一点清理，以处理不建议使用的 API。

各种服务器设置依赖于安全性、数据库、组织和编程语言而具有非常大的差别。秘诀 13-2 演示了实现这种功能所需的最少量的代码，无论你自己的服务器实现可能采取什么形式，它都可以充当一个模板。

13.6.1 沙盒和生产

pple 为推送通知提供了沙盒（开发）和生产（分发）环境，必须为每种环境创建单独的 SSL 证书。在把应用程序提交到 App Store 之前，沙盒有助于对它进行开发和测试。它使用一组小型的服务器，并且不打算用于大规模的测试。生产系统预留用于部署的应用程序，它们已经被 App Store 所接受。

- 沙盒服务器位于 gateway.sandbox.push.apple.com 上，端口 2195。
- 生产服务器位于 gateway.push.apple.com 上，端口 2195。

注意：
秘诀 13-2 打算进行编译并作为一个命令行实用程序运行在 Macintosh 上。

秘诀 13-2　把有效载荷推送给 APNS 服务器

```
// Adapted from code by Stefan Hafeneger
- (BOOL) push: (NSString *) payload
{
    otSocket socket;
    SSLContextRef context;
    SecKeychainRef keychain;
    SecIdentityRef identity;
    SecCertificateRef certificate;
    OSStatus result;

    // Ensure device token
    if (!deviceTokenID)
    {
        fprintf(stderr, "Error: Device Token is nil\n");
        return NO;
    }

    // Ensure certificate
    if (!certificateData)
    {
        fprintf(stderr, "Error: Certificate Data is nil\n");
        return NO;
    }

    // Establish connection to server.
    PeerSpec peer;
```

```
result = MakeServerConnection("gateway.sandbox.push.apple.com",
    2195, &socket, &peer);
if (result)
{
    fprintf(stderr, "Error creating server connection\n");
    return NO;
}

// Create new SSL context.
result = SSLNewContext(false, &context);
if (result)
{
    fprintf(stderr, "Error creating SSL context\n");
    return NO;
}
// Set callback functions for SSL context.
result = SSLSetIOFuncs(context, SocketRead, SocketWrite);
if (result)
{
    fprintf(stderr, "Error setting SSL context callback functions\n");
    return NO;
}

// Set SSL context connection.
result = SSLSetConnection(context, socket);
if (result)
{
    fprintf(stderr, "Error setting the SSL context connection\n");
    return NO;
}

// Set server domain name.
result = SSLSetPeerDomainName(context,
    "gateway.sandbox.push.apple.com", 30);
if (result)
{
    fprintf(stderr, "Error setting the server domain name\n");
    return NO;
}
```

```objc
// Open keychain.
result = SecKeychainCopyDefault(&keychain);
if (result)
{
    fprintf(stderr, "Error accessing keychain\n");
    return NO;
}

// Create certificate from data
CFDataRef data = (__bridge CFDataRef) self.certificateData;
certificate = SecCertificateCreateWithData(
    kCFAllocatorDefault, data);
if (!certificate)
{
    printf("Error creating certificate from data\n");
    return nil;
}

// Create identity.
result = SecIdentityCreateWithCertificate(keychain, certificate,
    &identity);
if (result)
{
    fprintf(stderr, "Error creating identity from certificate\n");
    return NO;
}

// Set client certificate.
CFArrayRef certificates = CFArrayCreate(NULL,
    (const void **)&identity, 1, NULL);
result = SSLSetCertificate(context, certificates);
if (result)
{
    fprintf(stderr, "Error setting the client certificate\n");
    return NO;
}

CFRelease(certificates);

// Perform SSL handshake.
```

13.6 秘诀：发送通知

```
do {result = SSLHandshake(context);}
    while(result == errSSLWouldBlock);


// Convert string into device token data.
NSMutableData *deviceToken = [NSMutableData data];
unsigned value;
NSScanner *scanner = [NSScanner
    scannerWithString:self.deviceTokenID];
while(![scanner isAtEnd]) {
    [scanner scanHexInt:&value];
    value = htonl(value);
    [deviceToken appendBytes:&value length:sizeof(value)];
}

// Create C input variables.
char *deviceTokenBinary = (char *)deviceToken.bytes;
char *payloadBinary = (char *)[payload UTF8String];
size_t payloadLength = strlen(payloadBinary);

// Prepare message
uint8_t command = 0;
char message[293];
char *pointer = message;
uint16_t networkTokenLength = htons(32);
uint16_t networkPayloadLength = htons(payloadLength);

// Compose message.
memcpy(pointer, &command, sizeof(uint8_t));
pointer += sizeof(uint8_t);
memcpy(pointer, &networkTokenLength, sizeof(uint16_t));
pointer += sizeof(uint16_t);
memcpy(pointer, deviceTokenBinary, 32);
pointer += 32;
memcpy(pointer, &networkPayloadLength, sizeof(uint16_t));
pointer += sizeof(uint16_t);
memcpy(pointer, payloadBinary, payloadLength);
pointer += payloadLength;

// Send message over SSL.
size_t processed = 0;
```

```
    result = SSLWrite(context, &message, (pointer - message),
        &processed);
    if (result)
    {
        fprintf(stderr, "Error sending message via SSL.\n");
        return NO;
    }
    else
    {
        printf("Message sent.\n");
        return YES;
    }
}
```

获取这个秘诀的代码

要查找这个秘诀的完整示例项目，可以浏览 https://github.com/erica/iOS-6-Advanced-Cookbook，并进入第 13 章的文件夹。

13.7 反馈服务

应用程序不会永远存在，用户一直都会添加、删除和替换他们设备上的应用程序。从 APNS 的角度讲，把通知递送给不再宿主应用程序的设备是无意义的。作为推送提供者，你有义务从活动支持列表中删除不活动的设备令牌。如 Apple 所指出的："APNS 将监测提供者是否在细致彻底地检查反馈服务，并且制止把推送通知发送给设备上不存在的应用程序。""老大哥"正盯着呢！

Apple 提供了一种简单的方式来管理不活动的设备令牌。当用户从设备上卸载应用程序时，推送通知将开始失败。Apple 将跟踪这些失败，并通过它的 APNS 反馈服务器提供报告。APNS 反馈服务将列出无法接收通知的设备。作为提供者，你需要定期获取该报告，并清除你的设备令牌。

反馈服务器将宿主沙盒和生产地址，就像通知服务器一样。可以在 feedback.push.apple.com （端口 2196）和 feedback.sandbox.push.apple.com（端口 2196）上找到它们。可以利用生产 SSL 证书联系服务器，并以与发送通知相同的方式进行握手。在握手之后，读取你的结果。服务器将立即发送数据，而无需你提供任何进一步的明确命令。

反馈数据包含 38 个字节，其中包括时间（4 字节）、令牌长度（2 字节）和令牌本身（32 字节）。时间戳指示 APNS 何时第一次确定应用程序不再存在于设备上，这使用一个标准的 UNIX 时间戳，即从 1970 年 1 月 1 日子夜起经过的秒数。设备令牌以二进制格式存储。如果使用字符串存储令牌数据，则需要把它转换成十六进制表示法，以使之匹配你的设备令牌。在编写本书时，可以忽略长度字节，它们总是 0 和 32，指设备令牌的 32 字节的长度：

```
// Retrieve message from SSL.
size_t processed = 0;
char buffer[38];
do
{
    // Fetch the next item
    result = SSLRead(context, buffer, 38, &processed);
    if (result) break;

    // Recover Date from data
    char *b = buffer;
    NSTimeInterval ti = ((unsigned char)b[0] << 24) +
        ((unsigned char)b[1] << 16) +
        ((unsigned char)b[2] << 8) +
        (unsigned char)b[3];
    NSDate *date = [NSDate dateWithTimeIntervalSince1970:ti];

    // Recover Device ID
    NSMutableString *deviceID = [NSMutableString string];
    b += 6;
    for (int i = 0; i < 32; i++) [
        deviceID appendFormat:@"%02x", (unsigned char)b[i]];

    // Add dictionary to results
    [results addObject:
        [NSDictionary dictionaryWithObject:date
        forKey:deviceID]];
} while (processed > 0);
```

> 注意：
> 在 Xcode Organizer Console 上搜索 "aps"，定位 APNS 错误消息。

13.8 设计推送

在设计推送时，要牢记规模问题。正常的计算不需要考虑规模。在编码完成时，应用程序将使用本地 CPU 运行在设备上。如果开发人员部署了另外 1 万份副本，除了增加计算支持之外，将不会涉及更多的投资。

推送计算确实需要考虑规模，无论你是有 1 万、10 万还是 100 万个用户事件都要如此，这是由于开发人员必须提供服务层，为销售的每个单元处理操作。支持的用户越多，成本将越高。要考虑到这些服务需要完全可靠，并且消费者将不会容忍延长的停机时间。

在可靠性之上，还要加入安全考虑。许多轮询的服务需要安全的凭证，必须把这些凭证上传给服务以便远程使用，而不是只存储在设备上。即使使用的服务没有用到那种身份验证，允许服务联系特定设备的设备令牌本身也是敏感的。如果那个标识符被窃取，它可能让垃圾邮件发送者发送不请自来的提醒。涉足这个领域的任何开发人员都必须严肃地对待这些可能的威胁，并且为存储和保护信息提供安全的解决方案。

第三方提供者 Urban Airship（urbanairship.com）提供了准备好使用的推送基础设施，它广泛用在 iOS 开发人员社区中。而它的雄心勃勃的竞争对手 Newcomer Parse（parse.com）还简化了推送的组合方式和部署。

13.9 小结

在本章中，从客户构建的角度以及从提供者的角度探讨了推送通知。你学习了可以发送的通知的种类，以及如何创建有效载荷，把那些通知转移到设备上；还了解了注册和取消注册设备，以及用户怎样决定加入和退出服务。

量的推送故事都不属于本章介绍的范畴。由你自己决定建立服务器以及处理安全、带宽和规模问题。部署的现实情况是：有许多平台和语言可以使用，它们超越了这里显示的 Objective-C 示例代码，通常可以使用 Ruby、Python 和 PHP。无论如何，本章中讨论的概念和显示的秘诀给你提供了一个良好的起点。你知道问题是什么以及事情必须如何工作，现在由你自己把它们投入良好的应用。

- 通知的长处在于它们的即时更新和立即展示。像 SMS 消息一样，当它们到达你的设备上时，将难以忽略它们。如果应用程序不需要那种即时性，则可以决定退出推送机制，而不会引起任何错误。
- 看管好你的 SSL 证书和设备令牌。Apple 响应安全漏洞的方式会有所不同，但是经验表明它将是棘手的并且令人不快。
- 当你承诺给用户提供服务时，就要信守诺言。在业务计划中构建一张时间表，预计它要采用什么以便随着时间的推移保持递送通知，以及你将怎样投资于它。消费者将不会容忍延长的停机时间，你的服务必须可靠。
- Apple 为推送通知提供了一个繁琐的登录工具，在它的开发人员论坛站点上可以了解该工具。
- 不要给用户发送垃圾邮件。推送通知并不打算用于销售产品或促销，要尊重你的用户群。
- 在构建应用程序时要考虑到规模。尽管应用程序最初可能没有成千上万的用户，要预计成功的应用程序启动以及一种适度的规模。创建一个可以与你的用户群一起增长的系统。